Palgrave Studies in Education and the Environment

Series Editors
Alan Reid, Faculty of Education, Monash University, Melbourne, VIC,
Australia
Marcia McKenzie, College of Education, University of Saskatchewan,
Saskatoon, SK, Canada

This series focuses on new developments in the study of education and environment. Promoting theoretically-rich works, contributions include empirical and conceptual studies that advance critical analysis in environmental education and related fields. Concerned with the underlying assumptions and limitations of current educational theories in conceptualizing environmental and sustainability education, the series highlights works of theoretical depth and sophistication, accessibility and applicability, with critical orientations to matters of public concern. It engages interdisciplinary and diverse perspectives as these relate to domains of policy, practice, and research. Studies in the series may span a range of scales from the more micro level of empirical thick description to macro conceptual analyses, highlighting current and upcoming turns in theoretical thought. Tapping into a growing body of theoretical scholarship in this domain, the series provides a venue for examining and expanding theorizations and approaches to the interdisciplinary intersections of environment and education. Its timeliness is clear as education becomes a key mode of response to environmental and sustainability issues internationally. The series will offer fresh perspectives on a range of topics such as:

- curricular responses to contemporary accounts of human-environment relations (e.g., the Anthropocene, nature-culture, animal studies, transdisciplinary studies)
- the power and limits of new materialist perspectives for philosophies of education
- denial and other responses to climate change in education practice and theory
- place-based and land-based orientations to education and scholarship
- postcolonial and intersectional critiques of environmental education and its research
- policy research, horizons, and contexts in environmental and sustainability education

More information about this series at
https://link.springer.com/bookseries/15084

Maria F.G. Wallace · Jesse Bazzul · Marc Higgins ·
Sara Tolbert
Editors

Reimagining Science
Education
in the Anthropocene

Editors
Maria F.G. Wallace
Center for Science and Mathematics
Education
University of Southern Mississippi
Hattiesburg, MS, USA

Jesse Bazzul
Department of Science and
Environmental Education
University of Regina
Regina, SK, Canada

Marc Higgins
Secondary Education
University of Alberta
Edmonton, AB, Canada

Sara Tolbert
Department of Science and
Environmental Education
University of Canterbury
Christchurch, Canterbury, New Zealand

ISSN 2662-6519 ISSN 2662-6527 (electronic)
Palgrave Studies in Education and the Environment
ISBN 978-3-030-79624-2 ISBN 978-3-030-79622-8 (eBook)
https://doi.org/10.1007/978-3-030-79622-8

PRAISE FOR *Reimagining Science Education in the Anthropocene*

"Situated in the era of the Anthropocene, this book volume recognizes the political urgency of re-envisioning science education *with* and *for* the community while dismantling the taken-for-granted deficit narratives of what science [education] is. Transcending disciplinary and geographical boundaries, the book calls us to reimagine science education in a more-than-human world, which places ecojustice, critical pedagogies, solidarity, and collectivity at the forefront. A timely, morally courageous, and indispensable reading."

—Lucy Avraamidou, *Associate Professor and Rosalind Franklin Fellow, University of Groningen, The Netherlands*

"This inspiring collection showcases the kind of creative thinking-without-borders we would need to prepare our students to meet the challenges of the Anthropocene. It makes me wish I were back in grad school to begin my research career afresh with the help of the wonderful assortment of ideas, insights, and perspectives that this book so generously offers. A much-needed and long-overdue contribution to the field!"

—Ajay Sharma, *Associate Professor, University of Georgia, USA*

"This book takes seriously the charge that current ways of thinking-doing in science education are terribly inadequate for addressing the complexities of the Anthropocene. In response, the editors bring together global scholars across disciplines who generate creative, relational, anti-racist, decolonizing, and speculative alternatives offering tools to map different futures. As such, this volume is an exciting development for the field that will be of interest to a range of educational scholar-practitioners looking to reimagine science education."

—Kathryn (Katie) Strom, *Assistant Professor, California State University, East Bay, USA*

"This volume is a collection of diverse (post)critical analyses, dialogues, and practices that address reconceptualizations of science education in the Anthropocene. A timely and significant contribution that takes justice-oriented science pedagogy beyond issues of representation by interrogating conventional episteme (know-what) and technē (know-how). Wallace, Bazzul, Higgins, and Tolbert have curated a phenomenal transdisciplinary exhibit for morally guided and praxis-oriented science education that can inform scholarship and transform teaching and learning."

—Rouhollah Aghasaleh, *Assistant Professor, Humboldt State University, USA*

CONTENTS

Notes on Contributors

Chessa Adsit-Morris is a curriculum theorist, environmental educator, and assistant director of the Center for Creative Ecologies housed in the Visual Studies Department at the University of California, Santa Cruz. Her research and teaching interests include curriculum studies, science education, philosophy of science, feminist science studies, art activism, and environmental justice. She is the author of *Restorying Environmental Education: Figurations, Fictions, and Feral Subjectivities* (Palgrave Macmillan, 2017).

Adilene Aguilera is a chemistry teacher and 10th grade level team lead at George Washington High School, a Chicago Public School. She is passionate and focused on a student-led curriculum which includes developing student leadership, research, and self-advocacy in science following a social justice framework. Her main goal is to guide students through the research and development of their passions through science.

Steve Alsop is a professor in the Faculty of Education at York University. He teaches courses and supervises graduate students in the fields of education, science and technology studies, environmental sustainability and interdisciplinary studies. Professor Alsop's research explores the personal, social, political and pedagogical articulations of scientific knowledge and technologies in educational settings and contexts.

Jesse Bazzul is associate professor of science and environmental education at the University of Regina. His work currently centres on ethics and how diverse beings inhabit shared worlds. Since ethics lies at the heart of education, it is important that the discipline of education be studied as one of the arts and humanities, as well as a social science and science.

Jessie Beier is a teacher, artist, writer, and conjurer of strange pedagogies for unthought futures. Working from Amiskwaciwâskahikan/ ᐊᒥᐢᑲ᙮ ᑭᐧ ᐣ ᑳᐦ ᐊᑲᐣ

(Edmonton, Alberta, Canada), Jessie's artistic and academic practices experiment with the potential for weird pedagogies to mobilize a break from orthodox referents and habits of repetition towards more ecological modes of thought. Jessie is currently a Ph.D. candidate at the University of Alberta, where she also teaches undergraduate courses in the Faculty of Education and in the Department of Women's and Gender Studies. When she is not trying to wrap (or better, warp) her head around black holes, you can find her making mixtapes, watching bad reality TV, or cooking for friends and family.

Nicole Bowers (Ph.D. Candidate, Arizona State University) is a graduate student in the Mary Lou Fulton Teachers College at Arizona State University. Her work interlaces philosophy and methodology to explore socioecological issues with a focus on educating for uncertain futures. Her interests have led her to promiscuously engage across disciplines. Her most current research explores approaches to inquiry past epistemology in the context of environmental and sustainability education in the Anthropocene.

Christie C. Byers is former elementary teacher and current Ph.D. candidate at George Mason University with specializations in science education research, qualitative inquiry, and teaching and teacher education. Her dissertation research involves following the phenomenon of wonder around and attempting to write about some of its complex ontology and promise for a more life-affirming approach to education.

Karen Canales Salas is the Environmental Justice (EJ) Education Coordinator with the Little Village Environmental Justice Organization (LVEJO). Through her role at LVEJO, Karen has engaged with thousands of youth all over the City of Chicago around environmental justice issues. Her work has concentrated on youth-led community science efforts throughout the city's EJ communities. She believes grassroots research and community outreach are critical in our fight toward a just transition.

Mindy J. Chappell is a native of East Saint Louis, Illinois. She currently serves as a science teacher in Chicago Public Schools. She is a science education doctoral candidate at the University of Illinois at Chicago. Her research focuses on the experiences of Black youth as they navigate science through the lens of arts-based research. Specifically, she explores how ethnodance informs the study of Black students' science identity.

Giani Clay will graduate from George Washington High School in June 2021. Giani plans to attend Eastern Illinois University and study electrical engineering. Giani's hobbies are playing sports, reading mangas, and watching science fiction television shows.

Darrin Collins is an author, educator, graduate student, and international martial arts competitor. Born and raised in Chicago, Darrin is a global traveler and pulls inspiration for his characters from his many international adventures. As a Chicago high school science teacher, he is committed to integrating themes (identity, social justice, judicial policy) into his writing that directly relate to and inspire the youth with whom he works. His aim is to intensify and increase the provocativeness of the conversation and evaluation of these themes by incorporating elements of science fiction.

Adam Devitt is an assistant professor of science education at California State University, Stanislaus. He engages in researching how preservice, elementary teachers develop STEM-specific pedagogies, in developing curriculum and instructional practices for supporting students with disabilities in STEM education, and developing methods and methodologies for understanding how teacher practitioners promote their own professional learning in math and science teaching. In his free time he enjoys road cycling and exploring the mountains and wilderness areas around the country.

Priyanka Dutt (she/they), though preferring to go by Pri, is the oldest child of three to Sunil and Promila Dutt, first-generation immigrants to Canada. Pri is a student, teacher, and an artist, attempting to contort the lines between science and culture. They hold a B.Sc. from Mount Royal University, a B.Ed. from the University of Alberta, and will hold an M.Sc. from the University of Lethbridge where they study aging in duckweed. Pri has a fascination with wasp dynamics and painting strange creatures of the world.

Colin Hennessy Elliott, Ph.D. is a postdoctoral research fellow at the Utah State University Department of Instructional Technology and Learning Sciences. His research interests include broadly exploring STEM educational spaces with a critical eye to imagine what it would take for a more just STEM educational future. He previously taught high school physics and continues to volunteer supporting youth in Newark, New Jersey.

Anastasya Fateyeva is a queer prairie artist, activist, musician, and early career teacher with Edmonton Public Schools. When she's not teaching kids about climate change and how to engage in organized action for social justice, she's creating art, food, and music for her community. Anastasya has a broad range of interdisciplinary academic interests, encompassing the visual arts, biological sciences, humanities, and critical perspectives on K–12 education.

Miranda Field is an educator in the fields of education and psychology and is currently completing her Ph.D. at the University of Regina. As a scientist, practitioner, advocate, and ally, she is currently focusing her work on Indigenous youth mental health. Miranda works in the area of academic and behaviour inclusive student support for an urban public school division in Saskatchewan, Canada.

Alejandra Frausto Aceves is the project based learning manager for Chicago Public Schools. Prior to this role, she has been a middle school and high school science teacher, a curriculum coach, and a school administrator. She has presented nationally and internationally at conferences, schools, universities, and other educational spaces about her work in context and culturally responsive curriculum and critical pedagogy. She believes in teaching and learning that goes beyond the classroom in order to position youth to be transformative intellectuals.

Michelle Gabereau is a Mi'kmaw educator at Ehpewapahk Alternate School. The foundation of her pedagogy is Wakotowin, and she is guided by Nehiyaw Pimatisiwa. Michelle is the best Auntie, a bead artist, and a novice moccasin maker.

Dr. Jane Gilbert is a professor of education at Auckland University of Technology (in Auckland, New Zealand). She was previously chief researcher at the New Zealand Council for Educational Research (NZCER). She has been involved in research and teaching in science education for nearly thirty years, focussing in particular on equity issues. Her current work is mainly in the area of educational futures. Recent projects have focussed on knowledge's changing meaning, science education's future, complexity thinking, and climate change education. In the last 10–15 years she has published two books and 25 refereed journal articles/book chapters on these topics.

Annette Gough is professor emerita of science and environmental education in the School of Education at RMIT University in Melbourne, Australia, and a Life Fellow of the Australian Association for Environmental Education (since 1992). She has been an adjunct or visiting professor at universities in Canada, South Africa, and Hong Kong and worked with UNESCO, UNEP, and UNESCO-UNEVOC on research and development projects. Her research interests include curriculum policy and development in science and environmental education; feminist, posthuman, critical, and poststructuralist research; research methodologies; and, most recently, child-centred disaster resilience education. She has written over 150 publications including books, chapters, reports, essays, articles, and curriculum materials.

Noel Gough is professor emeritus in the School of Education at La Trobe University, Melbourne, Australia. His teaching, research, and publications focus on research methodology and curriculum studies, with particular reference to environmental education, science education, internationalization, and globalization. He coedited and contributed to *Curriculum Visions* (Peter Lang, 2002), *Internationalisation and Globalisation in Mathematics and Science Education* (Springer, 2007) and *Transnational Education and Curriculum Studies: International Perspectives* (Routledge, 2021) and is founding editor of *Transnational Curriculum Inquiry*.

Simon Gough recently completed doctoral research in Japanese studies, with a particular emphasis on the production and consumption patterns that surround anime, manga, and other forms of popular media and their reception in Western fan and consumer culture. He is currently engaged in a number of independent projects which continue these investigations, with particular reference to the concept of "grand non-narratives"—aggregates of information, coded elements, and other fragments which combine in conceptual "wholes," but lack a single unifying narrative. He is book reviews editor for the *Electronic Journal of Contemporary Japanese Studies*.

Andreia Guerra is a senior researcher and full professor at Graduate Program Science, Technology and Education from CEFET/RJ, Rio de Janeiro, Brazil. She has a Ph.D. in History of Science for Science Education and is granted as CNPq (Brazilian Research Council) associate researcher. Her main research themes include science education; history, philosophy, and sociology for science education; and cultural studies. She is the president of IHPST Group (2019–2022) and co-editor of a Brazilian journal on physics teaching (CBEF).

Bronwyn Hayward is a professor of political science at Te Whare Wānanga o Waitaha University of Canterbury in Christchurch, New Zealand. Bronwyn's scholarship focuses on the intersections of youth, sustainability, and climate change. She is director of the University of Canterbury Hei Puāwaitanga Sustainable Citizenship and Civic Imagination research group and co-principal investigator for the University of Surrey's Centre for Understanding Sustainable Prosperity (CUSP). She is lead author on two reports for the Intergovernmental Panel on Climate Change (IPCC). Her most recent publication is *Children, Citizenship, and Environment #SchoolStrikeEdition*.

Marc Higgins is an assistant professor in the Department of Secondary Education at the University of Alberta, Canada, where he is affiliated with the Faculty of Education's Aboriginal Teacher Education Program (ATEP). His research labours the methodological space between Indigenous, post-structural, and post-humanist theories in order to respond to contested ways-of-knowing and -being, such as Indigenous science.

Xia Ji is a mother, daughter, a striving human being on Turtle Island, and a dedicated environmental educator and teacher educator. For over 20 years Xia's teaching, research, and community engagement have centered on science & environmental education, teacher professional development, and lifelong learning. Since 2008, Xia's work has greater focus on eco-social justice, Earth democracy, and wellbeing through civic discourse, contemplative practices, integrative curriculum making, and regenerative pedagogies.

Vicki Kirby is emeritus professor of sociology and anthropology in the School of Social Sciences, The University of New South Wales, Sydney. She is also a visiting professorial fellow, Institute of Art and Architecture, Academy of Fine Arts Vienna, where she holds a Peek Grant. She is a prominent figure in feminist and new materialist debates and a member of the international think tank, Terra Critica. Most recent books include *What If Culture Was Nature All Along?* (2017, Edinburgh UP) and *Quantum Anthropologies: Life at Large* (2011, Duke UP). The motivating question behind her research is the puzzle of the nature/culture, body/mind, body/technology division because so many political and ethical decisions are configured and defended in terms of this opposition.

Delani Lopez is a senior at North-Grand High School in the Chicago Public Schools.

Catherine Milne is a professor in science education and chair of the Department of Teaching and Learning at New York University. Her research interests include the role of material culture in teaching and learning, socio-cultural elements of teaching and learning science, the role of the history of science in learning science, and models of teacher education. She is the author of *The Invention of Science: Why History of Science Matters for the Classroom* (2011). Her co-edited volumes include *Sociocultural Studies and Implications for Science Education: The Experiential and the Virtual* (2015) and *Material Practice and Materiality: Too Long Ignored in Science Education* (2019). She is co-editor-in-chief for the international journal *Cultural Studies of Science Education* and co-editor of two book series, one for Springer Nature and the other, Brill Sense Publishers.

Daniel Morales-Doyle is an assistant professor of science education in the Department of Curriculum and Instruction at the University of Illinois at Chicago. He was a teacher in the Chicago Public Schools for more than a decade before joining the faculty at UIC. His work focuses on the science question in social justice education: How can teaching and learning science contribute to efforts to build communities that are more just and sustainable?

Cristiano B. Moura is a professor of the Graduate Program in Science, Technology and Education and High School Teacher at CEFET/RJ. He is part of the NIEHCC Research Group at the same institution. His research interests span from the history, philosophy and sociology of science in science education and curriculum theories to cultural history and cultural studies of science education. Cristiano is deeply interested in politics and increasingly concerned with social injustices at the global level. He loves poetry (and sometimes writes poems) and loves to teach his high school students.

Fikile Nxumalo has a scholarship that focuses on reconceptualizing and resituating place-based and environmental education within current times of ecological precarity. She is interested in curricular and pedagogical possibilities for disrupting anti-Black and settler colonial educational frames that erase and/or enact deficit constructions of Black and Indigenous (and their intersections) childhoods, knowledges, and place relations. As an interdisciplinary scholar, Dr. Nxumalo's research brings together early childhood education, childhood studies, Indigenous and Black studies, children's geographies, and the environmental humanities and sciences.

Molly Quinn endowed professor at Louisiana State University, is the author of *Going Out, Not Knowing Whither: Education, the Upward Journey and* *the Faith of Reason* (2001), and *Peace and Pedagogy* (2014); and editor of *Complexifying Curriculum Studies:Reflections on the Generative and Generous Gifts of William E. Doll Jr.* (2019). Avidly embracing Haraway's (2016) "making kin," "odd kin" call in all her wondering and worlding for the earth's healing; much of her scholarship engages "spiritual" and philosophical criticism towards embracing a vision of education that cultivates beauty, awareness, compassion, creativity, community and social action.

Tomasz G. Rajski is a teacher in the Chicago Public Schools. He hopes to encourage learners of science to question the sustainability of our communities and develop the tools needed to advocate for change. He is driven to explore the elements of history and anthropology which impact the ways in which we view and utilize scientific knowledge and natural resources.

Aswathy Raveendran is a faculty member with the Department of Humanities and Social Sciences at the Birla Institute of Technology and Science Pilani, Hyderabad Campus. Her areas of interest include critical studies of science education and feminist science and technology studies. At BITS Pilani, she teaches courses that situate science and technology in the sociopolitical context and gender studies.

Dr. Carolina Castano Rodriguez is a biologist and a science and environmental educator. Carolina is an internationally recognised scholar for her work with marginalised and socio-economically disadvantaged communities in the field of critical pedagogies, transformative learning, peace education and pedagogies of care. She has a particular interest in the link between science education, social justice and ecojustice. Carolina has led several projects in Australia, Ecuador, Argentina and Colombia.

Kathryn Scantlebury is a professor in the Department of Chemistry and Biochemistry at the University of Delaware. Her research interests focus on gender issues in science education from a material feminist perspective. Scantlebury is a guest researcher at the Center for Gender Research at Uppsala University, co-editor-in-chief for *Gender and Education*, and co-editor of two book series for Brill Publishers.

Himanshu Srivastava is a graduate student at the Homi Bhabha Centre for Science Education in Mumbai who believes that the mainstream science and environment education in Indian schools exerts a kind of symbolic violence on the students from marginalized communities and is disempowering for them. His doctoral work is focused on problematizing (and reimagining) the educational discourse on selected themes from the standpoint of one such community that lives close to India's largest landfill site.

Sara Tolbert is associate professor of science and environmental education at Te Whare Wānanga o Waitaha University of Canterbury in Aotearoa New

Zealand. Her current work focuses on critical/feminist praxis in teacher education and education for regenerative and justice-oriented environmental present futures.

Briony Towers is a research fellow in the Centre for Urban Research at RMIT University. She lives and works on the unceded lands of the Boonwarrung, Woiwurrung and Dja Dja Warrung people of the Kulin Nation. Her Ph.D. in sociocultural psychology involved an in-depth investigation of children's knowledge of vulnerability and resilience to wildfire in south-eastern Australia. For the last 10 years, she has been collaborating with emergency service agencies, educators, children, and families to develop effective and sustainable models of wildfire education for Australian schools.

Anna L. Tsing is professor of anthropology at the University of California in Santa Cruz. Her transdisciplinary work straddles the fields of ecology, history, anthropology, and political economy. Notable publications include *The Mushroom at the End of the World: On the Possibility of Life in Capitalist Ruins* and the recently released *Feral Atlas*. *Feral Atlas*, which Tsing discusses, in this volume, maps the feral effects of human infrastructures from multiple perspectives and "nonscalable" scales.

Dr. Blanche Verlie is a climate change education researcher and practitioner and currently a postdoctoral researcher at the Sydney Environment Institute at the University of Sydney. Her work explores more-than-human pedagogies of learning with and as climate, and the ways that we become (with) climate change. She is currently writing a book, *Learning to Live with Climate Change*, to be published with Routledge in 2021.

Maria F.G. Wallace is an assistant professor of elementary STEM education and Women & Gender Studies Program faculty affiliate at the University of Southern Mississippi. Earning a Ph.D. in curriculum and instruction with specializations in curriculum theory, science education, and a graduate minor in women's and gender studies from Louisiana State University, Dr. Wallace's research and teaching aim to deterritorialize beginning (science) teachers' subjectivities and practice. Drawing on conventional qualitative research methods and post-qualitative modes of inquiry, Dr. Wallace's research (re)imagines ways beginning science teachers are "known," named, and re/produced.

Travis Weiland is currently an assistant professor of mathematics education at the University of Houston and a former high school and collegiate mathematics instructor. His work focuses on considering how the disciplinary practices of statistics and data science translate to K–12 education through a statistical and critical literacy perspective. In particular, Travis is interested in thinking about how statistics can be taught in schools through the interrogation of sociopolitical issues in order to support students in developing as critical citizens in democratic societies.

List of Figures

CHAPTER 1

Introduction

Maria F.G. Wallace, Jesse Bazzul, Marc Higgins, and Sara Tolbert

This edited volume invites transdisciplinary scholars to re-vision science education in the era of the Anthropocene. The collection encompasses the works of educators from many walks of life and areas of practice together to help reorient science education toward the problems and peculiarities associated with the geologic times many call the Anthropocene. It has become evident that science education, the way it is currently institutionalized in various forms of school science, government policy, classroom practice, educational research, and public/private research laboratories, is *ill-equipped* and *ill-conceived* to deal with the expansive and urgent contexts of the Anthropocene. Paying

M. F.G. Wallace (✉)
Center for Science and Mathematics Education, University of Southern Mississippi, Hattiesburg, MS, USA
e-mail: maria.wallace@usm.edu

J. Bazzul
Science and Environmental Education, University of Regina, Regina, SK, Canada
e-mail: Jesse.Bazzul@uregina.ca

M. Higgins
Secondary Education, University of Alberta, Edmonton, AB, Canada
e-mail: marc.higgins@ualberta.ca

S. Tolbert
Science and Environmental Education, University of Canterbury, Christchurch, New Zealand
e-mail: sara.tolbert@canterbury.ac.nz

© The Author(s) 2022   1
M. F.G. Wallace et al. (eds.), *Reimagining Science Education in the Anthropocene*, Palgrave Studies in Education and the Environment,
https://doi.org/10.1007/978-3-030-79622-8_1

homage to myopic knowledge systems, rigid state education directives, and academic-professional communities intent on reproducing the same practices, knowledges, and relationships that have endangered our shared world and shared presents/presence is not where educators should be investing their energy. For example, the forces and flows of science education render the Anthropocene an epistemological object to be learnt *about* rather than *with* or *through* such that it might implicate the learner (Gilbert, 2016). Science education does not (and cannot not in its current forms) meet the needs of the post/human moment(s) in which we find ourselves.

This work continues the transdisciplinary project of transforming the ways communities *inherit* science education. Specifically, authors were invited not to fit questions of the Anthropocene *into* science (or) education but rather attend to their cross product(ion). In other words, authors attended to the proliferation of possibilities and (re)orientations made possible through reading these dialogically rather than dialectically. Not unlike de Freitas et al. (2017), "we hope this cross product... amplifies the philosophical insights from each, stretching scholarship in new directions and across disciplines" (p. 551). Throughout the book, authors nurture productive relationships between science education and fields such as science studies, environmental studies, philosophy, political science, the natural sciences, Indigenous studies, feminist studies, critical race studies, and critical theory in order to provoke a science education that actively seeks to remake our shared ecological and social spaces in the coming decades and centuries. After Stengers (2018), we exclaim that "another science [education] is possible!"—but also necessary in rethinking and regenerating our world yet-to-come.

Our understanding of the Anthropocene is necessarily open and pluralistic, as different beings on our planet experience *this* time of crisis in different ways. Notably, the Anthropocene threatens large swaths of the Global South, animal and plant species, Indigenous peoples, and marginalized communities of color (both rural and urban), in ways that affluent, privileged/colonizing communities of the Global North have purposefully ignored (see Davis & Todd, 2016; Whyte, 2018; Yussof, 2018). These inequalities and differential effects of the Anthropocene are inextricable from the scourge of late capitalism and its consumption of the natural/social commons (Tsing, 2015; Moore, 2015). Further, the pluralistic orientations offered within this book also challenge the notion that the Anthropocene is singular to *this* time, when considering the figure of Man (i.e., the masculine subject of Western modernity): that this *end of the world* is premised on the end of multiple "worlds whose disappearance was assumed at the outset of the Anthropocene" (de la Cadena & Blaser, 2018, p. 2). Authors also consider how the naming of this shared catastrophe in a way that holds the *"anthropos"* responsible (e.g., vs. unfettered capitalism, settler colonialism, and patriarchy) perhaps masks more than it reveals (Moore, 2015; Kirby, 2018), while also masking the efforts of new and old human and more-than-human communities and collectives that are working toward hopeful, regenerative present futures (Haraway, 2016; Hayward & Tolbert, this volume).

Working with authors situated across several different continents since 2018, the publication of this text has never seemed so timely. From landscapes and communities (of all kinds) ablaze, the COVID-19 pandemic, and pervasive racial and environmental injustice, it has become quite apparent that science education sits at the nexus of reconfiguring our human and more-than-human relationships. Centering (for just a moment), the COVID-19 pandemic (among several simultaneous pandemics), we invite readers to witness the power of a non-human agent, SARS-CoV-2 (the virus that causes COVID-19), to completely reconfigure our more-than-human relationships with each other. SARS-CoV-2, alongside the social, political, and cultural systems in which it is implicated, demands that we do several things: (a) recognize the multifaceted challenge of teaching (i.e., intellectual, emotional, skillful, physical, and, and, and across all contexts and roles); (b) witness and act on the grave material and immaterial inequities that persist across time and space; (c) reconfigure our more-than-human social relationships with land, water, and other non-human living organisms; and (d) center care, empathy, and patience over productivity, efficiency, and haste. Like SARS CoV 2 and COVID-19, we invite readers to pay attention to our shared non-human teachers. Might we *think-with* the Anthropocene in such a way that renders us response-able (Haraway, 2016)? Might attuning to the Anthropocene (re)open our ability to respond, and in turn, produce forms of actionable responsibility toward ourselves, other(ed) humans, more-than-human kin, the planet, and institutions?

What the COVID-19 pandemic makes clear is that the time to argue for a more socially, politically, and ecologically rich science education is effectively over. Education, perhaps especially science education, must play a role in nurturing the kinds of new relationships and modes of living that will carry life through the Anthropocene. This realization is less of an ethical stance than it is a necessity. The COVID-19 pandemic affirms the role of science education in bringing about well-being, health, and community vibrance. More specifically, science educator fields find themselves caught up in battles against right-wing populism and political attacks that seek to delegitimize all collective institutions (including the sciences). In these divisive times, we need collectivity more than ever. The sciences, and their related fields, owe their life-beyond-the-laboratory to science *connoisseurs*, those who are neither expert nor amateur, who "appreciate the originality or the relevance of an idea but also pay attention to questions or possibilities that were not taken into account in its production, but that might become important in other circumstances" (Stengers, 2018, p. 9).

These are people such as activists, citizen scientists, and community members who support things like action on climate change, environmental protection, access to STEM education, and ethical research in the natural sciences. Part of science and science education's legitimacy in the early-mid twenty-first century is the result of those who have expanded science participation, cultivating a "science" that thrives outside of its intended environments. Coalitions of activists, educators, community members, etc., have played key roles in seeding and growing public resistance to climate and pandemic denial positions, some invoking slogans such as "Believe science." Yet, it is important to highlight that the very logics through which science is defined can also work against possible coalitions. While these logics can help delegitimize dangerous pseudoscientific claims, they can also marginalize and work against the important coalition work we have just described. For example, as Bang and colleagues (2018) poignantly ask, "if Indigenous peoples stand with the sciences—as we will—will scientists also stand with us?" (p. 151). Lastly, and in turn, the COVID-19 pandemic makes clear that life, with all its activity and production, exists in a web of relations that requires the utmost sensitivity and care.

The chapters herein, while written largely before (and revised during) the COVID-19 pandemic, recognize the contexts of political urgency and the need for care and sensitivity in education. One thing educators can be sure of is that education in the Anthropocene will need to engage the grand disturbances that will come to define the trajectory of community life. From relations with fire/water and magical realisms to political manifestos and Anthropocenic detonators, this volume takes loving, but bold, steps into different worlds—and the pedagogical relations that might widen these worlds. Of course, it does so cautiously, knowing full well that the trappings of things like colonialism, capitalism, patriarchy, and white supremacy still operate at the edges of these texts. While science education in the Anthropocene works at the margins to dismantle these forms of oppression, we are dismayed at how so many of our institutions, and more broadly, our field(s), still continue to give only cursory attention to entrenched and often unexamined systemic inequities. Yet, we are inspired by the creative and transformative work of the authors in this collection, and remain optimistic that indeed, another science education is possible (it is already happening!).

PART I: KINSHIP, MAGIC, AND THE UNTHINKABLE

The first section of *Reimagining Science Education in the Anthropocene* doesn't necessarily ease the reader into this "changed space" for science education. As we (the editors) have said above, the time for arguing for a sociopolitically engaged and transdisciplinary science education for multispecies survival is effectively over. The book therefore proceeds under assumption that different pedagogical and ethical pathways and openings are needed. There is now less of a need to justify these pathways within the institutional norms and policy statements of the recent past, as it is abundantly clear that different kinds of thinking and institutional arrangements are needed today. The COVID-19 pandemic is a prime example of how modern institutions and community investment, as they currently exist, are inadequate to the tasks at hand. The chapters in this section bend toward futurity, strange possibility, and new conceptions of kinship. Here's a little preview of what's to come, given in the order that the chapters appear.

In Chapter 2 ("'Trees Don't Sing!... Eagle Feather Has No Power!' Be Wary of the Potential Numbing Effects of School"), Xia Ji begins with intimate conversations about science education with her own children. These conversations, like all important dialogues, contain ruptures of possibility and meaning. Among these ruptures, is the idea that science education could be so much more meaningful for students if only it opened itself to other ways of thinking and experiencing the world. The first chapter reminds educators that those moments of possibility and rupture will inevitably occur in daily life. In a time where scientific ways of knowing the world seek ethical, spiritual, and ecological guidance from Indigenous ways of living and knowing, is it still possible to say that "*trees don't sing*," or that "*feathers have no power*"? What do educators stand to lose by taking a narrow stance on such questions with their students?

With the tone for different kinds of encounters and relations set, Jessie Beier introduces the idea of unthinkability into our pedagogical lexicon (at least for this volume) in Chapter 3 ("Tracing a Black Hole: Probing Cosmic Darkness in Anthropocenic Times"). Using black holes as an image of our current existential predicament, as opposed to the iconic image of earth from space, Beier suggests that our time of the Anthropocene demands we embrace uncertainty and unknowability as important aspects of existence. In this way, Beier introduces a dimension of obfuscation into our view of science education. That is, the idea that there are things in our world that cannot be known here and now, and this should give educators pause as they go about portraying the world as knowable and thinkable.

In Chapter 4 ("The Waring Worlds of H. G. Wells: The Entangled Histories of Education, Sociobiology, Post-Genomics, and Science Fiction"), Chessa Adsit-Morris demonstrates that science fiction is a powerful tool for orienting science education toward the future. Adsit-Morris's chapter goes back and forth between writings of H. G. Wells and the new sociobiologies and market-driven genomic research of our current time. Adsit-Morris not only points to new racisms on the horizon, but how possibilities for a "new science education" were missed a hundred years ago. One lesson being that science educators and students might turn to the work of writers and artists to expand the boundaries of what's possible and thinkable.

Like the other sections of this book, section one has a very strong affective dimension running through it. Nicole Bowers takes us on journey into the productive space of magical realism in Chapter 5 ("Creating Magical Research: Writing for a Felt Reality in a More-Than-Human World"). What if the existence of extra-worlds, virtual possibilities, and a suspension of "natural" laws only served to make our shared actual worlds more vibrant and comprehensible? Does opening our world to strange possibilities provide a new space for pedagogical experimentation in science education? This question is not only relevant for Bowers' chapter directly, but all the chapters in section one.

Lastly, in Chapter 6 ("Fire as Unruly Kin: Curriculum Silences and Human Responses"), Annette Gough, Brony Towers, and Blanch Verlie open the question of non-human or more-than-human kinship by positioning elemental fire as teacher, healer, and unruly relation. This repositioning of kin, and what counts as kin, shifts education toward new relations that are absolutely necessary to avoid climate disaster and ecological collapse. The multiplicitous relations that fire-as-kin creates allow educators to recast the historical, geologic, and spiritual relations essential for recreating the world today.

PART II: DECOLONIZING ANTHROPOCENE(S)

Just as the first section does not ease the reader into the central question of how we might respond to this contemporary moment in science education, this second section begins by troubling the notion that the Anthropocene is singular. As de la Cadena and Blaser (2018) state, what distinguishes the contemporary moment is that "the colonizers are threatened as the worlds they displaced and destroyed when they took over what they called *terra nullius*" are at risk (p. 3, emphasis in original). In turn, this section endeavors to set and expand the ethical context for the book by recognizing the ways in which *the* Anthropocene is predicated on and preceded by a multiplicity of Black, brown, Indigenous, and other-than-human Anthropocenes that are marked by genocides and mass extinction, as well as the ongoing and lasting effects of colonization.

More than strictly a (re)thinking of science education, this section takes seriously the notion that perhaps "modern academic literacies and technologies can make what has been made invisible by colonialism visibly absent, but they cannot make it present" (Ahenakew, 2017, p. 89): the Anthropocene(s) need new ways to be felt.

In Chapter 7 ("Redrawing Relationalities at the Anthropocene(s): Disrupting and Dismantling the Colonial Logics of Shared Identity through Thinking with Kim Tallbear"), Priyanka Dutt, Anastasya Fateyeva, Michelle Gabereau, and Marc Higgins trouble the ways in which the naming of the Anthropocene is at once an admission of guilt that simultaneously also works to mask some of the culpability. Particularly, thinking with Dakota scholar Kim Tallbear's work on Indigenous genetics, parallels are drawn to explore the ways in which the production of *an* identity meeting a shared crisis defers and diminishes responsibility for ongoing colonialism: dispossession of Indigenous lands, disenfranchisement of Indigenous peoples, and the genocide of Indigenous ecologies. Rather than prescribe *a* response (as *a* meaning, such as *an* identity, singular, can be problematic), Dutt and colleagues invites science educators to consider *why* science education cannot or has not been able to respond to these uneven inheritances and effects through a series of artful provocations. As Plains Cree scholar Cash Ahenakew (2017) states, "the work of decolonization is not about what we do not imagine, but what we cannot imagine from our Western ways of knowing" (p. 88): we need new ways to (re)open what we can even imagine within science education as we respond to the Anthropocene(s).

In Chapter 9 ("Still Joy: A Call for Wonder(ing) in Science Education as Anti-Racist Vibrant Life-Living"), Christie C. Byers continues the trend of making the Anthropocene(s) felt otherwise by poetically disrupting sense-making. This is done as a means of interrupting already received and embodied notions of wonder and the ways in which they are leveraged toward re-asserting science education as usual either through quelling curiosity or re-directing it towards an already knowable "nature." By theorizing and attending to wonder as a more-than-individual affective flow, Byers focuses specifically on anti-blackness as a significant way in which the life-giving prolif-eration of possibilities that is wonderment is rendered inert within and beyond science education. Creatively juxtaposing her own work as science teacher educator with that of Black poets and critical theorists, Byers invites consid-eration of the ways in which science education disciplines Black bodies (and, by extension, the ways in which Black bodies are literally policed) through the (re)construction of a scientific subjectivity which is premised on the othering of nature, but also those "closer" to nature (through their other(ed) construc-tion). But this is nothing new: as Haraway (1988) states, for science "Nature is only the raw material of culture, appropriated, preserved, enslaved, exalted, or otherwise made flexible for disposal by culture in the logic of capitalist colonialism" (p. 592).

In Chapter 8 ("Decolonizing Healing through Indigenous Ways of Knowing"), Miranda Field takes seriously the notion that, after Mi'kmaq scholar Marie Battiste (2018), "we all must become critical learners and healers within a wounded space." Presenting psychology as pharmakon—as both poison and panacea—Field disrupts and displaces the ways in which the field is entangled within longstanding and ongoing colonial violence toward Indigenous lifeways, while attending to the healing possibilities to come by extending already present openings within medicine and psychology. In the affirmative, she suggests that psychology can and should move beyond the pathologizing of Indigenous peoples toward strength-based approaches that are rooted in Indigenous land-based practices and work to regenerate other significant relationships (e.g., community, more-than-humans) fragmented by colonial logics and practices. Importantly, the chapter functions as more than an ethical injunction to do no (colonial) harm as it also proposes other-ways-of-being-in-relation. Of note, Field highlights the ways in which healing, like learning (in wounded spaces), can and must be a journey whose pathway cannot be wholly prescribed or predicted: for Nature to be teacher or healer, one must learn to slow down and attune differently.

In Chapter 10 ("The Salt of the Earth: Inspired by Cherokee Creation Story"), Darrin Collins recasts a Cherokee origins story in order to make Black and Indigenous Anthropocene(s) felt otherwise. Jumping right into the *doing* rather than the *theorizing about* (re)storying how we might engage *at* and *with* the Anthropocene(s), Collins' adaptation of the traditional story engages with Blackness and Indigeneity in ways that take seriously the colonial harm done by the figure of "Man" (i.e., Western, modern, white, settler, "rationalist") through erasure and othering without making the harm the story. Rather, given the ways in which settler colonialism pits those othered by it against each other, this is not only a tale of resilience, strength, and survivance, but also of allyship. Further, the storying blurs the lines between Indigenous storywork and traditional storytelling, as well as historical and science fiction, becoming a rich way of speculating pasts and futures to-come, as well as a story in which Indigenous science is enacted (rather than spoken about). This has the effect of not only making it unclear as to whether the story told is occurring prior to our contemporary moment or after, but also, in honoring Indigenous temporalities, disrupting linear notions of time such that there is a spiral quality to time, in which it is a creation story whose (re)occurrence signals that time has circled back. Lastly, the notable absence of particular dominant figures speaks to the ways in which there is no future for "Man" given the ways in which the damage to the Earth is not evenly spread, and the need for some to learn, in humility, from ways-of-knowing-in-being that have been practiced in place since time immemorial.

Part III: Politics and Political Reverberations

An engagement with politics, eco-politics, and new/old forms of solidarity and collectivity are essential to science education in the Anthropocene. This third section includes contributions that interrogate and illustrate historicity in science education, critical pedagogical interventions, and dynamics of power in STEM fields. The four chapters in the section help us envision different political and pedagogical visions/theories/forces for future worlds.

In Chapter 11 ("The Science of Data, Data Science: Perversions and Possibilities in the Anthropocene through a Spatial Justice Lens"), Travis Weiland presents us with dualities of data science, illustrating how data (and data science) are not inherently objective and have been used as tools for oppression—but also have significant and necessary potential as tools for social and ecological justice. He turns his attention as well to how mapping can also have oppressive as well as justice-oriented goals and outcomes. Using North Carolina gerrymandering as a case study of racial and spatial injustice, he outlines possibilities for a critical data curriculum that supports students' development of data literacy (and critical data literacy) while engaging them in socially transformative learning. He concludes with a very personal and important reflection—and a call to action for all of us—about how we read, who we read, and who we cite as we continue to engage in justice-oriented scholarship.

In Chapter 12 ("Science and Environment Education in the Times of the Anthropocene: Some Reflections from India"), Aswathy Raveendran and Himanshu Srivastava reflect through metalogue on how Anthropocenic discourses that privilege growth as "sustainable development" are identifiable in Indian policy, curriculum and education standards, textbooks, and students' subjectivities. They explore through their reflective dialogue how they came to question their own training as scientists (Aswathy) and engineers (Himanshu) and found their place in critical science education. They delve into the ways that textbooks and national political movements in India have come to include environment-related themes but in ways that privilege technoscientific "innovation" and marginalize issues of political inequality. Finally, they highlight the need for coalition building, solidarity, political emotion, and political love as key to a more justice-oriented educational, ecological, and political future. As Aswathy concludes, "The challenge for educators working with marginalized communities is to find ways to inculcate political emotions that have the power to alter their living conditions."

In Chapter 13 ("Rethinking Historical Approaches for Science Education in the Anthropocene"), Cristiano B. Moura and Andreia Guerra challenge the "single story" often communicated about Western modern science in the history of science fields. Using botany as a case study, they critically analyze how European scientists and capitalists extracted, appropriated, and/or erased knowledge they "collected" as "data" from the Americas. Moura and Guerra then reflect on how science educators and historians of science must challenge this single story and seek to tell new histories, ones that honor and attribute marginalized knowledges and their place in the "becoming" of Western modern science, or what we now often refer to as "science." By doing so, we can (re)generate new possibilities for living well together in the Anthropocene.

In Chapter 14 ("Reflections on Teaching and Learning Chemistry through Youth Participatory Science"), Daniel Morales-Doyle, Alejandra Frausto Aceves, Karen Canales Salas, Mindy J. Chappell, Tomasz Rajski, Adilene Aguilera, Giani Clay, and Delani Lopez provide a window into work they presented at a town hall meeting session of the Science Educators for Equity, Diversity, and Social Justice (SEEDS; http://seedsweb.org) conference in Norfolk, Virginia, 2019. From the perspective of university researchers, teachers, teacher educators, high school students, and community organizers on Chicago's Southside, they grapple with what it means to *really do* science for social justice in formal school settings. Their work in engaging youth in analyzing heavy metal contamination in the soils of their communities reveals that chemistry education can be a powerful vehicle for transformative learning. They also comment on the ongoing tensions of their endeavors. They share, for example, their thoughts on the politics of regimes of evidence—what it means to have to "prove" that your soils are contaminated or to find evidence to support why you do not want a polluter to be located in your community. As Daniel points out, "Why is the burden on the largely Mexican, working class community to have to find evidence of harm? Why doesn't the multinational corporation that owns this plant have to provide evidence to the surrounding community that what they're doing is safe and sustainable?".

Part IV: A Science Education
for a World-Yet-to-Come

The fourth section includes chapters that offer alternative entry points into science education as it has come to be known. Several of the chapters in this section clarify and critique interpretations of "the Anthropocene" head-on as it pertains to curating alternative futures for the field of science education. From curriculum studies to reorientating relationships with the more-than-human, the authors in this section invite readers to engage new ways of attuning to the multifaceted experience of science education.

In Chapter 15 ("Learning from Flint: How Matter Imposes Itself in the Anthropocene and What that Means for Education"), Catherine Milne, Colin Hennessy Elliot, Adam Devitt, and Kathryn Scantlebury invite readers to take a deeper dive into the Flint, Michigan, water crisis, an ontological disturbance worthy of deeper attention within science education. Using new materialist analyses, the authors present a compelling case for re-examining the "ontological underpinnings of the Earth as an outcome of bio, geo, chemo intra-actions" and thus also foundational ideas around "hands-on science."

In Chapter 16 ("Resurrecting Science Education by Re-inserting Women, Nature, and Complexity"), Jane Gilbert offers a feminist critique of science in and of the Anthropocene to suggest that science-as-we-know-it cannot provide solutions to the issues it (i.e., science), education, and our collective mo(ve)ments are confronted. Gilbert maps a pedagogical approach for deconstructing science and science education for our new geologic epoch.

In Chapter 17 ("Watchmen, Scientific Imaginaries, and the Capitalocene: The Media and Their Messages for Science Educators"), Noel and Simon Gough, a father-and-son duo, take inspiration from their own cross-generational experiences with the popular graphic novel and recent film remake, *Watchmen*. Upon troubling the Anthropocene and deconstructing the science of *Watchmen*, the Goughs illuminate a fourth dimension, *simultaneity*, visible for rethinking the spacetime configuration(s) of science education.

In Chapter 18 ("Curricular Experiments for Peace in Columbia: Re-imagining Science Education in Post-Conflict Societies"), Carolina Castano-Rodriguez, Steve Alsop, and Molly Quinn invite readers to wonder: How could science education contribute to empower marginalized societies and, particularly, how could it contribute to create lasting peace in Colombia? Using a short personal narrative set in 2050 Columbia, feminist standpoint theory, and Escher's *Relativity* (1953), the authors explore how science education might be central element of critical reconciliation. Thinking peace, as an enduring verb, a way of being and becoming together, a science education for and of peace is made thinkable.

PART V: COMPLICATED CONVERSATIONS

One of the early features we adopted for this book were interviews with scholars that have inspired transdisciplinary work across a variety of fields, such as ecology, environmental studies, education, science studies, political science, and social theory. As a writing form, the interview introduces an element of interlocution, and there is a small sense that we (the readers) are the ones asking the questions. How will the interviewee answer the pressing questions put to them? Will there be some kind of answer that we (the readers) can hold on to? The interviews contain a small amount of pleasure, either in their investment (in the process or conversation), or the capture of a definite moment in time where a comprehensive exchange took place. In the spring of 2019, the editors met via video conferencing to discuss who might be able to expand the confines of our edited collection and allow us to think differently. We consider ourselves very fortunate to have been able to speak directly with Anna L. Tsing, Fikile Nxumalo, Vicki Kirby, and Bronwyn Hayward. Here is a preview of what's to come in section five.

In Jesse's interview with Anna L. Tsing, the Anthropocene is discussed through conceptual lenses like "Empire," Capital, and Acceleration. *Feral Atlas*, Anna L. Tsing's latest project, is discussed as a transdisciplinary approach to talking about the feral effects of human infrastructures. How do we think about livability and environment in the ruins of capitalism? Next is Maria's in-depth conversation with Fikile Nxumalo, which explores a refiguration of the universalizing discourses around Anthropocene that might account for the inequalities hidden by such concepts. Early on, Maria asks Fikile to talk about the kinds of things educators "inherit" and how these figure into what is possible and doable—with the hope of doing and thinking otherwise. Marc's conversation with Vicky Kirby is a fast-paced discussion about the long-established boundaries between science and other disciplines (to name just one thing). How are we as (post)critical scholars to mediate these boundaries, and how might we be mindful of how our questions are bound-up and produced alongside a plethora of aporias, exclusions, and strange relations? Lastly, Sara's conversation with professor of political science and IPCC member/writer Bronwyn Hayward explores possibilities for critical hope and new solidarities across local communities, nations, and disciplinary boundaries. Bronwyn also shares encouraging perspectives from her research with youth, including climate change activists and students in New Zealand and around the world, about what we can learn from their hopes and visions for the future. What does it mean to educate in a changing climate and ongoing ecological and political uncertainty? We hope your interest is piqued by these questions and topics, and if so, please feel free to begin reading this book at the interviews (or any section that speaks to you).

References

Ahenakew, C. R. (2017). Mapping and complicating conversations about indigenous education. *Diaspora, Indigenous, and Minority Education, 11*(2), 80–91.

Bang, M., Marin, A., & Medin, D. (2018). If indigenous peoples stand with the sciences, will scientists stand with us? *Daedalus, 147*(2), 148–159.

Battiste, M. (2018, May 27). Decolonizing and indigenizing the academy: Achieving cognitive justice. Plenary talk at 2018 Canadian Society for the Study of Education (CSSE) Annual Meeting: Gathering Diversity, Regina, SK. https://www.youtube.com/watch?v=XKQyvcOvAPI.

de Freitas, E., Lupinacci, J., & Pais, A. (2017). Science and technology studies × educational studies: Critical and creative perspectives on the future of STEM education. *Educational Studies, 53*(6), 551–559.

de la Cadena, M., & Blaser, M. (Eds.). (2018). *A world of many worlds*. Duke University Press.

Gilbert, J. (2016). Transforming science education for the Anthropocene—Is it possible? *Research in Science Education, 46*(2), 187–201.

Haraway, D. (1988). Situated knowledges: The science question in feminism and the privilege of partial perspective. *Feminist studies, 14*(3), 575–599.

Haraway, D. J. (2016). *Staying with the trouble: Making kin in the Chthulucene*. Duke University Press.

Hayward, B., & Tolbert, S. (2021). Conversations on citizenship, critical hope, and climate change: An interview with Bronwyn Hayward. In M. Wallace, J. Bazzul, M. Higgins, & S. Tolbert (Eds.), *Reimagining science education in/for the Anthropocene*. Palgrave Macmillan.

Kirby, V. (2018). Un/limited ecologies. In D. Wood, M. Fritsch, & P. Lynes (Eds.), *Eco-deconstruction: Derrida and environmental ethics* (pp. 121–140). Fordham University Press.

Moore, J. W. (2015). *Capitalism in the web of life: Ecology and the accumulation of capital*. Verso Books.

Stengers, I. (2018). *Another science is possible: A manifesto for slow science*. Polity.

Todd, Z. (2016). An indigenous feminist's take on the ontological turn: 'Ontology' is just another word for colonialism. *Journal of Historical Sociology, 29*(1), 4–22.

Tsing, A. L. (2015). *The mushroom at the end of the world: On the possibility of life in capitalist ruins*. Princeton University Press.

Whyte, K. P. (2018). Indigenous science (fiction) for the Anthropocene: Ancestral dystopias and fantasies of climate change crises. *Environment and Planning E: Nature and Space, 1*(1–2), 224–242.

Yussof, K. (2018). *A billion black Anthropocenes or none*. University of Minnesota Press.

Part I

Kinship, Magic, and the Unthinkable

"Trees Don't Sing! … Eagle Feather Has no Power!"—Be Wary of the Potential Numbing Effects of School Science

Xia Ji

INTRODUCTION

Let's start in the middle. (Aamodt & Bazzul, 2019)

Three family conversations prompted the writing and the title of this essay. Each of the conversations happened at different times with each of my three children when they were 8, 11, and 13 year old, respectively. Although having little firsthand experience of my children's science classes or their experiences at school in general, I could not help wondering about how much their formal education has played a part in the responses they gave me that have somewhat alarmed me not just as a parent, but also as a science educator. Let me first share the three conversations.

Conversation 1

A mid-summer afternoon, 2018, wind was gusting at about 30 mph, swinging branches and crowns of trees, a typical day in the prairie part of Saskatchewan; mother and her 8-year-old son were walking side by side in the neighborhood. We haven't done this for a while for some reason, so we were both excited to be with one another combing the sidewalk and chatting now and then. After

X. Ji (✉)
University of Regina, Regina, SK, Canada
e-mail: xia.ji@uregina.ca

© The Author(s) 2022
M. F.G. Wallace et al. (eds.), *Reimagining Science Education in the Anthropocene*, Palgrave Studies in Education and the Environment,
https://doi.org/10.1007/978-3-030-79622-8_2

one round around the block, we came under a quaking aspen, one of the many stimulating trees in our neighborhood, and paused...

> Mother: *"Jun! Listen—the quaking aspen is singing . . . very loud!"*
> Jun looked up: *"What do you mean?"*
> Mother: *"Remember, we used to stop under each tree and listen to them sing?"*
> Jun, with a bewildered look: *"No, trees don't sing!"*
> Mother's heart sank and wasn't sure what had happened to her son.

My thoughts went far back to when Jun was about 3 years old and before he started any formal schooling. We did the neighborhood walk quite regularly—almost daily. At that time Jun and I carried our conversations in mandarin Chinese—the language I have spoken/sung to him since he was born—well, actually while he was still in the womb. We referred to all the trees as "大树公公 Da Shu Gong Gong"—a Chinese phrase meaning "Big Tree Grand One." That time we rarely walked side by side but more often Jun was either ahead of or behind me—examining something new or exciting, such as a leaf or small branch, a feather, a puddle. ... On one of those windy days in late spring Jun was way ahead of me running with his rainbow-colored pinwheel raised above his head to see how fast it could spin. Then suddenly he slowed down and backed up in front of our neighbor Rick's sidewalk and stopped there while looking up around and about. Curious about what happened I quickened my steps and caught up with him, asking in Chinese:

> *"Tell me, Jun, tell me, what did you find?"*
> With the pinwheel in front of his chest, his mouth in an exaggerated "O" shape, Jun quietly said to me: *"Listen! Da Shu Gong Gong is singing!"*
> *"Oh—what is Da Shu Gong Gong singing about?"*
> *"Happy song!"* Jun said with a big, radiating smile.

I listened more intently under the tree. With the coming and going of each gust of wind, I heard the somewhat lower pitched yet cheerful rustling sounds crescendo and decrescendo above and around me. I recognized the tree from my naturalist training years and knew right away from its greenish powdery bark, its distinctive heart shaped leaves, and their fluttering movement that it is a quaking aspen! We both were mesmerized by the chorus of this tree mixing from time to time with that of the nearby Dutch Elm trees, as well as the quieter songs of the spruce tree. My naturalist training taught me that quaking aspen is the largest living organism, growing in clones that reproduce primarily by sending up sprouts from their roots. One clone of this aspen tree in Minnesota was estimated to be 8,000 years old! So the Chinese phrase "Big Tree Grand One" indeed makes sense! We both closed our eyes, stood there under the quaking aspen, and listened! I was grateful that Jun invited me to listen, and was thrilled to see the trees through his little being! From then on we made it a habit to pause under each tree during our walk and listen to them sing—especially when the wind was stirred up! Now five years later Jun was telling me that "Trees don't sing!".

Conversation 2

May, 2018. Returning home at the end of a day after attending the Canadian Society of Studies in Education (CSSE) conference session on indigenous medicinal plants and healing practices with a Cree Elder, I could not contain my excitement. Hovering over the kitchen table, mother and 13-year-old daughter Li started a chat:

Li: *"Where were you mom?"* (usually she could read my face—which was now spilling over with apparent exhilaration!)

Mother: *"I just attended a most interesting session with a Cree Elder named George. He showed us some of his ancient tools for healing."*

Without a pause, mother kept rambling: *"One of the things he used in healing practice was a fan of eagle feathers. It is almost one hundred years old—the Elder told us! The feathers looked plain and rather torn in many parts, but he said it has healing power while the other kind—the modern one with glittering decorations—doesn't. He even passed the eagle feather fan around and let us hold it!"*

Li's face grew in impatience and disbelief as she snapped: *"Mom, what are you talking about?! Eagle feather has no power!!"*

My heart crinkled upon hearing this certain conviction from my 13-year-old daughter.

Mother: *"How come eagle feather can't have power?!"*.

Li: *"Are you kidding, Mom?!"*.

Mother, trying to stay calm: *"Well, tell me how come an eagle feather can't have power. The elder uses his eagle feather fan as part of his healing practice for people. How can you be so sure that it has no power?"*

Li: *"A feather is not even alive. You can't believe whatever the elder tells you."*

Mother: *"You have a point that we should not believe whatever others say. But what if in that context the eagle feather has healing power? You don't know for sure it has no power!"*

I do not remember much of the rest of our conversation other than that Li eventually rolled her eyes and walked away to her room. Whatever explanation or invitation I tried to offer her was only met with deaf ears and a closed mind. She seemed to have a hard time putting side by side the possibility that an eagle feather can be dead as determined by the Western science criteria, but can also carry healing power in another context, such as in this Cree Elder's healing practice, and that one view does not have to exclude the other to be valid.

Conversation 3

End of school year in June 2019. Getting ready for bed, mother and 11-year-old daughter Ann were at the bathroom—brushing teeth together. Out of nowhere, the following exchange just happened:

> Ann: *"You know ... mom ... I am very good at science ... but I hated it!"*
>
> Mother: *"What do you mean?"* apparently alarmed and concerned, not just as a mom, but also as a science educator.
>
> Ann: *"Well, I got all As because I am good at doing what she (the teacher) asks us to do, but science is just ... ughhh."*
>
> Mother: *"Oh ... I am sorry to hear that. How come you sound so disgusted by school science!?"*
>
> Ann: *"I don't know ... all we do is ... read the textbook, the handouts, answer questions on the worksheets! Then repeat!!!"*
>
> Mother: *"That is NOT science! That is reading comprehension!!"*

By the end of Grade 6, Ann already determined that she does not like school science. The hopeful side is that at least she was aware enough to know that she was disgusted by school science. All my three children went to the same elementary school and had mostly the same teachers. In no way I am here to suggest their teachers are to blame, but I could not help wondering about how science education is carried out in their elementary school. As a science teacher educator, I am aware of the challenges of science education at the elementary school level, such as minimized time given to science teaching and learning due to more emphasis on literacy and numeracy in most schools. Teachers in elementary school in general have taken a limited number of science courses and tend to by and large shy away from teaching through the sciences. As Ann informed me in the above conversation, it is not uncommon for science to be approached as a reading exercise, which can be mostly content driven instead of process driven. Very little hands-on/experiential inquiry based learning takes place to truly develop students' scientific literacy and competency. There is no way for me to verify, but I wonder if the responses from Jun and Li in Conversations 1 & 2 have something to do with what they have been taught at school. It would be very concerning if my children have learned through school science education to have to choose either this or that when it comes to seeing, understanding, and experiencing our world. It would be as concerning as if they were taught to choose the Chinese or Nepalese or Canadian side of their identity while they are all three, because they were born to a Chinese mother, a Nepalese Father, and are growing up in Canada.

While Ann was aware of how she felt about school science, Jun and Li were not even conscious that school science may have impacted them in ways that might limit their future experiences of what is possible. At such tender ages of 8 and 13, with only a few years of formal schooling behind them, they both have already become crusted and closed to other possibilities of experiencing the world—and the worlds still in the making. I wonder how "educative" my children's science learning experiences have been at school. Dewey (1938) proposed a set of criteria for signifying "educative experience" (pp. 33–50), one of which is the principle of continuity of experience. It covers the "formation of attitudes, attitudes that are emotional and intellectual; it covers our basic sensitivities and ways of meeting and responding to all the conditions which we meet in living." One exemplification of the principle of continuity of experience is growth, or growing as developing, not only physically but intellectually and morally (p. 36). School science might have contributed to my children's attitudes toward science or other possibilities of experiencing the world that are not "conducive to continued growth," but instead can be detrimental.

What's also alarming is how similar my three children's experiences of school science were to mine from decades ago when I was schooled out of the openness to or even tolerance of any other possibilities aside from scientific ways of knowing and truth defined by so-called science teachers. My formal introduction to science started in middle school (grade 6) chemistry and physics classes in the People's Republic of China during the late 1980s. China, then under Deng Xiaoping's leadership, was going through a nation-wide economic reform and opening up to the outside world. As stated in the 863 Program or State High-Tech Research & Development Program, China aimed to stimulate the development of advanced technologies in a wide range of fields for the purpose of rendering China independent of financial obligations to foreign technologies. We secondary school students were mostly Youth League members (团员) and were regarded as the heirs of the Chinese Communist Party. Studying the sciences and technology became our obligation in order to better serve society and to serve China. The influence of school science on my worldviews and beliefs on truth and what is possible is subtle, gradual, yet profound—just like the metaphor of a frog sitting in a pot of cold water being heated up slowly to boiling point. The frog never jumped out, and would eventually be boiled to death. These learned ways of seeing and legitimizing have to some extent alienated me from my mother, who has very little formal schooling and never took any science class but has much experiential knowledge through her active living and experiential knowing. For instance she truly sees plants as persons with feelings and believes that our ways of working with them can influence their (the plant's) feelings, and consequently how well they grow. I must have rolled my eyes in my teenage years when mom tried to teach me how to treat plants right so as to not upset them. Here is a conversation I had with my mom after I harvested some 韭菜 (jiucai, *Allium tuberosum*) from our backyard garden.

Mom: (after taking a look at the handful of jiucai in my left hand) *"How did you cut them?"*

Xia: *"I only cut those that looked good—tall and broad-leaved. Aren't these great?!"*

Mom: *"Yeah! But you mustn't cut them like that anymore!"*

Xia: *"Why?"*

Mom, paused for a few seconds as if trying to find the right words to communicate her concern, *"The way you cut the jiucai would upset the ones you didn't cut! They would be so upset that they would not grow well. You should have cut them roll by roll—leaving none behind."*

Xia: *"Jiucai has feelings?! How funny you believe that they would be upset!"*

As an adolescent I could be very insensitive to my mom and often laughed at her unusual remarks from time to time—attributing her off-key beliefs to her lack of schooling and lack of science education. Now 30 some years later as I recalled this conversation about harvesting jiucai, I was reminded of the experimental study with sweet grass by Robin Kimmerer and her graduate student Laurie (Kimmerer, 2013, p. 156). In their research over two years to compare the effects of two different ways of harvesting on the growth of sweet grass, they challenged a dominant view held by mainstream academic ecologists that "to protect a dwindling species was to leave it alone and keep people away" (p. 163). Their research verified the theory known to Kimmerer's ancestors that "If we use a plant respectfully it will stay with us and flourish. If we ignore it, it will go away." Kimmerer (2013) reminded us that "we are all the product of our worldviews—even scientists who claim pure objectivity" (p. 163). I have to admit that my learning in the school sciences have greatly contributed to the alienation I developed toward my mother's worldviews and ways of knowing/doing/being. The following poem highlights some of my own experiences as a student of science in middle school and high school in mainland China.

As a Student of Science

Science is for the smart students,
But who are not talented enough
To go into the creative arts!

Science track will surely lead you
To some sort of professional job,
So that you don't have to repeat
The back-breaking lifestyle of your parents
Who were frozen in value (if any)
As farmers and factory workers.

Science is the new life force,
Science and technology is the
Primary productive power,
Sought after by every sector of society!
Indoctrinated by such slogans as
Scientification/modernization of agriculture,

Scientification/modernization of industry,
Scientification/modernization of defense,
Modernization of science and technology!

I repeated these pledges
Day after day, year after year:
"Love the motherland,
"Love the people,"
"Love the Communist Party,
"Serve the people!"
These energizing words
Echoed throughout my impressionable years.

Science is about the absolute truth,
It is amoral, apolitical, acultural.
Science has been put on a pedestal,
Believing itself entitled to be enthroned!
All should yield in the advance of science;
No one shall, nor can be in his way!

A very sad and confusing thing is that my years of formal schooling have gradually steered me away from my mother and her ways of knowing, being, and relating. I learned from my school subjects (especially the so-called natural sciences, such as physics, chemistry, and biology) that anything that does not align with the established sciences is primitive, backward, and even superstition, and should be despised and abolished. The alienation from my mother and her worldviews and ways of knowing grew with my accumulated years of formal schooling. I remember the feeling of shame for having a mother like mine.

Also, in school science, we learned about the glorification of many scientists who in serving their country with the mission to advance China's science and technology development sacrificed their lives and families. Many more scientists lived incognito until years after retirement from their research career in the nuclear science and military/defense technologies. For instance, from 1961 to 1989 a Chinese scientist named 于敏 (Yu Min) concealed his identity so as to devote himself to the research on hydrogen bombs, leading to the success of China's hydrogen bomb development! According to the Baidu website, many more scientists dedicated their lives to science and technology research work classified as state top secret (implicating national security). Most of the time even their families did not know where they were working and what they were working on. Behind the success of the China's first atomic bomb many more scientists lived/worked incognito. I remember having mixed feelings and reactions in high school as we watched the documentaries of these scientists held as national heroes. On the one hand I admired their selflessness and dedication to their work, on the other I wondered about how any scientist with good consciousness would willingly contribute to the creation of deadly weapons of mass destruction.

After six solid years of being a student of science in junior and senior high school in mainland China, somehow I did not want to choose science as the field of study for my post-secondary education as most of my peers did. I was drawn to the creative arts instead, even though I did not feel I had the "talents" for them. However, like many children from working-class families, I was confronted by the dilemma of "what job prospects can the arts bring you?" as questioned by my parents, and reminded by my high school teachers. So, there I was—accepted into the English Department at the West China University of Medical Sciences and Technology, which was a compromise with the daunting forces trying to leave a mark in my adolescent and young adult life. I studied English for Medical Science and Technology—really not sure what job prospect or career this degree would lead me to. As the first person in my family and in my neighborhood to go to university I had no idea that an English degree from a university of medical sciences and technology would not lead me to become a medical doctor. Most of us (in an 18-student cohort) were very disappointed to learn this harsh reality! Two students actually switched to the medical school after one year of waiting. I also contemplated switching to medicine—although I did not have the financial or social means.

A lab incident at the medical school in the winter of 1995 totally turned me away from the idea of pursuing a career in medicine. My cohort of 16 students pursuing an English major were divided into groups of four to learn and carry out the tracheotomy procedure. For this lab, our instructor brought four lab rabbits in a cage. We learned that after undergoing these medical procedures, the rabbits would eventually die and be disposed of, which did not sound necessary or right to my group of four young women. We tried our best to comply by getting the anesthesia drug ready and injecting it through the vein on the rabbit's ear. As we waited for our rabbit to be anesthetized, there was a commotion in the lab when a couple of students were trying to catch a rabbit from the cage. Some yelled in excitement: "she is pregnant… looks like she is about to labor… any moment!!" Well, the lab instructor could have turned this into a teaching/learning moment—guiding us through the birthing process of the rabbit, comparing that with the human experience. The four members in my group all spoke up—requesting our instructor to spare this rabbit from the tracheotomy procedure as she had young ones coming and needed to nurse and take care of them. We even offered to share our rabbit with the other group so that they can learn the procedure as well without harming this pregnant one. What a relief when our request was tentatively approved by the instructor: "Well, we will see. Let's leave her alone for now." So we went on with our lab work, while the mother rabbit gave birth to four babies. However, by the end of the lab session, we learned that another group of students got hold of the mother and carried out the procedure on her anyway, obviously with the instructor's silent approval. I felt disgusted, powerless, and indignant as I painfully recalled the well-rehearsed slogan of my minor years that "In the pursuit of science, life will be sacrificed!" Throughout history (distant and recent) and globally so many lives have perished in the name of scientific research and progress.

Upon graduation from the West China University of Medical Sciences and Technology, I continued to Graduate School at the University of Minnesota (St. Paul/ Minneapolis) in the United States. As a graduate student there from 1997 to 2007 I had the good fortune to take classes from a variety of colleges. There was no restriction as to what I could take as long as they were graduate courses. At the turn of this century I sat in one class on humanist geography, and learned that one way to understand aesthetics is its opposite, which is anesthesia (Tuan, 1993). According to Tuan, the aesthetic is "not merely one aspect of culture but its central core—both its driving force and its ultimate goal." He wisely noted that "the pervasive role of the aesthetic is suggested by its root meaning of feeling—not just any kind of feeling, but shaped feeling and sensitive perception… and the opposite of aesthetic is anesthetic, which is a lack of feeling—the condition of living death" (p. 1). Is this what we want to school the next generation of learners into—"living death?"

The good news is that the lack of aesthetics in my science learning experiences was compensated by the required courses in English literature as I majored in English. My five years of immersion in British and American literature kept me human/humane and well aware of the recursive "bad faith and cruelty" and "savage inequalities" (Kozol in Greene, 1993, p. 211) operating in all societies (Baldwin, 1963, p. 17). Literature also provides ample examples of heroes and heroines who are "not bound by the expediencies of any given administration, any given policy, any given morality, and who exercise their birth right to examine everything and come to their own views of the world and make their own decisions" (Baldwin, 1963, pp. 18–19). In a way, immersion in literature helped to prevent me from becoming the "living death." I kept searching for and remained open to the next adventures in life. After all, as Kenko (1967) wisely said, the most precious thing in life is its uncertainty.

The following significant life experience (Ji, 2011) got me questioning what I have been learning (and not learning) from my formal science classes and opened a new path for me in the field of environmental education.

Encountering Living Water: A Turning Point in My Science Learning Journey

Serendipity is the one word I can think of to express how I remember the experience in the summer of 1995. It was my third year of university, at the end of the spring semester, the week after all the final exams. I was relaxed and enjoying my mid-day nap when I felt shaken side to side and stirred up by a familiar voice:

"Amy, Amy, wake up, wake up!"
  "What?!!"

I opened my eyes, unhappy about the fact that Helen (my best friend) disrupted my nice and well-deserved nap. We all had English names then given our major in English.

Amy: *"What's the matter that you had to wake me up from my nap!?"*.

I was still rubbing my eyes—trying to stay awake. We were the only two left in the dorm room of five women. The rest of them had all left for their homes in other provinces for the summer holidays.

Helen: *"You said you want to stay here to work this summer, right!? Now is our chance! Come, they are interviewing students for summer jobs, interpreter job! It's said to be very well paid! Come on!"*

Amy: *"Really?! When do we start?"*.

Helen: *"Don't know. We will find out."*

I rolled out of bed, ran to the public washing area, hand washed my face with cold tap water, skipped back to the dorm, combed my hair, and checked my clothes—which weren't obviously wrinkled from lying in bed. Off we went—Helen and I along with three other students who were one year ahead of us, toward the Foreign Faculty's Residence of the West China University of Medical Sciences and Technology.

This is how I got my first summer job—to work as a Chinese-English interpreter for the Keepers of the Waters project in the city of Chengdu, China. That is where and when I met Betsy Damon, an eco-feminist, performance artist, and environmental and civil rights activist, and Jill Jacoby—a water resource scientist and environmental educator, both from the United States. Keepers of the Waters is a community-based water activism initiative, which was founded and directed by Betsy Damon in 1991. It brings scientists and artists together to stimulate community action on water quality issues. As Damon (2016) shared in her lecture titled "I Am Water," she intended to "create a language of consciousness," with which she has certainly reached a deeper dimension/consciousness of me as I worked as a language interpreter initially and later as a volunteer for the Keepers of the Waters projects in Chengdu and Lhasa. A couple of the photographic images of water drops (see Fig. 2.1) she shared with us totally challenged my established and unquestioned scientific view on water, which is made of three atoms—two hydrogen and one oxygen, and is believed to be the same everywhere (Fig. 2.1).

Betsy shared Schwenk's research, which has discovered that "If water is of good quality or tested from a natural source it will express rosette and vortical patterns. If water is of poor quality or contaminated by pollutants it will lack expression." This view and understanding of water did not register with me until I drew and painted the living water drop image again and again alongside Betsy Damon, trying each time to make my drawing of the water drop more alive and expressive!

Fig. 2.1 Photo & caption credit to Theodore Schwenk, used with permission

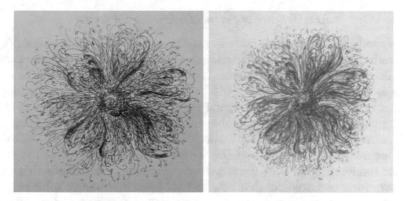

Fig. 2.2 Artwork and photo credit to Xia Ji

Writing this chapter I tried again to draw a picture of the living water drop image (see Fig. 2.2) as this practice has had a profound impact on my perception and consciousness. Drawing and painting is a process that forced me to attend to my attention as the energy of my attention often resulted in different things I observed and what I could replicate or interpret as a result (Fig. 2.2).

For both summers in 1995 and 1996, I worked and volunteered as an interpreter and assistant to Betsy Damon, and was immersed in public performances and installation arts created by a diverse group of artists to raise awareness of source water protection, particularly the Fu and Nan Rivers and Lhasa Rivers. Details of these two summer community-based art events were recently included in the Asia Art Archive (AAA-Hong Kong, 2017). Through these experiences I learned that "the arts can serve to liberate human perception, challenge human creativity, and stir the human soul" (Jacoby & Ji, 2010, p. 135). I was reminded of Eisner's understanding that "artists invent fresh ways to show us aspects of the world we had not noticed; they release us from the stupor of the familiar" (Jacoby & Ji, 2010, p. 135). I also came to the realization that "In the face of our collective daunting global environmental and health challenges, we do not just need science and engineering, but also other forms of knowledge and ways of knowing. As the 'creative force' in our society artists can serve as awakening, educating, and provoking agents as they themselves learn about the various social and environmental challenges of our time" (Jacoby & Ji, 2010, p. 143).

So far in the above sections I have shared some of my observations and concerns regarding potential numbing effects of school science on students through the conversations I had with my three school-age children and my own experiences as a student of science. I have also shared my experience with the living water garden and the Keepers of the Waters projects as a major awakening stage in my life and work. Next, I would like to turn to my experience as a teacher of science/science educator.

As a Teacher of Science/ a Science Educator

It is beyond the scope of this essay to provide a detailed remedy to the potential numbing effects of school science as shown above in the experiences of my children and my own experience. However, I want to share a few commitments and strategies that I have tried over the last ten years of teaching in the area of science and environmental education. Science education as status quo in public schools is not acceptable and cannot meet the complex demands and challenges of our time or the future. Darling-Hammond (2015) called for "new learning for a rapidly changing world." She reminded us of the reality we face today, that "2/3 of today's young people will enter jobs that don't yet exist, using technologies that have not yet been invented to solve complex problems we have not solved—such as climate change, water quality and security, food insecurity, [and] poverty." Six decades earlier Baldwin prophetically wrote the following text.

> ... we are living through a very dangerous time... we are in a revolutionary situation, no matter how unpopular that word has become in this country. The society in which we live is desperately menaced . . . from within. To any citizen of this country who figures himself as responsible—and particularly those of us who deal with the minds and hearts of young people—must be prepared to go for broke. (p. 17)

Baldwin pointed out that "the whole process of education occurs within a social framework and is designed to perpetuate the aims of society.... The paradox of education is that as one begins to become conscious one begins to examine the society in which he is being educated;... at the point when you begin to develop a conscience, you must find yourself at war with your society" (p. 19). If a country does not "find a way to use that energy of the young generation, it will be destroyed by that energy" (p. 20). We educators are left with no choice but to take up the challenge of our time. If we do not, we are part of the problem.

In my own practice as a science and environmental educator working alongside pre-service teachers mostly in their early 20s, I have certainly come to recognize the gravity of our responsibilities. In my practice as a science teacher educator I consciously invite my students to engage with me in currere. Currere is a concept and method suggested by Pinar (1975, 2004, 2012), which is an infinitive form of curriculum and encourages educators and learners to engage in an autobiographical examination of themselves and their lived experiences. As Pinar states, "The method of currere reconceptualized curriculum from course objectives to complicated conversation with oneself (as a 'private' intellectual), an ongoing project of self-understanding in which one becomes mobilized for engaged pedagogical action—as a private—and—public intellectual—with others in the social reconstruction of the public sphere" (Pinar, 1978, p. 318). Currere encourages participants to "confront difficulty in order to loosen its grip" (Pinar, 2004). One exercise I always introduce to my students is an autobiographic writing/drawing/rumination about their lived experiences as a student of science. Aside from treating science as a process of sustained disciplined inquiry I have also been exploring the following strategies to truly open up possibilities for both my students and myself.

Science Education as a Healing and Restorative Experience

If we truly believe in the interconnectedness of all as featured by the David Suzuki Foundation's Declaration of Interdependence (https://davidsuzuki. org/about/declaration-of-interdependence/), we must feel deeply the pain and suffering of the millions for our scientific progress and for our pursuit of comfort, profit, pleasure, power, and dominance! The amount of damage, suffering, and trauma that have been inflicted upon so-called "others," including our fellow human beings, has been accumulating historically, and is ongoing. Therefore, we need to deepen our consciousness of the need for healing, healing of ourselves, and healing of all. Greene beautifully advocated for a "curriculum required to help provoke persons to reach past themselves and to become" (Greene, 1993, p. 220). Science curriculum and pedagogy should and can champion a commitment to a healing and restorative experience for all, starting with whomever we have in a class community, and from "the world that touches us" (Remen, 2005).

What I learned from medical school—"Do no harm"—a pledge we made on day one of attending the West China University of Medical Sciences and Technology, is a minimum ethical standard and commitment we must uphold as educators. I have also tried contemplative practices with my students in the science and environmental education classes, including journaling, music and song writing, visualization, deep listening, insight meditation, loving kindness, walking meditation, and circle of trust. These practices have been healing and restorative for me and for many of my students. As the ongoing COVID-19 pandemic and the revitalized anti-racism and global climate justice movements remind us, we have much collective trauma to confront and an increasing need for collective healing. Our classroom spaces have to not just pose no harm to anyone, but also need to be healing and restorative places for all who inhabit these spaces. If not, what else should they be?

Teaching Science as Humanities and as Narrative Knowing

Greene described "women's ways of knowing" as something that is "concrete, transactional, narrative in form." With this in recognition, approaches to science can be affected—"engagements *with* the objects of study rather than analytic work *on* them" (Greene, 1993, p. 219). Following Greene's conception of curriculum as something emerging out of "an interplay among conceptions of knowledge, conceptions of the human being, and conceptions of the social order" (p. 216), we can really make science learning into diverse human experiences of various combinations of possibilities! The aforementioned currere method is one of the ways to approach science education as humanities and narrative learning. Storying and story-telling of our lived experiences as students and teachers of science can shed some light, so does engagement with the history of science and stories of scientists of diverse cultures and backgrounds. Storying can be "an endeavor that is oriented towards liberation and transformation" (Goodson & Gill, 2014, cited in Bazzul & Siry, 2019, p. 6); one in which we re-visit/re-evaluate the past as we work to attend to the present (Ji, 2014), and construct the future (Bazzul & Siry, 2019, p. 6). Science educators/scholars/researchers from dominant groups should "stop trying to know the Other or give voice to the Other, and listen, instead to the plural voices of those Othered, as constructors and agents of knowledge" (Scott, 1991, as cited in Fine, 1994, p. 75). Teaching science as humanities and as narrative knowing/learning is an area I have immense interest in exploring further to see how to actualize this possibility for science education.

Restoring the Centrality of the Arts/Aesthetics in (Science) Education

Educational philosopher and social activist Maxine Greene called for the "centrality of the arts" (1993, p. 214) to education at all levels decades ago, yet the aesthetic core of education has been often negated to the margin or completely forgotten in public education, as shown in recent drastic school measures to respond to the COVID-19 pandemic situation by only offering literacy and numeracy instruction. What is desperately needed in formal education is what Greene termed as "wide-awakeness" or "being attentive to the beauty and cruelty of life," to "aesthetic encounters" and "living in the world esthetically" (2001, 2005). Maxine Greene seeks to define aesthetic education for us in a number of ways. For example, she wrote that aesthetic education is defined as "an intentional undertaking designed to nurture appreciative, reflective, cultural, participatory engagements with the arts by enabling learners to notice what is there to be noticed, and to lend works of art their lives in such a way that they can achieve them as variously meaningful" (Greene, 2001, p. 6).

Engagement with the creative arts and artists can bring out more diversity, which includes diverse participants and perspectives, diverse worldview and experiences, diverse forms of knowing and knowledge, diverse way of defining and solving problems, and diverse possibilities and ways to realizing what is possible. I was fortunate to experience these transformative learning opportunities decades ago as an undergraduate university student, which is documented in the Asia Art Archive project (AAA-Hong Kong, 2017). I look forward to collaborating more with eco-social justice-minded and community-based artists in this area of work. I have tremendous hope for the possibilities in centering the arts and collaborating with artists in our work as (science) educators.

CONCLUSION

This essay was drafted in the mist of the COVID-19 pandemic. At this point I am still not sure how I feel about the very term "Anthropocene" considering the tiny novel coronavirus COVID-19 (less than 0.5 micron in size) and not even classified as life (in the strict sense of Western science) has been wreaking havoc across the globe and bringing humanity to our knees in such a short period of time. Could we then say that we are now into a corona virulence after a fleeting kiss with Anthropocene? Maybe the need to focus on Anthropocene is not that urgent or even relevant—considering what we know and what we do not know. This chapter is written from the orientation championed by Bazzul and Siry (2019) that "science education needs to be for the wellbeing of communities and justice for our shared planet... and for the creation of environmentally and socially just futures" (p. 3). Collectively we need to find "openings for transformation" (p. 5). In the face of the pandemic and climate crisis we might as well focus more on the opportunities they can bring—such as collective awakening to what is essential to life/livelihood, collective healing from past and present intergenerational traumas, and greater public demand and alliance for justice. The Chinese characters for "crisis"—危机 (weiji)—illustrate, 危 (wei) means danger, and 机 (ji) means chance/opportunity. Maybe what is required is "a profound reversal in our perspective of ourselves and the universe around us" (Berry, 1999, p. 159). The few strategies I suggested above can hopefully contribute to restoring science education onto a more life-affirming and life-sustaining path that would enable all to not only survive, but also to thrive. After I engage more with what I have proposed above as a teacher of science/a science educator I hope to share my experiences and findings. I have the conviction that the aesthetics which the creative arts are best at can provide possible antidote to the "living death" plague, and can counter the forces which have been trying to enclose and foreclose the experiences we can have in this world and beyond.

REFERENCES

计划 (863 Program). https://baike.baidu.com/item/%E5%9B%BD%E5%AE%B6%E9%AB%98%E6%8A%80%E6%9C%AF%E7%A0%94%E7%A9%B6%E5%8F%91%E5%B1%95%E8%AE%A1%E5%88%92%EF%BC%88863%E8%AE%A1%E5%88%92%EF%BC%89/9529335?fromtitle=863%E8%AE%A1%E5%88%92&fromid=114257.

那些为制造核武器而隐姓埋名的科学家 (Nuclear Scientists in Cognito). https://wapbaidu.baidu.com/theme/%E9%82%A3%E4%BA%9B%E4%B8%BA%E5%88%B6%E9%80%A0%E6%A0%B8%E6%AD%A6%E5%99%A8%E8%80%8C%E9%9A%90%E5%A7%93%E5%9F%8B%E5%90%8D%E7%9A%84%E7%A7%91%E5%AD%A6%E5%AE%B6/46456912763?bk_fr=lemma.

Aamodt, A., & Bazzul, J. (2019). In the middle of treaty walking: Entangling truth, ethics, and the risky narratives of two settler(colonial)s. In J. Bazzul & C. Siry (Eds.), *Critical voices in science education research: Narratives of hope and struggle* (pp. 179–187). Cultural Studies of Science Education (Vol. 17). Springer.

Asia Art Archive (AAA). (2017). Betsy Damon archive: Keepers of the waters (Chengdu and Lhasa). 貝特西達蒙檔案: 水的保衛者 (成都與拉薩). https://aaa.org.hk/en/collections/search/archive/betsy-damon-archive-keepers-of-the-waters-chengdu-and-lhasa-11525.

Baldwin J. (1963). A talk to teachers. *Yearbook of the National Society for the Study of Education, 107*(2), 15–20. https://onlinelibrary.wiley.com/doi/pdf/10.1111/j.1744-7984.2008.00154.x.

Bazzul, J., & Siry, C. (2019). *Critical voices in science education research: Narratives of hope and struggle.* Cultural Studies of Science Education (Vol. 17). Springer.

Berry, T. (1999). *The great work: Our way into the future.* Bell Tower, New York.

Damon, E. (1995, 2012). *Keepers of the waters.* https://vimeo.com/22747563.

Damon, E. (1995, 1996, 2018). *Betsy Damon archive: Keepers of the waters (Chengdu and Lhasa).* https://aaa.org.hk/en/collection/search/archive/betsy-damon-archive-keepers-of-the-waters-chengdu-and-lhasa-11525.

Damon, E. (2016). *I am water.* https://vimeo.com/148517336.

Darling-Hammond, L. (2015). *New learning for a rapidly changing world.* https://www.youtube.com/watch?v=f14g0OG4mz0.

David Suzuki Foundation. (1992). *Declaration of interdependence.* https://davidsuzuki.org/about/declaration-of-interdependence/.

Dewey, J. (1938). *Experience and education.* Kappa Delta Pi.

Fine, M. (1994). Working the hyphens: Reinventing self and other in qualitative research. In N. K. Denzin & Y. S. Lincoln (Eds.), *Handbook of qualitative research* (pp. 70–82). Sage.

Greene, M. (1993). Diversity and inclusion: Toward a curriculum for human beings. *Teacher College Record, 95*(1), 211–221.

Greene, M. (2001). *Variations on a blue guitar: The Lincoln Center Institute lectures on aesthetic education.* Teachers College Press.

Greene, M. (2005). *Releasing the imagination: Essays on education, the arts, and social change.* Jossey-Bass.

Jacoby, J., & Ji, X. (2010). Artists as transformative leaders for sustainability. In B. W. Redekop (Ed.), *Leadership for environmental sustainability* (pp. 133–144). Routledge Studies in Business Ethics. Routledge.

Ji, X. (2011). Environmental education as the mountain: Exploring Chinese-ness of environmental education. *Australian Journal of Environmental Education, 27*(1), 109–121.

Ji, X. (2014). The heart of education and well-being is spiritual: Autoethnographic inquiry as an educational practice for sustainable well-being. In F. Deer, T. Falkenberg, B. McMillan, & L. Sims (Eds.), *Sustainable well-being: Concepts, issues, and educational practices* (pp. 121–137). ESWB Press. https://www.eswb-press.org/upl oads/1/2/8/9/12899389/sustainable_well-being_2014.pdf.

Keepers of the Waters. (n.d.). *Living water garden.* https://www.keepersofthewaters. org/projects/living-water-garden.

Kenko, Y. (1967). *Essays in idleness: The Tsurezuregusa of Kenko* (D. Keene, Trans.). Columbia University Press (Original work published ca. 1330–1332).

Kimmerer, R. W. (2013). *Braiding sweetgrass: Indigenous wisdom, scientific knowledge, and the teachings of plants.* Milkweed Editions.

Pinar, W. F. (1975, April). *The method of currere.* Paper presented at the Annual Meeting of the American Research Association. Washington, DC. https://files.eric. ed.gov/fulltext/ED104766.pdf.

Pinar, W. F. (1978). Currere: A case study. In G. Willis (Ed.), *Qualitative evaluation: Concepts and cases in curriculum criticism* (pp. 316–342). McCutchan.

Pinar, W. F. (2004). *What is curriculum theory.* Routledge.

Pinar, W. F. (2012, 2019). *What is curriculum theory?* Routledge.

Remen, R. N. (2005, August 11). *Dr. Rachael Naomi Remen: The difference between curing and healing.* On Being with Krista Tippet. https://onbeing.org/programs/ rachel-naomi-remen-the-difference-between-fixing-and-healing-nov2018/.

Schwenk, T. (2013). *Images of living water vs. dead water.* https://truespring.files. wordpress.com/2013/01/dropphotography.jpg https://stroemungsinstitut.de/ins titute/.

Tuan, Y. (1993). *Passing strange and wonderful: Aesthetics, nature, and culture.* Island Press.

CHAPTER 3

Tracing a Black Hole: Probing Cosmic Darkness in Anthropocenic Times

Jessie Beier

Seeing the Unseeable

On April 10, 2019, the team at the Event Horizon Telescope (EHT) project released an unprecedented image of a supermassive black hole at the centre of galaxy Messier 87. The image, which shows a dark disc outlined by swirling hot gas circling the black hole's event horizon, exhibits a 55-million-year-old cosmic event in the Virgo galaxy cluster—a void of stellar mass measuring some 6.5 billion times that of Earth's sun. After almost a decade of work and the collaboration of an international team of scientists, a network of radio telescopes synched through atomic clocks and the creation of a powerful algorithm capable of correlating and calibrating huge amounts of data while sifting through "noise," the image produced by the EHT has been offered as unprecedented "visual evidence" of what has, until now, been invisible—evidence of the event horizon of a black hole (Drake, 2019). As EHT director and astrophysicist Shep Doelman put it when introducing the image, "[w]e are delighted to be able to report to you today that we have seen what we thought was unseeable. We have seen and taken a picture of a black hole" (cited in Chappell, 2019) (Fig. 3.1).

While the news of this extraordinary image was but a blip on newsfeeds at the time, its significance has reverberated throughout scientific communities, where the picture has been hailed as a "precious" techno-scientific discovery

J. Beier (✉)
University of Alberta, Edmonton, AB, Canada
e-mail: jlbeier@ualberta.ca

© The Author(s) 2022
M. F.G. Wallace et al. (eds.), *Reimagining Science Education in the Anthropocene*, Palgrave Studies in Education and the Environment,
https://doi.org/10.1007/978-3-030-79622-8_3

Fig. 3.1   Radio image of the black hole located in Messier 87, EHT Collaboration (2019)

that affirms some of the most important theories subtending the field of astrophysics (Chappell, 2019). For instance, Einstein's theory of general relativity and its related calculations were used to predict the size and shape of the black hole's event horizon, which was then confirmed with the image, resulting in a renewed confidence in the physics that is said to drive the large-scale structure of the universe (Chappell, 2019). While there have been extensive simulations created in order to speculate on how the laws of physics might extend to the outer limits of deep space, black holes have never been directly observed, and thus have not, until this image, been empirically "proven." With this in mind, the 2019 image has been deemed precious because, as Heino Falcke, chair of the EHT Science Council put it, "this one is finally *real*" (my italics, cited in Chappell, 2019).

The image has also been deemed significant based on the technical ingenuity required to visually represent such a distant and compact cosmic event. In order to reach into the depths of spacetime, the globally dispersed EHT team needed to create an extremely large and powerful telescope, one at least 10,000 kilometres in diameter or almost as big as planet Earth. Due to the material impossibility of such an endeavour, the team responded by developing the EHT, a computational telescope, which, combined with a powerful algorithm, is capable of resolving structure on the scale of a black hole's event horizon.

Beyond affirming fundamental scientific theories and demonstrating what is possible through well-funded and technologically innovative international scientific collaboration, the image has also been celebrated for the cosmological presumptions it seemingly affirms, that is, presumptions about humanity and our role and place in the cosmos. For instance, upon its release, the image was dubbed a beacon of hope, a "ray of light" in these dark times (O'Hagan, 2019). As *Guardian* journalist Ellie May O'Hagan (2019) put it on the day of the image's publication:

> On our own turquoise speck in the cosmos, we're living through ecological breakdown, the rise of authoritarianism and the appallingly unequal distribution of resources. We're devoting a lot of time and energy to destroying our home and to hurting one another. It can be hard not to feel like the human race has become trapped in its own event horizon, and we're inevitably and inexorably being pulled towards the darkness. The grainy image those scientists released on Wednesday reminds us of something different.

Counter to the melanoheliophobic narratives that often circle around discussions of black holes (for instance, at the press conference where the image was introduced the black hole was described as "[t]he gates of hell, the end of space and time,") O'Hagan highlights a more optimistic narrative, one wherein this scientific and technological feat is a testament to the ingenuity, curiosity, and resourcefulness of the human species. In this narration, the black hole image is not only celebrated for its techno-scientific implications, but, importantly for this investigation, also for the optimistic narratives of human progress and futurity it seemingly affirms.

In what follows, I continue to trace the image produced by EHT in terms of such narratives, but also in terms of the counter-narratives that might be speculatively deployed when questions of human progress and futurity are put in contact with the strange logic of black holes and their imaging. This tracing is situated within what has been labelled (contentiously, as we will see) the Anthropocene, a geologic era wherein the magnitude, variety, and longevity of human-induced transformations have emerged as a geological force now altering the planet's climate and environment (Lewis & Maslin, 2015). More specifically, this tracing is situated within the milieu of (science) education today and takes off from the increasingly common, if uncomfortable, claim that contemporary educational domains are ill-conceived and ill-equipped to deal with the urgent, if illusive, concept of the Anthropocene and thus what is required is a "substantial rethinking [of education]—of its content, its purposes and its relationships" (Gilbert, 2016, p. 188).

With this call for "substantial rethinking" in mind, this tracing brings the black hole image in contact with the site of (science) education today in order to experiment with how questions of sustainability and, ultimately, educational futurity might be *resituated* given the planetary shifts on the horizon (or that are, in many cases, already here). Through a series of brief forays into black hole physics, historical examples of cosmic imaging and further exploration of the image created by EHT, this tracing outlines the black hole and its apparent horizons in order to propose a strange vantage point from which pedagogical problem-posing might be interrupted, mutated, and relaunched.

By turning to that which lies outside of the traditional science classroom—beyond the school, beyond curriculum, beyond state-imposed policy, national rankings and international benchmarks, indeed, beyond the planet itself, deep into the cosmos—this tracing seeks to probe this black hole event in terms of its weird and weirding pedagogical trajectories so as to speculate on unthought possibilities for resituating (science) education in the age of the Anthropocene.

Apparent Horizons: Cosmological Shifts, Pedagogical Resituation

A black hole is typically described as what remains after a star is unable to resist gravity and collapses inwards. This process, sometimes called the "death" of a star, is more accurately the result of what happens when high-mass stars use up their fuel at a pace that creates a supernova explosion, resulting in an altered gravitational state. As a more common naming of what had been previously called Totally Gravitationally Collapsed Objects, the name "black hole" references both light and gravity: black holes are considered "black" because they are a place in space where gravity pulls so much that light cannot get out, and they are considered "holes" because the dense inner region of a black hole, known as its singularity, permanently warps spacetime, thus creating a "hole" in the fabric of the universe. As such, a black hole does not have a surface, but is instead defined by a special boundary called an event horizon. The pedagogical example often used to simulate this strange cosmic boundary is the one where an astronaut shines a flashlight on either side of the black hole's event horizon. If the flashlight is outside of the event horizon, then the light rays are able to escape the pull of gravity and thus can be seen from a distance. If the flashlight is at or inside of the event horizon, however, the light cannot escape the gravitational pull produced by the collapsed star, and thus any light emitted is trapped inside of the black hole. It is this threshold between (human) visibility and invisibility that defines the event horizon of a black hole.

Of course, it is not exactly the case that the light is "trapped," but rather that it is unobservable from the vantage of the astronaut. This realization shines light on, so to speak, one of the most interesting, if vexing, scientific debates surrounding black holes and their cosmological implications. That is, physicists remain largely undecided as to whether the prediction of black holes and their singularities, supported as they are by fundamental theories about the "laws" and structure of the universe, actually exist (or existed at some point), or if it is the case that current knowledge and theories of the cosmos are simply insufficient to describe what happens at such extremely dense points in spacetime. As such, the very concept of the event horizon—that special boundary beyond which events cannot affect a human observer—is still up for debate, catalyzing difficult questions about, for instance, the nature of photon spheres, black hole thermodynamics, Hawking radiation, and information-loss paradoxes, to name but a few research trajectories.

Stephen Hawking (2014), for instance, suggested that the very idea of an event horizon should be replaced with what he called "apparent horizons," an assertion founded on the claim that quantum effects around a black hole cause spacetime to fluctuate too wildly for a sharp boundary surface to exist. In contrast to an event horizon, which refers almost exclusively to the possibility of an absolute horizon, that is, one defined teleologically in terms of an asymptotically flat spacetime, an apparent horizon is instead dependent on the "slicing" of spacetime. Put another way, unlike an event horizon, which relies on an absolute horizon—one that is very geometrical and requires the full history (all the way into the future) of spacetime in order to be known—the very location and even existence of an apparent horizon instead depends on the way that spacetime is itself divided into space and time. Put yet another way, which is sometimes necessary when grappling with astrophysical concepts (at least for me), apparent horizons are not invariant and immutable properties of spacetime that can be known in advance, but are instead local and observer-dependent—the boundary at *this* instant—whereas an event horizon is the boundary of a black hole for light in the future, where the future is defined teleologically by the laws of physics.

What is important to note here is not how this distinction works to deny the existence of black holes and their singularities, but rather, how the definitions and narrative contingencies of scientific phenomena are interpolated within a set of broader cosmological presumptions. This is not to say that there are no particular facts or regularities between certain certainties, that everything can be reduced to social constructions and cultural tropes, but rather, that the "laws" and definitions that chase these regularities are always contingent. In the case of black holes, for example, the very definition of a black hole varies radically, often in conflicting ways, based on the disciplinary background and cosmological milieus from which such definitions emerge (Curiel, 2019, p. 27). As Erik Curiel (2019) suggests, this uncertainty around black hole definitions is not so much about the lack of a single, canonical answer but rather that "there are too many good possible answers to the question, not all consistent with each other" (p. 27). For Curiel, then, the multitude of definitions is an important virtue of research related to black holes, not something to be resolved or unified. What matters, then, is how such definitions are purposefully interpolated into scientific practice across domains and for different theoretical, observational, and foundational contexts (Curiel, 2019, pp. 33–34).

In her essay *Platform Cosmologies: Enabling Resituation*, artist, writer, and designer Patricia Reed (2019) provides a provisional framework for understanding such interpolations, honing in on the relationship between techno-scientific development and cosmological shifts. In this essay, Reed works to think through the conditions for enabling resituation given today's planetary realities by asking how techno-scientific change manifests, how it is narrated, how it is oriented by purposes, and, ultimately, how such purposes might be reoriented towards a desired otherworld (p. 27). Drawing on the work of political scientist Bentley Allan, Reed highlights how "despite the import of scientific development as a catalyst for novel understandings of the world, it's not until these developments become interpolated into a cosmological order that they begin to influence general purposefulness within it" (p. 28). In this way, Reed draws attention to the role and import of the narrative contextualization of techno-scientific developments and how such narrations work to operationalize new knowledge, both socially and politically, in turn impacting broader cosmological presumptions. For Reed, this focus on narrative contingency marks an important difference between merely acquiring new knowledge about the world and *existing in* that knowledge, which, she notes, "entails working out *how* one navigates the world anew with this knowledge, as well as its instrumentalization" (p. 28).

In the case of a black hole and its unprecedented imaging, then, Reed might say that the importance of such a discovery lies not only in what it can tell us about ourselves and our place in the universe, but also in how such narrations are purposefully extended into particular domains so as to affirm (or negate) cosmological presumptions. In this way, the question of enabling resituation, that is, enabling the interpolation of new knowledge and cosmological shifts adequate to the pressing challenges on the horizon today, is not just about knowing more or even knowing differently, but instead, the "cosmological stakes lie in making claims on the construction of social, political, and ethical narrations to ramify the meanings we ought to extrapolate from them, influencing a cosmological milieu we need for their just deployment" (p. 29). Returning back to the specific cosmological milieu of this investigation, to the site of (science) education in/and the Anthropocene, I draw upon Reed's platform for enabling resituation in order to experiment with what it might mean for educational thought to *exist in* an encounter with a black hole. It is through this encounter that I aim to grapple with the difficult question of pedagogical resituation today.

While calls for rethinking, re-orienting, re-imagining, etc.... (science) education are becoming more and more pronounced against the backdrop of transformed and transforming planetary realities, responding adequately to such calls remains incredibly difficult. This is not only due to the immense effort, energy, and time required, but because such thinking is itself limited by what it cannot think, or as educational theorist Jane Gilbert (2016) notes, "the conceptual categories that structure our thinking are themselves part of the problem" (p. 188). For Gilbert (2016), this means any sort of educational re-imagining must confront its own assumptions and limits: "we need to look at ourselves to dig up some of our assumptions about science education—what it *is* and what it is *for*—and [thus] our assumptions about science, education, society and the future" (p. 190). Such a confrontation not only involves grappling with the difficult task of interrogating cosmological presumptions and their pedagogical ramifications, but importantly, I wager, this confrontation must experiment with putting educational thought in contact with what it is currently unable to think—in contact with the (apparent) horizons of thinking itself—so as to put pressure on the conceptual categories that delimit thinking in the first place.

As Reed notes, it is through the transformative activity of thinking itself, of constructing concepts in relation to objects, that we might be resituated (p. 30). That is, through the mutual formation between thought and its (apparent) object, "the situatedness of all thought (located within a particular cosmological milieu) always contains the possibility of overflowing its situation" (Reed, 2019, p. 30). It is through this spillage that alien milieus and yet unthought pedagogical trajectories open up. In what follows, I endeavour towards such encounters and experimentation by bringing the black hole image produced by EHT in contact with this question of pedagogical resituation, and specifically, with some of the conceptual categories and narrative contingencies that undergird the (Good) Anthropocene and its visions for a sustainable education.

Messages to Humanity: From Earthrise *to* Pōwehi

Upon its release, the image produced by the Event Horizon Telescope team was quickly placed into the lineage of those other iconic environmental photographs—for example, *Earthrise* (1968), *The Blue Marble* (1972), *The Pale Blue Dot* (1990)—that have "both captured the public imagination and offered scientists insight into how the universe works" (Reuell, 2019). As EHT science council member Dan Marrone noted at the time of the image's release, "photos can change the way we think about ourselves and our place in the universe." To support this claim, Marrone likened the black hole image to *Earthrise*, a photo taken by Apollo 8 astronaut Bill Anders in December of 1968 (Fig. 3.2).

Earthrise is often cited as both an inspirational and aspirational image, widely credited for helping to spur the environmental movement and for influencing scientific and ecological theories, such as Lovelock and Margulis' Gaia Hypothesis.[1] Captured during the first manned mission to the moon, *Earthrise* shows a small and delicate Earth rising over a rugged grey lunar surface, "a magnificent spot of colour in the vast blackness of space" (Anders, 2018). Upon encountering the wondrous view during lunar orbit, part of which was broadcasted live into homes across America on December 24, 1968, Apollo 8's mission astronauts expressed feelings of humility, gratitude, and admiration as they described the unprecedented sight. Command Module Pilot Jim Lovell, for instance, stated that "the vast loneliness is awe-inspiring and it makes you realize just what you have back there on Earth" (cited in Williams, n.d.) In short, and as adventure photographer Galen Rowell has asserted, *Earthrise* is

---

[1] The Gaia Hypothesis proposes that the Earth and its organisms form a synergistic complex system that self-regulates so as to maintain and propagate the conditions for life. By providing a view of this complex system as a whole from the vantage of space, Earthrise has been attributed to influencing this hypothesis.

Fig. 3.2   Earthrise, NASA (1968)

considered "the most influential environmental photograph ever taken" due to the way in which it has inspired contemplation about our fragile existence and our place in the cosmos (cited in Coulter, 2009). Since its release, *Earthrise* has not only continued to rank high on lists of era-defining images, but has also become a common representational stand-in for environmental protection and planetary stewardship. Showing up on NASA-endorsed stamps and influential magazines such as *Life* and *Time*, as well as in countercultural texts ranging from Stewart Brand's *Whole Earth Catalogue* to Joni Mitchell's 1976 song "Refuge of the Roads," this view of Earth from a distance has been called a "message for humanity," a message of both our cosmic insignificance and the precious nature of this planet we call Home.

It is to this lineage that Marrone was perhaps referring, one wherein the black hole image not only provides evidence of an unprecedented encounter with the mysteries of the cosmos, but, more importantly, acts as a catalyst for bringing about broader social, cultural, and cosmological shifts. Fast forward just over 50 years, however, and it is clear that while *Earthrise* has become a common trope within environmental communications, the notion that this image has been an impetus for more care-full planetary stewardship is today dubitable, particularly given the increasingly disastrous state of the planet. As crises converge and proliferate across the globe—i.e., uncontrollable fires and historic flooding but also unprecedented droughts, the end of farmable land, food scarcity, climate plagues and global pandemics, unbreathable air, poisoned oceans, the list goes on and on and on—it appears that *Earthrise's* "message to humanity" has gone all but unanswered.

Resituated within the cosmological milieu of the Anthropocene, *Earthrise* no longer tells a story of cosmic humility and planetary protection, but instead spins a tale of anthropic omnipotence, one narrated from a God-like perspective, a "view from nowhere" as Donna Haraway (1988) might have it, that unifies vision across time and space from but a single (all-too-human) perspective. Returning to the Apollo 8 Christmas Eve broadcast, it is perhaps interesting to note that after sharing pictures of the Earth and Moon as seen from lunar orbit, the live broadcast (the most watched TV programme at the time) ended with the crew taking turns reading from the book of Genesis (Williams, n.d.). Refracted through this God-like view, with its fantasies of both dominion and salvation, the Earth in *Earthrise* is transformed from a precious "blue dot" in need of protection to a mutable object over which "we" can exert control and domination, extracting and exploiting for anthropocentric ends. Not unlike the image itself, which was edited so as to exaggerate the Earth's presence (NASA flipped the photo and cropped it in order to make the Earth a bigger focal point[2]), from the vantage of the Anthropocene, *Earthrise's* "message to humanity" is manipulated to satisfy anthropocentric ends. From this vantage, *Earthrise* no longer reads as a symbol of environmental stewardship, but instead, as a stand-in for assumptions about human control, manipulation, and ultimately, mastery over planetary life.

With this in mind, the case may be made that the black hole image follows in the lineage of *Earthrise* based on how it too tells a story of human mastery, cosmic frontierism, and an understanding of progress defined by "the recursively institutionalized, neoclassical economic ideal that 'human progress [is equal to] the unleashing of scientific and technological progress itself'" (Reed, 2019, p. 29). Indeed, this narration was characteristic of the mainstream dispatches that circulated upon the image's release. As O'Hagan (2019) wrote in her optimistic report, "[a]s far as we know, we are alone in the infinite darkness—yet we have not succumbed to despair of this terrible possibility. Instead, we have looked out into the darkness, transfixed by its mystery. The universe is spellbinding and miraculous, *but so are human beings*" (my italics). Like *Earthrise* before it, the narrative contingencies emanating from the black hole image tell a story of an ingenious human species, one that is discrete and separate from the cosmos, but nevertheless able to image its deep mysteries. It is this characterization of humanity, situated as it is within a broader set of cosmological presumptions, through which the Anthropocene itself is often narrated, particularly in its "Good" iterations.

The idea of the "Good Anthropocene" originated with Erle Ellis, a landscape ecologist and senior fellow at The Breakthrough Institute, who champions what he calls "postnatural environmentalism" and the assertion

[2] see Moran (2018).

that we should "forget Mother Nature" and recognize that "this is a world of our making" (Ellis, 2009, 2011). According to Ellis, if we first "stop trying to save the planet," then we can embrace the Anthropocene and our new role in it as "the creators, engineers, and permanent global stewards of a *sustainable* human nature" (my italics, Ellis, 2011). In many ways, Ellis' "Good Anthropocene" is perhaps just a more explicit articulation of how the Anthropocene has been narrated more generally, that is, as the expansion and extension of yet another epic story where humanity—*anthropos*—is not only capable of transforming planetary realities, but more importantly, "if 'we' discover ourselves to be an agent of destruction, then 'we' must re-form, re-group and live on" (Cohen & Colebrook, 2016, p. 9).

Whereas the introduction of the Anthropocene might otherwise be positioned as a conceptual "shock," "a point of bifurcation in the history of the Earth, life and humans [that] overturns our representations of the world" (Bonneuil & Fressoz, 2016, p. 29), a "Good Anthropocene" is "one in which humans can be proud of their achievements rather than lose too much sleep over the side effects" (Ellis cited in Zylinska, 2018, pp. 16–17). As Joanna Zylinksa (2018) writes in her proposal for a feminist counter-apocalypse, the Anthropocene not only signals our presently unfolding planetary emergency, but also offers a "new epistemological filter through which we humans can see ourselves" (p. 3). Zylinska discusses how dominant Anthropocene narratives fail to recognize how their own discursive tropes and points of reference "bring forth a temporarily wounded yet ultimately redeemed *man* who can conquer time and space by rising above the geological mess he has created!" (my italics, Zylinksa, 2018, p. 12). Evidenced through, for instance, the immortality projects of Silicon Valley and the interstellar colonization projects of Elon Musk's SpaceX, the Anthropocene ushers in narratives wherein something like anthropogenic climate change merely requires a "technical fix," and thus, *anthropos* himself is also "fully fixable" through projects aimed at human upgrade and redemption (p. 18). As Zylinska highlights, in the tale of the (Good) Anthropocene the human is reinstalled as the protagonist, the ultimate hero of the story, the moral of which is that "we made this" and so too will we overcome it.

Viewed through this rose-coloured filter, the image of the black hole might be positioned as support for this narrative, one wherein humanity is not only capable of imaging the Earth from a distance, but of controlling and transforming its geology, climate, and future trajectories. However, if I have learned anything from black holes so far, it is that this is but one way of approaching its apparent horizons. As Reed (2019) might remind us, "the meaning and comprehensive significance of scientific development is not as matter-of-fact as the concrete discovery itself" (p. 29). There are alternative narrations emanating from this stellar object, albeit ones that have been necessarily obfuscated so as to affirm and reproduce dominant cosmological orders. After all, what was produced by the EHT does not technically show visual evidence of a black hole, for as we saw (or perhaps didn't see), black holes are, by definition, completely dark incredibly dense objects from which even light cannot escape. Instead, the image shows a shadow, a boundary of perceptibility that manifests in darkness, forever exceeding human representational schema. This sense of darkness has been expressed through the designation of the black hole itself, which was bestowed with the name Pōwehi, a Hawaiian word that comes from the *Kumulipo*, an eighteenth-century creation chant, and means "the adorned fathomless dark creation" or "embellished dark source of unending creation" (Mele, 2019). In contrast to images of human curiosity and wonder, of enlightenment and insight, in the narrative of Pōwehi, the black hole image instead signals a horizon—a limit case—for human sense-making apparatuses.

This limit is evidenced through the creation of the EHT image itself, which not only required an international team of human scientists, but, perhaps more importantly, necessitated a complex meshwork of inhuman and non-human sense-making assemblages. In order to capture the image, the EHT project needed to create an incredibly large telescope capable of reaching galaxy Messier 87, which involved the help of a powerful machine-learning algorithm that could stitch together data collected from telescopes scattered around the globe while also discerning "good" images from "noise" and calculating for hidden delays in astronomical signal processing. While the working parts of this inhuman/non-human assemblage are too extensive to detail here, a starting list includes: 8 radio telescopes, atomic clocks (a.k.a. hydrogen masers), thousands of terabytes of data, highly specialized supercomputers, three different algorithmic imaging methods, and 13 stakeholder institutions (plus over 60 affiliated institutions). And, if we were to unravel each of these elements, the list would, of course, grow exponentially, revealing a complex network of partial objects and machinic flows that challenge the very notion of creating a bulleted list.

By interpolating the black hole image through its technical apparatus, the narrative contingencies that support the cosmological presumption that this technological feat reflects an all-too-human ingenuity is made vulnerable. Taking this into account, the 2019 image of the black hole may, once again, follow in the footsteps of *Earthrise* as an iconic environmental image—a representation of these anthropocenic times—albeit one that pushes back against (Good) Anthropocene impulses and desires. In contradistinction to fantasies of anthropic omnipotence and planetary control, the black hole image, in both its form and content, instead affirms a cosmological milieu characterized by rhythms and patternings of inhuman and non-human cosmic forces far beyond our perception, let alone control. In this narrative, the human species is resituated as but a partial component in a dynamic network of connections that resist God-like, totalizing views and tidy causal determinations. Resituated in this way, the 2019 black hole image might indeed be put forth as an important "message to humanity," alerting us, for instance, to the ways in which "[t]he 'human' *is* not, except as an effect of complex, shifting, becoming relations among a multiplicity of actants" (Weaver & Snaza, 2016, p. 1063).

From this resituated position, the pedagogical question is not just about how to better explain or theorize or understand this knowledge, but how to *exist in* this knowledge, that is, how to navigate the alien milieus that are opened up when the situatedness of thought exceeds the commonsense categories and delineations that have come to undergird a given cosmological milieu. As Reed writes, "situated thought may always stem from a partial position embedded within a particular (cosmological) milieu, but neither that position nor that milieu is fixed absolutely; it is subject to the forces of thought's speculative mobility to permeate the given and render its subtending mythical grounds temporary" (p. 30). In the case of the black hole, the question is thus how to *sustain* encounters with this force of thought and the risks and uncertainties that come with it.

Alien Territories: Thwarting Laplacean Dreams, Resituating Sustainability

It is here where the question of (science) education returns to the forefront. Of the narrative contingencies traced above, it might be wagered, as I do here, that dominant educational discourses today conform to (Good) Anthropocene vibes, in turn dejecting and obscuring some of the more unfathomable aspects that might otherwise be raised by the Anthropocene designation. While ecological catastrophes gather force and speed around the globe, alongside a range of distressing social, political, and psychological crises, education (and its reasons) has yet to fully confront the bleak realization that the planet is undergoing uncertain and catastrophic transformations. Indeed, one of the main, if obscured, stakes raised by the current moment of ecological degradation concerns the very existence of humans on planet Earth, or to put it bluntly, "the emerging path that we are collectively clearing that leads towards the end of human earthly habitation" (Heimans, 2018, p. 6). As Jason Wallin (2017) notes, the significance of educational research today should therefore be predicated on its ability to engage with the ecological, economic, and political challenges of the Anthropocene. As he writes: "to deny that education must be fundamentally rethought in relation to such ecological complexity marks a failure to engage not only the challenges to human and non-human life intimate to the [A]nthropocene, but a reluctance to forge a speculative encounter with the quite real potential of human extinction" (p. 1100).

This non-confrontation with today's planetary realities is not only evidenced by education's more conservative approaches, but is also made manifest by well-intentioned "progressive" approaches, many of which take cues from (Good) Anthropocene narratives and impulses, perhaps unknowingly. For instance, education is often nominated as an important player in "sustainable development" and, as such, it is positioned as one of the most "powerful tools for transformation, in order to make the Anthropocene long-lasting, equitable, and worth living" (Leinfelder, 2013, p. 10). Here, education is oriented towards "envisioning a sustainable human presence on Earth in which humans would no longer be 'invaders' but rather participants in shaping the natural environment" (Leinfelder, 2013, p. 9). The cosmological presumptions of this orientation are in line with those of broader (Good) Anthropocene narratives wherein humans are positioned as geological agents that must "take this agency upon themselves and direct it better than it has thus far been directed" (Snaza, 2018, p. 339). Within this narrative, the Anthropocene is not seen as a "'crisis to end all crises,' the catalyst needed to provoke real change" in educational domains, as Gilbert (2016, p. 188) puts it, but is instead positioned as an opportunity to reform and responsibilize education towards more sustainable practices.

As such, this narrative not only downplays and denies the current reality of anthropogenic climate change and its coming (or in many cases, current) annihilations, but provides a breeding ground for a "new" education to emerge, an education *after* education, wherein our current planetary predicament is simply an aberration in need of rehabilitation and re-engineering. Within this redemptive narrative, it is simply assumed that education will forever be founded on the project of (human) sustainability and an ultimately positive, or "Good," future for "us." In turn, potentials for conceiving educational futurity from the unfathomable perspective of, for instance, human extinction and ecocatastrophic abolition, are disappeared and occluded in the name of the (Good) Anthropocene's imperative to sustain current cosmological orders, no matter the cost. Put bluntly, a confrontation with the pressing issues of insufficiency, finitude, diminishment, divestment, and extinction that might otherwise re-orient educational futurity in these anthropocenic times is disappeared and replaced with a vision of times-to-come that are always-already given in adequation to a present of which the future is a mere perpetuation.

Caught amidst this perpetual present, attempts to rethink (science) education must therefore not only interrogate current educational situations, but must actively experiment with speculative pedagogical trajectories that involve processes of alienation that are able to "denaturalize the givenness of a certain worldpicture, our position, and the understanding of our agencies within it" (Reed, 2019, p. 28). As Reed (2019) asserts, knowing about something like anthropogenic climate change is very different than *existing in* that knowledge, raising the important, if uncomfortable, point that the necessary illusions and obfuscations through which the (Good) Anthropocene is realized do not suddenly dissolve and transform because we are shown, scientifically or otherwise, differently. With this in mind, the task of rethinking education is not just about knowing more or even knowing differently, but instead about creating speculative methods and hypothetical probes that might catalyze processes of alienation from one world to a desired otherworld. It is here where the image of Pōwehi—that image of "adorned fathomless dark creation"—returns to the frame.

In addition to providing an index of anthropic limits and the power of inhuman and non-human planetary sensing, the black hole image produced by EHT offers one site where practices of speculative alienation might be exercised so as to interrupt and mutate some of the inheritances that have come to define and delimit possibilities for rethinking (science) education. As just one example of such inheritances, and to bring us back into the world of black hole physics once again, we might look, for instance, to something like causal determinism. Causal determinism, also known as "Laplace's demon," is the theory that all physical events in the universe are determined completely by previously existing causes, or the idea that every event is necessitated by antecedent events and conditions alongside the "the laws of nature" (Weaver & Snaza, 2016, p. 1055). The pedagogical example used to explain this goes something like this: if some omnipotent being (the demon) could account for the precise location and momentum of all matter in the universe, both the past and future state of this mattering could be predetermined in a causal manner through the formulas and calculations provided by the "laws" of physics. Within this narrative of the cosmos, the future is not inherently unknown, but rather, its unknowability is the result of inadequate information. As such, causal determinism hypothesizes that if we could gather enough data alongside the ability to sort through and manage that information, we would be able to account for all of time. Thinking back to the distinction between event and apparent horizons, it is partly through the narrative contingencies of casual determinism that the very existence of absolute horizons, wherein the future is defined teleologically, is supported.

Causal determinism undergirds much scientific theory and practice, but also extends and intersects with a larger tapestry of philosophical, social, economic, political, and cultural significations. Alex Garland's (2020) timely science-fiction miniseries *Devs* provides one example of such intersections, in this case presenting a speculative account of what might happen if Laplace's demon were to be actualized by harnessing the power of big data and quantum computing. (Spoiler alert, it doesn't end well.) Theories of causal determinism also have educational significations and implications. As educational theorists John Weaver and Nathan Snaza (2016) assert, for instance, the "Laplacean dream"—one wherein reality can be ordered around the "the certainty of facts, the predictability of the future, the stability of isolated phenomenon, the universalization of mathematical thinking [and] the necessity of hypothetical/deductive thinking" (p. 1055)—has come to undergird the very practice of educational research. Further, it is, in part, the "Laplacean dream" that foregrounds postures of anthropic omnipotence and thus the possibility of seeing the Earth from a distance, seeing it as a known "object" over which an all-knowing transcendent "subject" (called humanity) is able to capture, manipulate, and control planetary realities. This deterministic dream not only prioritizes consistency over contingency, standardized and standardizing techniques, and a posture of anthropocentrism that erases phenomena that cannot be accounted for in deterministic ways, but also positions the future as something that can be predicted and controlled, so long as we have adequate information and the correct calculations. Within this narrative, questions of sustainability are driven by the cosmological presumption that "we" humans will be able to overcome the unsustainability of the present world order through careful accounting, strategic risk-management, and mitigation strategies so as to sustain present political, social, and ecological orders.

The black hole image, however, provides an unsettling reminder of how the conceptual categories to which we have become accustomed—be it "Laplacean dreams," illusions of well-intentioned geologic agency or fantasies of a redeemed and redeeming educational future—are narratively contingent and thus vulnerable to alternative speculations. By thwarting the "Laplacean dream" of isolated phenomena, objective stability, and the universalization of mathematical calculation, for instance, the black hole image provides an encounter with a potential limit case for scientific intelligibility, rendering epistemology itself mute and in need of new means of hypothetical expression. Put another way, the black hole image provides "proof," albeit paradoxically, of a physical phenomenon that puts at risk what, and on what basis, we can claim to "know" things in the first place. At the same time that the image provides "visual evidence" of a cosmic event that affirms fundamental cosmological presumptions about the universe, it simultaneously provides evidence of an actual limit case for testing such hypotheses. As such, no matter how complex the technological mediations and interpretations, no matter how nuanced the theorizations of, for instance, quantum states and apparent horizons might be, it cannot be determined once and for all whether black holes do, in fact, conform to the "laws" of science.

After all, and to paraphrase Steven Hawking (2008), if an object were to enter the event horizon of a black hole, we wouldn't be able to predict the future of that object because there is no telling what occurs at the event horizon. As Hawking (2008) quips: "[i]t could emit a working television set, or a leather bound volume of the complete works of Shakespeare, though the chance of such exotic emissions is very low. It is much more likely to be thermal radiation, like the glow from red hot metal" (para. 14). What is significant here is that if causal determinism, and thus the very "laws" of science, do not hold in this instance then there is always the possibility they might not hold in other situations. As Hawking (2008) puts it:

> [t]here could be virtual black holes that appear as fluctuations out of the vacuum, absorb one set of particles, emit another, and disappear into the vacuum again. Even worse, if determinism breaks down, we can't be sure of our past history either. The history books and our memories could just be illusions. It is the past that tells us who we are. Without it, we lose our identity. (para. 14)

With Hawking's words in mind, the black hole image not only provides "evidence" of a limit case for scientific intelligibility, but, importantly for this tracing, conjures a weird encounter with an unthinkable thought—with that which we cannot even begin to think—an encounter with an unfathomable event wherein the cosmos itself ties a knot that cannot be untied. What this encounter reveals is not only the limits of human exploration and intelligibility, but also how the commonsense categories, delineations, and correlations through which we make sense of the world on a more daily basis are perhaps much, much stranger than we like to think. Or as theoretical physicist often attributed to popularizing the term "black hole," John Wheeler (2000), puts it, "[the black hole] teaches us that space can be crumpled like a piece of paper into an infinitesimal dot, that time can be extinguished like a blown-out flame, and that the laws of physics that we regard as 'sacred,' as immutable, are anything but" (p. 298). In short, in this weird narration, the black hole image provides an encounter with the perplexing thought that "[n]othing can be said, here and now, to be impossible or to be closed down or determined once and for all [...] the only impossibility is the determination in advance that certain events would be impossible" (Colebrook, 2016, p. 103).

Returning to the question of rethinking (science) education in/and the Anthropocene, the pedagogical task is how to *exist in* such a thought, that is, how to *sustain* an encounter with the alien milieus that open up, "milieus that are both bristling with the possibilities and fraught with the risks and uncertainties that come with it" (Reed, 2019, p. 30). For me, this is a very different notion of sustainability than that which is most often proffered by (Good) Anthropocene narratives and their advocates. In this narrative of sustainability the goal is not to maintain and perpetuate current cosmological orders, but instead, to sustain the active practice of confronting and re-working the pain and pleasure, the urgency and exhaustion, the deep discomfort and existential uncertainty that comes with *existing in* knowing, or rather, *existing in* not knowing. With this resituated notion of sustainability in mind, and as a way to conclude, the black hole image offers a site wherein the promise of a "new," more sustainable (science) education *after* education, is dislodged and made vulnerable to revision and, perhaps, rethinking. As Reed (2019) concludes in her own investigation of enabling resituation, while it is "indispensable to map the diagnostic terrain we find ourselves in, including the cosmological predispositions that have legitimized these processes, [...] it is equally crucial to speculate on the locations, means and alternative narrations to make the current, entirely destructive path we are on an object of history" (p. 34). Taking cues from Reed one last time, the 2019 black hole image is not offered up here as a new framework or metaphor for (science) education, but instead provides just one experimental site (we might also look to reality TV, financial derivatives, or glacial melt) for navigating alien territories and developing sustainable practices capable of speculating on education and its future(s) otherwise.

REFERENCES

Anders, W. (Photographer). (1968, December 24). *Earthrise.* http://www.hq.nasa.gov/office/pao/History/alsj/a410/AS8-14-2383HR.jpg.

Anders, B. (2018). 50 years after 'Earthrise,' a Christmas Eve message from its photographer. https://www.space.com/42848-earthrise-photo-apollo-8-legacy-bill-anders.html.

Bonneuil, C., & Fressoz, J. (2016). *The shock of the Anthropocene: The earth, history, and us.* Verso.

Chappell, B. (2019). Earth sees first image of a black hole. *NPR.* https://www.npr.org/2019/04/10/711723383/watch-earth-gets-its-first-look-at-a-black-hole.

Colebrook, C. (2016). What is the anthropo-political? In T. Cohen, C. Colebrook, & J. H. Miller (Eds.), *Twilight of the Anthropocene idols* (pp. 81–125). Open Humanities Press.

Coulter, D. (2009). Exploring the moon, discovering Earth. *NASA Science.* https://science.nasa.gov/science-news/science-at-nasa/2009/17jul_discoveringearth.

Cureil, E. (2019). The many definitions of a black hole. *Nature Astronomy, 3*(1), 27–34.

Drake, N. (2019). First-ever picture of a black hole unveiled. *National Geographic.* https://www.nationalgcographic.com/science/2019/04/first-picture-black-hole-revealed-m87-event-horizon-telescope-astrophysics/.

EHT Collaboration. (2019, April 10). *First image of a black hole.* https://www.eso.org/public/images/eso1907a/.

Ellis, E. (2009). Stop trying to save the planet. *Wired.* https://www.wired.com/2009/05/ftf-ellis-1/.

Ellis, E. (2011). Forget mother nature: This is a world of our making. *New Scientist.* https://www.newscientist.com/article/mg21028165-700-forget-mother-nature-this-is-a-world-of-our-making/.

Garland, A. (Director). (2020). *Devs* [TV miniseries]. DNA TV.

Gilbert, J. (2016). Transforming science education for the Anthropocene—Is it possible? *Research in Science Education, 46*(2), 187–201.

Haraway, D. (1988). Situated knowledges: The science question in feminism and the privilege of partial perspective. *Feminist Studies, 14*(3), 575–599.

Hawking, S. (2008). *Into a black hole.* http://www.hawking.org.uk/into-a-black-hole.html.

Hawking, S. (2014). Information preservation and weather forecasting for black holes. arXiv:1401.5761v1.

Heimans, S. (2018). The world is gone, I must carry you: A provocation for doing post-critical educational research with the Anthropocene. In V. C. Reyes (Ed.), *Educational research in the age of Anthropocene.* Information Science Reference.

Leinfelder, R. (2013). Assuming responsibility for the Anthropocene: Challenges and opportunities in education. In H. Trischler (Ed.), *Anthropocene: Exploring the future of the age of humans* (pp. 9–28). RCC Perspectives.

Lewis, S. L., & Maslin, M. A. (2015). Defining the Anthropocene. *Nature, 519*(7542), 171–180.

Mele, C. (2019). That first black hole seen in an image is now called Pōwehi, at least in Hawaii. *The New York Times.* https://www.nytimes.com/2019/04/13/science/powehi-black-hole.html.

Moran, J. (2018). Earthrise: The story behind our planet's most famous photo. *The Guardian*. https://www.theguardian.com/artanddesign/2018/dec/22/beh old-blue-plant-photograph-earthrise.

O'Hagan, E. M. (2019). Why the black hole is a ray of light in these dark times. *The Guardian*. https://www.theguardian.com/commentisfree/2019/apr/11/black-hole-science-humanity.

Reed, P. (2019). Platform cosmologies: Enabling resituation. *Angelaki: Journal of Theoretical Humanities, 24*(1), 27–36.

Reuell, P. (2019). A black hole, revealed. *The Harvard Gazette*. https://news.har vard.edu/gazette/story/2019/04/harvard-scientists-shed-light-on-importance-of-black-hole-image/

Snaza, N. (2018). The earth is not "ours" to save. In J. Jagodzinski (Ed.), *Interrogating the Anthropocene: Ecology, aesthetics, pedagogy, and the future in question*. Palgrave Macmillan.

Wallin, J. J. (2017). Pedagogy at the brink of the post-Anthropocene. *Educational Philosophy & Theory, 49*(11), 1099–1111.

Weaver, J. A., & Snaza, N. (2016). Against methodocentrism in educational research. *Educational Philosophy and Theory, 49*(11), 1055–1065.

Wheeler, J. A., & Ford, K. W. (2000). *Geons, black holes, and quantum foam: A life in physics*. Norton.

Williams, D. R. (n.d.). The Apollo 8 Christmas Eve broadcast. https://nssdc.gsfc.nasa.gov/planetary/lunar/apollo8_xmas.html.

Zylinska, J. (2018). *The end of man: A feminist counterapocalypse*. University of Minnesota Press.

CHAPTER 4

The Waring Worlds of H. G. Wells: The Entangled Histories of Education, Sociobiology, Post-genomics, and Science Fiction

Chessa Adsit-Morris

EDUCATION AND CATASTROPHE

In 1937 the so-called father of modern science fiction, author H. G. Wells (1966, p. 1063), presented a paper to the British Association for the Advancement of Science's (BAAS) Educational Science Section (Section L), in which he declared that:

> [O]ur schools lag some fifty years behind contemporary knowledge. The past half-century has written a fascinating history of the succession of living things in time and made plain all sorts of processes in the prosperity, decline, extinction and replacement of species.[1]

Wells envisioned an education system that taught children about the world by providing a foundation in scientific knowledge through natural experience, informed by the latest advances in science and technology.[2] Wells rejected the existing nationalistic curriculum focused on "the scandals and revenges" that made up the "criminal history" of British royalty which "once passed

---

[1] Wells was referred to as the "father of science fiction" (along with Jules Verne) as well as the "Shakespeare of science fiction" (see Wagar, 2004).

[2] As Wells (1966, 1062) states: "we ought to make the weather and the mud pie our introduction to what Huxley christened long ago as Elementary Physiography. We ought to build up simple and clear ideas from natural experience."

C. Adsit-Morris (✉)
University of California, Santa Cruz, CA, USA
e-mail: cadsitmo@ucsc.edu

© The Author(s) 2022
M. F.G. Wallace et al. (eds.), *Reimagining Science Education in the Anthropocene*, Palgrave Studies in Education and the Environment, https://doi.org/10.1007/978-3-030-79622-8_4

as English history" (Wells, 1966, p. 1062). Instead he proposed a transnational scientific curriculum that taught students about the diverse historical and current ways humans and nonhumans occupy and negotiate their way through the world, including subjects such as social organization, politics, comparative religion, sociology, biology, chemistry, and natural history. Unbeknownst to many contemporary readers, Wells' first job after attending the Normal School of Science, where he learned biology under the tutelage of T. H. Huxley, was as a science teacher at Holt Academy, and his first book publication was the *Text-Book of Biology* (1893), which remained in print for thirty years.[3] Wells even attempted to become a Member of Parliament, running on a campaign platform of educational reform.[4] Wells was a socialist and revolutionary; he was a member of the Fabian Society who believed strongly that educational reform was fundamental to the social evolution of man, so much so that he declared in his popular publication *The Outline of History* (1920, p. 594): "human history becomes more and more a race between education and catastrophe."[5]

Wells presented his address to the British Association at a time of social, political, and intellectual ferment, in the wake of the first Great War, the industrial revolution and subsequent reform movements, the rebellious roaring twenties, and the burgeoning surrealist movement. It was a time of scientific and industrial innovation, advancement, and progress: the first X-ray machine was created by Wilhelm Conrad Röntgen in 1895; Svante Arrhenius derived the basic principles for the greenhouse effect in 1896; the Wright brothers took flight in the first modern powered aircraft in 1903; and in 1911 Ernest Rutherford provided evidence for his atomic model of the atom, leading to a reconceptualization of the modern atom by Niels Bohr in 1913 and subsequent quantum revolution. Between 1895 and the early 1920s scientists and engineers created faster and more efficient ways to extract, process, produce, transport, and sell consumer products. By the mid-1920s most cities had reliable transportation and communication networks that included refrigerated railroad cars, radios, public telephones, mail-order catalogues, and affordable silent motion pictures. By 1937, it was clear (at least to Wells) that humans were on a crash course toward catastrophe unless socio-cultural systems—including economic systems, social structures, political organizations, and educational systems—were reformed in ways that enabled humans to adequately grapple with the implications and impacts of the coming "great acceleration" of human activity. Wells' science fiction novels *War of the Worlds* (1897) and *The World Set Free* (1919) not only predicted the coming world

---

[3] Wells wrote a revised biology textbook, *The Science of Life*, in 1934 in collaboration with Julian Huxley (T. H. Huxley's grandson) and his son George P. Wells.

[4] See Adam Roberts (2019) "Education," which describes Wells growing interest in education echoed in his scholarship *Joan and Peter* (1918), *The Undying Fire* (1919), *Socialism and the Scientific Motive* (1923), and *The Story of a Great Schoolmaster* (1924).

[5] For more detailed explanation of Wells' critique of the history curriculum and its nationalistic underpinnings in the interwar period—as well as a more general overview of what has been termed a "Wellsian education"—see Ken Osborne (2014).

wars—including the use of tanks, chemical warfare, and the atomic bomb—but also provided a prophetic warning for the Anthropocene, which he viewed as the inevitable result of man's complacent and yet negligent mastery of nature.[6]

Almost a century later we find ourselves at a similar juncture in history, on the precipice of a global catastrophe fueled by extractive capitalism and poorly regulated technological advancements, which have accompanied and exacerbated social, cultural, racial, economic, political, and militaristic inequalities.[7] Wells would have been disappointed but not surprised to hear of the continued political and military conflicts, and even more disheartened by the institutionalized necropolitics that shape the Anthropocene. As T. J. Demos (2018, p. 1) observes:

> The Anthropocene is proving to be an era of world war, or rather, worlds at war. Not that this is anything new. We are no doubt living in the continuation of longstanding onto-epistemological and politico-military conflicts set within (still unfolding) histories of colonial and global states of violence and dispossession.

Drawing on Wells' visionary texts, social critique, and revolutionary insights, this chapter revisits and recontextualizes questions raised by Wells almost a century ago around the adequacy of existing science curricula to grapple with the still unfolding Anthropocene. I deploy the methodological practice of "diffractive play" (see Adsit-Morris & Gough, 2020) by reading selected fictional, theoretical, scientific, and policy texts/discourses diffractively through each other in order to undo pervasive conceptions, trouble linear temporal logics, and foster collective imaginaries around what might yet still be possible. I begin by exploring the technological advances in molecular biology that have occurred over the last twenty years that have instigated an epistemological turn toward what many science studies scholars and social scientists are calling the *post-genomic era*, which not only entails a reconceptualization of scientific understandings of genes, genomes, and genetics, but the rise of biocapitalism, the commercialization of genetics, and the resurrection of a genetics of race (see Keller, 2015). I situate current education research and policy within the post-genomic era through new research in the field of sociobiology—what Robert Plomin (2018) calls the new genetics of intelligence—and further explore the gaps between current research in evolutionary biology and genomics, and the content of the contemporary science education curriculum.

---

[6] In the opening paragraph of *War of the Worlds* (1897, p. 1), Wells described man's relationship with nature: "With infinite complacency men went to and fro over this globe about their little affairs, serene in their assurance of their empire over matter."

[7] Of the many catastrophic threats humankind faces, catastrophic climate change—comprising anthropogenic global warming and sea-level rise, desertification and agricultural failures, ecosystem fragmentation and mass species extinction—is the most generally associated with the Anthropocene.

This diffractive narrative inquiry travels through various time periods and national borders as it jumps from pre-war Britain, to interwar Australia, to contemporary United States in order to explore the entanglements of science and education within the still unfolding histories of colonial and capitalistic processes of globalization. As Donna Haraway (1975, p. 446) notes: "It is a considerable leap from pre-war Britain to the recent past of the United States; there is danger that comparisons will be facile. But drawing parallels is tempting, and it might well be instructive in our own working out of the political nature of science and its pedagogy." So, proceeding with caution, I explore the possibility of a historically situated *transnational extended synthesis* within education and evolutionary biology that engages a "transknowledge" approach through *EcoEvoEdu*, allowing for the exploration of various gaps—knowledge and achievement gaps—which have been carefully constructed, maintained, and reified by existing onto-epistemological and socio-political systems.[8] Such a reconceptualization encourages critical examination of the impact of globalization, nationalism, and capitalism on science education and works to imagine how science education can be reformed, reimagined, and reconfigured to contribute to the radical actualization of a just, equitable, and sustainable world.

Education and Sociobiology

In 1901 Francis Galton, Charles Darwin's cousin, presented the Huxley Lecture of the Anthropological Institute titled "The Possible Improvement of the Human Breed under the Existing Conditions of Law and Sentiment,"

---

[8] Drawing on the recent shift in biology toward a more transdisciplinary approach—what has been termed EcoEvoDevo or ecological, evolutionary, and developmental biology—which is part of a larger "Extended Evolutionary Synthesis" (see Pigliucci & Müller, 2010; Jablonka & Lamb, 2020), Haraway (2017, M28) calls for such an extended synthesis to also include the arts in an attempt to draw together "human and nonhuman ecologies, evolution, development, history, affects, performances, technologies, and more." For example, in "The Camile Stories" (2016) Haraway puts forth a transknowledging approach called EcoEvoDevoHistoEthnoTechnoPsycho (Ecological Evolutionary Developmental Historical Ethnographic Technological Psychological). In this chapter I call for a "transnational extended synthesis" which engages the transknowledge approach EcoEvoEdu. Such an approach seeks to expand educational inquiry beyond national borders to encourage a more critical examination of the impact of globalization, nationalism, and capitalism on how science and education—including science education, education science, education research, etc.—is utilized to address (or reify) issues including gender and racial justice, human rights, the concerns of indigenous peoples, and poverty and social exclusion, drawing on current biological research at the intersection of ecology, evolution, and education (or behavioral and symbolic inheritance systems). My efforts in this chapter are specific to this inquiry and should not be interpreted as an attempt to solidify a particular field of inquiry, discipline, or general law of any kind. My diffractive narrative inquiry led me to explore the potential intersection between transnational curriculum inquiry and biology, but I encourage others to use my methods to explore other transknowledging approaches, including Haraway's EcoEvoDevoHistoEthnoTechnoPsycho.

which was republished a few months later in *Nature*. The paper built off his previous work on the heritability of reputation (eminence) and mental ability (genius), his research on nature vs. nurture, and his research on binomial distribution (i.e., the bell curve) (see Galton, 1869, 1876, 1889).[9] Known as the father of eugenics, Galton's research was instrumental in legitimizing eugenics, racial science, and establishing the conceptual foundation, methodological apparatus, and statistical tools for the future fields of behavioral, social, and educational psychology (i.e., psychobiology and sociobiology). Galton, along with his protégé Carl Pearson, had a lasting effect on education policy and practice, solidifying what Stephen Jay Gould (1981, p. 21) describes as "the theory of a measurable, genetically fixed, and unitary intelligence"—what became known as the "g" factor for "general intelligence" (Staub, 2018). They were also part of a larger shift in the eugenics movement, described by Ann Gibson Winfield (2007, pp. 59–75) as a move beyond simply "an adherence with science" to a more influential "social and public policy stance" as I.Q. testing "became a primary tool in efforts to limit immigration and create more efficient schools."[10] Indeed, Pearson and his influential associates, including Cyril Burt and Hans Eysenck in the UK, Charles B. Davenport, Robert Mearns Yerkes, Harry Hamilton Laughlin, and Carl Brinham (who created the SAT for the College Board in 1926) in the United States, established educational programs and laboratories at prominent universities, managed national records and archive offices, and held prominent positions on various governmental committees, influencing the establishment of national policies and laws managing immigration and implementing marriage restrictions, compulsory sterilization, and segregation (see Haraway, 1979).[11]

In response to Galton's Huxley Lecture, Wells (1903, p. 37) wrote an article for the *Fortnightly Review* describing Galton's system of classification for superior and inferior human types—based on race, class, and intelligence, evaluated using a normal distribution—subjecting it to considerable ridicule, stating that Galton "saturates the whole business in quantitative colour." Wells argued that traits such as ability, genius, and beauty were too complex to be subject to a "simple and uniform" quantitative evaluation as Galton had

[9] The First International Eugenics Congress held at the University of London in 1912, was held in Galton's honor as the founder of the "Science of Eugenics" (First International, 1912). Most of these methods and conceptual tools have been debunked as pseudoscience, including the nature/nurture debate (see Evelyn Fox Keller, 2010), twin studies (see Jay Joseph, 2017, 2018), and I.Q. testing (see Gould, 1981).

[10] Throughout the paper I will be using I.Q. to refer to the measurement of a person's intelligence quotient (per variations of the Simon–Binet Intelligence test) and IQ to refer to *Inuit Qaujimajatuqangit*.

[11] The direct intellectual influences between these men can be clearly traced: Galton taught Pearson, Pearson taught Burt, Burt taught Eysenck, Jenson was taught by both Eysenck and Burt, Davenport met Pearson in London and regularly contributed to his journal *Biometrika*, Davenport worked with both Yerkes and Laughlin through the Eugenics Records Office, and Bringham collaborated with Yerkes on I.Q. testing during WWI.

proposed. However, the cultural authority of "science" toward the end of the nineteenth and beginning of the twentieth century had grown significantly, as many academic disciplines sought to legitimize their research by turning to more "rigorous" and "objective" positivist methods. For many of the social sciences, including education (see Selleck, 1967), this also meant establishing a grounding in the physical and biological sciences (Cravens, 1985). As Kimberly Hamlin (2014, p. 59) notes, by the turn of the century, if one wanted to engage in debates about social reform, one needed to be armed with "scientific, ideally evolutionary, evidence." The importance of biology (and biological literacy) in the early 1900s can be seen in a review published in *Nature* (1931, p. 478) of Wells' updated three-volume biology textbook *The Science of Life* (1931), written in collaboration with his son G. P. Wells and T. H. Huxley's grandson Julian Huxley, as the anonymous *Nature* reviewer wrote:

> If, as we believe, mankind is at the dawn of a new era—the biological era, when an all-round appeal will be made to the biological sciences, as already to the physical, for guidance in the control of human life—then the big book of Wells, Huxley, and Wells will come to be regarded as an instalment of the relevant 'Law and Prophets.'

A century later, we find ourselves at the dawn of a new biological era—the post-genomic era—also fraught with social, political, and intellectual ferment. The post-genomic era began in the wake of the Human Genome Project (HGP), the first large-scale project aimed at sequencing all of the 3 billion nucleotides that make up the human genome. By the end of the HGP most molecular biologists agreed that much more was going on during transcription and translation of the genome than could be imagined, resulting in a conceptual shift from the genocentrism and reductionism of classical genetics, toward a focus on the complex mechanisms regulating gene expression.[12] The central dogma of molecular genetics (DNA → RNA → protein) was proven to be much more complex through alternative splicing, messenger RNA editing, and post-translational protein modification, through which multiple proteins can be produced from a single gene. Additionally, contemporary research illustrates that complex traits are influenced by multiple genes spread across the genome, referred to as *polygenic inheritance*. For example, ongoing research on human height has shown that there are hundreds of loci associated with height scattered across the genome in coding and noncoding regions that explain only 20% of the heritability of height (see Allen et al., 2010; Marouli et al., 2017). As Evelynn Fox Keller (2015, p. 10) describes, post-genomics has "turned our understanding of the basic role of the genome on its head, transforming it from an executive suite of directional instructions to

---

[12] Gene expression encompasses the processes by which functional gene products are created, including various protein and RNA products.

an exquisitely sensitive and reactive system that enables cells to regulate gene expression in response to their immediate environment."

The post-genomic era has also ushered in a surge of biotechnological innovation at the intersection of data-intensive informatics and genomic science, what Ben Williamson (2018a) calls "big biodata" with potential implications for education research and policy. It took scientists 13 years and cost a total of $3 billion dollars to sequence the first human genome; now it costs less than $1,000 (Gibbs, 2020). The HGP was the first "Big Science" project with an open access policy—later termed the "Bermuda Principles"—which required genome sequence data to be publicly accessible within 24 hours of its assembly, allowing the network of labs funded by the project (~200 labs) to collaborate. Genome data was also accessible to the public, initiating the development of privately funded labs and institutions headed by a "new breed of investigator, the scientific entrepreneur" (Jackson, 2015, p. 2). These labs worked tirelessly in a "race to the genome" (see Reardon, 2005) that spurred faster paced advances in genomic sequencing technologies and helped to significantly reduce the cost and time required for sequencing. These technologies included databases and biobanks, microarray chips, next-generation high-throughput genome sequencing technologies, and commercial genome kits. For Sarah Richardson (2015, p. 3) these technological and methodological shifts signify the advent of the post-genomic era in which whole-genome technologies have become "a shared platform for biological research across many fields and social arenas." One such platform is Genome Wide Association Studies (GWAS), which have become one of the main methods for making sense of genetic variation over the last decade by measuring the statistical correlations between single nucleotide polymorphisms (SNPs)—variations across the genomes in nucleotide bases—in thousands of genomes of individuals with a particular phenotypic trait (i.e., height, disease, etc.).

Education and Post-genomics

There has been a significant upsurge in the publication of research using GWAS to assess the correlation between race, socio-economic status, I.Q., and educational attainment (see Lee et al., 2018; Selzam et al., 2016; Kovas et al., 2016).[13] Through one of the largest genetic studies conducted to date, researchers generated a "polygenomic score" for educational attainment by comparing genomic data from the consumer genetics company 23andMe to self-reported educational data (EduYears) for one million individuals with

---

[13] One of the main researchers in the field of behavioral genetics, involved to some degree in all published studies, is Robert Plomin, who has recently published two books on the subject: *G is for Genes: The Impact of Genetics on Education and Achievement* (2014) and *Blueprint: How DNA Makes Us Who We Are* (2018). Plomin was an advocate of Herrnstein and Murray's *The Bell Curve* (1994), helping to initiate a statement of support for their book published in the *Wall Street Journal* and signed by 52 professors across the United States (Gottfredson, 1997).

European ancestry (Lee et al., 2018). Other research has been published on "the genetics of university success" (Smith-Woolley et al., 2018), DNA variants shared between personality traits and educational achievement (Smith-Woolley et al., 2019), and how "genetics affects choice of academic subjects as well as achievement" (Rimfeld et al., 2016). These studies have been touted for their ability to be used to "predict educational achievement for individuals directly from their DNA" (Selzam et al., 2016) and signals the emerging interest of biocapitalists in the development of "personal precision education" (see Williamson, 2018b). However, significant concerns have been raised around data bias, particularly disparities in the underlying ancestral diversity of genomic data, which has been dominated by participants of European ancestry (roughly < 80%) mainly from the United States, UK, and Iceland (see Mills & Rahal, 2019; Popejoy & Fullerton, 2016).[14] Additionally, concerns have been raised around the ambiguity of population categories via the use of "quasi-racial 'continental' terms" leading to fears around the "molecularization of race" (Panofsky & Bliss, 2017, p. 59).

Debates about the genetic (or hereditary) basis of race and I.Q.—which began with Galton's publication *Hereditary Genius* in 1869—have resurfaced time and time again, including the race and I.Q. debates (sparked by the *Brown vs. Board of Education* court case) in the 1960s and 1970s that pivoted around Hans Eysenck, Arthur Jensen, and Edward O. Wilson (founder of the field of sociobiology), provoking anti-racism protests and demonstrations at a number of universities including the University of Birmingham, the University of Sydney, the University of Melbourne, Harvard, and the University of California, Berkley; the publication of *The Bell Curve* in 1994 by Richard Herrnstein and Charles Murray; and the 2007 remarks made by James Watson (co-discoverer of the double helix structure with Francis Crick) stating that "people of African descent are not as intelligent as people of European descent" (Dean, 2007), igniting a critical uproar that prompted his resignation as chancellor of the Cold Spring Harbor Laboratory. Researchers such as Sarah Richardson (2011, p. 420) express concern that education research in the post-genomic age will "initiate a new era of scientific claims about the genetics of racial differences in I.Q." In a book review for *Nature* on behavioral geneticist Robert Plomin's book, *Blueprint: How DNA Makes Us Who We Are* (2018), Nathaniel Comfort (2018, para 1) begins:

> It's never a good time for another bout of genetic determinism, but it's hard to imagine a worse one than this. Social inequality gapes, exacerbated by climate change, driving hostility towards immigrants and flares of militant racism. At

---

[14] For example, studies on educational achievement at the college level exclude African Americans for several reasons; most important for this study is that African genomes have less linkage disequilibrium (LD) between alleles at different SNPs, and thus require a higher coverage platform in order to capture the variation across the African genome. Due to this, researchers found a "much lower predictive power in a sample of African-American individuals" (Lee et al., 2018, p. 116).

such a juncture, yet another expression of the discredited, simplistic idea that genes alone control human nature seems particularly insidious.

Indeed, Plomin (2018, p. 80) draws on Galton's research calling him the "nineteenth-century founder of behavioral genetics," arguing that environmental effects (including education) are "unsystematic, idiosyncratic, serendipitous events without lasting effects." Plomin falls unapologetically on the genetic side of the nature/nurture debate, what Comfort calls "vintage" genetic determinism, a stance that has historically adversely affected youth of color through the systematic exclusion of BIPOC students from adequate and equitable education through laws, policies, and social attitudes that result in resource disparities.

As educational theorists and practitioners, we know that intelligence, attainment, cognition, and learning are complex processes influenced simultaneously by socio-cultural, politico-economic, and biological factors. However, within this complex post-genomic landscape—in which genetic data is a shared central platform for researching human social, cultural, and biological difference within the wakes and waves of renewed biological determinisms—science education is essential to translating and mediating public health and educational applications of genomic research. Currently, there is a significant lag—as Wells warned against so long ago—in integrating genomic education with the foundational genetic principles taught in secondary and higher education classrooms (Reiss & Harms, 2019; Whitley et al., 2020; Zudaire & Napal Fraile, 2020). This leads students to a poor understanding of genomics based on reductive, deterministic, and gene-centered misconceptions; additionally, social media, movies, and television generally reinforce these outdated understandings of genomics (Stern & Kampourakis, 2017). Topics including the history and philosophy of science, bioethics, and feminist science studies are given little, if any, attention in the contemporary science classroom (Jones et al., 2010). Research has shown that teaching the history and philosophy of science helps students learn about genetics; better understand the nature of science; connect genetics to social, cultural, ethical, and political concerns; and enhance reasoning and critical thinking skills (Gericke & Smith, 2014).

Recently, the rapid growth of the #BlackLivesMatter movement in the United States in response to the unjust killings of Breonna Taylor, Rayshard Brooks, George Floyd, and countless others, has instigated radical calls for social reform including from Representative Dr. Eddie Bernice Johnson (2020, para. 3), who officially requested that the National Academies of Science, Engineering, and Medicine assemble a research committee to conduct a "rigorous and thoughtful analysis of the extent to which the U.S. scientific enterprise perpetuates systemic inequalities to the detriment of society as [a] whole, as well as how those inequities are manifest." As a fundamental component of the scientific enterprise, STEM education is implicated in this call for reform, particularly the complex relations between how scientific research informs education policy and practice, and how STEM education perpetuates and privileges the production of particular types of scientific knowledge, both of which help manifest the ethnic, racial, gender, disability, and income "achievement gaps" we are so fervently (and ineffectually) trying to close. As we face a future of accelerating global social, cultural, economic, political, and environmental crises fueled by complex histories of extractive capitalism and inequitable distributions of wealth, knowledge, power, and privilege, the question we urgently need to ask is: What role can, and should, science education play in creating a just, equitable, and sustainable future?

Reconceptualizing I.Q.

While studying at the Normal School of Science, Wells became involved in the British socialist revival of the 1880s, presenting and publishing his emerging ideas in school debates and as editor of the *Science Schools Journal* (Partington, 2016). He advanced his own particular brand of democratic socialism— captured in this futurological works *Anticipations* (1902), *Mankind in the Making* (1903), and *A Modern Utopia* (1905)—greatly influenced by late Victorian socialists including William Morris, George Bernard Shaw, and Graham Wallace, as well as the "ethical evolution" propounded by his former professor T. H. Huxley (1895) in his influential lecture "Evolution and Ethics." In 1903 Wells became an active member of the Fabian Society— a collective of socialist intellectuals established in 1884 who advocated for gradualist and reformist democratic policies—until Wells broke ties with the group in 1908 due to conflicts with fellow Fabians George Bernard Shaw and Annie Besant. Following William Morris, many socialists—also referred to as Socialist Darwinists—were drawn to Lamarck's theory of the inheritance of acquired characteristics, rejecting the Malthusian naturalization of competition and social inequality, and taking up Lamarckian presumptions about social change instead (Hale, 2010). As Piers Hale (2010, p. 60) describes: "In an era in which evolution touched all aspects of politics, and in which Lamarckism remained credible, it is unsurprising that socialists speculated upon how humanity might quickly adapt to a changed environment and to new ways of living, and upon how the inheritance of these adaptations might affect human evolution." However, the publication of August Weismann's germplasm theory in 1892, invalidating Lamarck, proved a fatal blow for many socialist reformers (Amigoni & Wallace, 1995). By 1908, Wells could no longer tolerate anti-Malthusian neo-Lamarckian socialists, claiming that their theories were unscientific and their practice ineffectual.[15] Although Wells' adopted more of a neo-Darwinian bent after reviewing Weismann's research on heredity, he believed (following T. H. Huxley) that culture had a greater impact on social evolution than biology, and that "*cultural* characters might

[15] For a more detailed description of Wells' critiques of the Fabian Society see John Partington (2016).

still be transmitted and compounded across the generations" (Hale, 2010, p. 39, emphasis in original). Believing that biological evolution would take generations to affect change, Wells' turned wholeheartedly to education as a means to influence social evolution through learned social and behavioral traits.

New research emerging at the cutting edge of post-genomic science has led to expanded views of inheritance, epitomized by Eva Jablonka and Marion Lamb's (2014, 2020) provocation to think through evolution in four dimensions—genetic, epigenetic, behavioral (i.e., social learning), and symbolic (i.e., cultural)—putting forth a developmental and ecological view of the origin of heritable variations and an expanded notion of transmission and selection. Such research has inspired biologists to call for an *Extended Evolutionary Synthesis* based on the concept of "inclusive inheritance" or non-genetic forms of inheritance such as epigenetic inheritance, inherited symbionts (i.e., microbiota), cultural inheritance, and ecological inheritance (or niche inheritance) blurring the boundaries between natural and cultural mechanisms of evolution (see Pigliucci & Müller, 2010; Jablonka & Lamb, 2020). The underlying molecular basis of non-genetic inheritance is called *epigenetics*—from the Greek ἐπί meaning "upon" or "on top of" genetic factors—which are essentially non-DNA elements that affect gene function, plasticity, and expression. The new field of epigenetics comprises the study of heritable phenotype changes not caused by changes to the underlying DNA sequence. Although there are several processes associated with epigenetics, the most common process drawn upon in social science fields is DNA methylation: the attachment of a methyl group ($-CH_3$) to one of the four nucleotides, Cytosine, creating 5-methycytosine. This results in deactivation of the gene by blocking the transcription process, therefore altering gene function and expression. What is most important about this research for the social sciences, and specifically education, is that methylation can be affected by social, cultural, and environmental factors—nutrition, stress, lifestyle, exposure to toxins, etc.—and these changes are heritable.

Following other STS post-genomic scholars who have contributed to a fundamental reconceptualization of the science of genomics, critically analyzing the social implications of genomic research by drawing on feminist, queer, critical race and decolonial theories, this paper takes up the task of reconceptualizing intelligence and I.Q. testing, tracing the history of culturally biased and racist mental testing practices, the results of which have been misused to further the colonial agendas of cultural genocide through assimilation—what Jasbir Puar (2017) calls "weaponizing epigenetics." Such a reconceptualization, I argue, also requires an engagement with the discourses and cultural resources of popular media and non-Western knowings, which tend to be ignored or devalued in mainstream science and philosophy. For example, the Inuit lifelong learning philosophy known as *Inuit Qaujimajatuqangit* (IQ)—from the verb *qaujima* "to know" used to describe Inuit epistemology—can be viewed as a powerful case study of social, cultural, and political resistance to historic eugenic policies including the use of I.Q. testing in compulsory sterilization practices and epigenetic violence inflicted through residential schooling.[16] IQ was formally adopted by the Government of Nunavut in 1999, and work to reform the Nunavut education system to incorporate IQ began in 2000 (Igloliorte, 2017). IQ as a philosophy is a complex indigenous cosmology that encompasses Inuit beliefs, values, worldview, social organization, life skills, language, and environmental knowledge (Canadian Council on Learning, 2007). IQ is based on the integration of three types of laws—natural laws (*maligarjuat*), cultural laws (*piqujat*), and communal laws (*tirigusuusiit*)—and guided by thirty-eight values, including cooperation, conservation, adaptability, consensus, endurance, generosity, respect, interconnectedness, and equality (to name a few). As a form of political resistance, IQ works to subvert the logics of Western science, the modern developmental state, and its colonial legacy, which continually attempts to reduce IQ to "traditional ecological knowledge" useful for informing scientific knowledge, instead of viewing IQ as an integrated holistic knowledge system (Tester & Irniq, 2008). As a holistic onto-epistemological system, IQ is a radical worlding practice that not only attempts to subvert Western logics and reductive epistemological practices, but advocates for—and *actualizes*—a collective vision for a more just, equitable, and sustainable world.

"We Know Better Now"

In 1939 Wells was invited to present at the Australian and New Zealand Association for the Advancement of Science (ANZAAS) conference to speak to the Education Section to discuss the role of science in world affairs, during which he stated:

> What spendthrift ancestors we have had! What wastrels we still are! And all because history teaches us no better. Man burns and cuts down forests, he destroys soil, he acclimatizes destructive animals. A map of the world showing the devastated regions, where devastation is due to mankind, would amaze most people. It ought to be put in every child's atlas. (Walkom, 1939)

Well's visionary insight, political philosophy, and advocacy for human rights and education make him an interesting figuration to think through in relation to contemporary discourses around the Anthropocene. As Liam Gearon (2018, p. 765) notes: "His early science fiction was—with its social and political commentary—invariably related to the future of the planet and humanity. With a characteristically pessimistic view he shared with many of

---

[16] The epigenetic and intergenerational violence—cultural genocide—inflicted on Indigenous children (First Nations, Métis, and Inuit) included separation of children from their communities, not allowing children to speak their first languages or practice cultural traditions, malnutrition, poor sanitation, lack of medical treatment, medical experimentation, and physical, sexual, and psychological abuse.

his 19th-century peers, Wells' vision has arguably even more resonance with environmental and related concerns today." Indeed, Wells' writing and ideas exemplify the rapidly growing discourse of the Anthropocene—albeit written a century earlier—by setting the human species (and human history) within cosmic, geological, and evolutionary scales; blurring the boundaries between science fiction and science fact; and exploring the speculative implications of the entanglements between human exceptionalism and techno-scientific capitalism. Along with *Men Like Gods*, Wells' influential and highly popular publication *The Outline of History* (1920)—which sold over two million copies, was translated into several languages, and had a significant impact on the teaching of history—was a pioneering work aimed at constructing a universal story of the "Anthropos" set within the history of the planet from its formation. Although the publication has been subject to critiques about its Euro-centric universalizing tendencies (Wrong, 1921)—as has similarly (and rightly) been launched against the concept of the Anthropocene (see Haraway et al., 2016; Malm, 2015; Todd, 2015)—it was an attempt to establish a transnational history that provided a foundation for people to discuss issues of human rights and social justice that arise from national, ethnic, religious, and colonial conflicts. For Wells *The Outline of History* (1920) was a history of *worlds at war* that echoes to the present as a series of onto-epistemological, nationalistic, politico-military conflicts set within the still unfolding Anthropocene. On the centenary anniversary of its publication, Wells' (1938, p. 41) words are more prophetic and poignant than ever, as he states: "We know better now. Now the consequences of this change of scale force themselves upon our attention everywhere."

We know, as Bruno Latour (2017) states, that the Anthropocene is an era in which nature and culture can no longer be studied independently, where science and capitalism touch everything, and (as a result) science education can no longer avoid engaging with the critical discourses of the social sciences, arts, and humanities. As Wells stated in a conversation with Stalin in 1934: "There can be no revolution without a radical change in the educational system" (Stalin, 1978, p. 40). The dominant science education and communication paradigm fails to adequately address the complex world-historical workings of power and privilege within (and outside of) the scientific enterprise. As Haraway (1975, p. 442) notes: "Science education for responsible political behavior naturally does not encourage a radical activist approach to environmental, population, or armament issues." For example, although the concept of the Anthropocene has spurred critical debate and received various conceptual challenges from the arts, humanities, and social sciences, in the main, the fields of science, technology, and engineering have remained uncritical, focusing efforts on so-called innovative technofixes: techno-scientific and engineering-based approaches to Anthropocenic survival. Haraway (2017) argues that the Extended Evolutionary Synthesis requires the entanglement of the arts and sciences through human and nonhuman ecologies, histories, technologies, and affects. In the last half century contemporary research has written a fascinating multispecies symbiotic history of the bumptious succession of life on earth, one that fosters new collective imaginaries around what might yet still be possible. I argue that science education needs a transnational and transknowledge extended synthesis (or *EcoEvoEdu*), one that encourages critical examination of the impact of globalization, nationalism, and capitalism on science education and works to imagine how science education can be reformed, reimagined, and reconfigured to contribute to the radical actualization of a just, equitable, and sustainable world. Wells' knew—as can be seen in his increasingly prophetic voice around the future uses and abuses of science and technology, and the importance of education for the future of mankind—that the Anthropocene era is an era of worlds at war with nothing less than the future of the planet at stake.

By the end of his career—having lived through two world wars and witnessed the dropping of atomic bombs on Hiroshima and Nagasaki—Wells grew less optimistic about the future and science's role in it. In *Mind at the End of Its Tether* (1945, p. 18), Wells' last published book, he put forth a bleak vision of the future, explaining that: "A series of events has forced upon the intelligent observer the realization that the human story has already come to an end and that *Homo sapiens*, as he has been pleased to call himself, is in his present form played out." Wells (1945, p. 1) expands on these logics:

> If his thinking has been sound, then this world is at the end of its tether. The end of everything we call life is close at hand and cannot be evaded. … [W]e are confronted with strange convincing realities so overwhelming that, were he indeed one of those logical consistent creatures we incline to claim we are, he would think day and night in a passion of concentration, dismay and mental struggle upon the ultimate disaster that confronts our species.

Upon reviewing *Mind at the End of Its Tether*, many critics claimed that, at 78 years old, it was Wells' mind that had become untethered from reality, as they were unable to comprehend the narrative of global catastrophe via human techno-scientific and politico-capitalist progress—*we know better now*. Wells was able to see that our (Western Euro-centric) mental agencies (i.e., science) had become *untethered* from cosmic processes, and there was no going back. Following many queer, feminist, posthuman, and decolonial scholars who view the Anthropocene as signaling the end of a particular Western Euro-centric colonial and capitalistic world, I believe the task at hand requires the radical actualization of another world. So, as we stand on the precipice of worlds at war, the question of what science education will become is still yet to be determined.

REFERENCES

Adsit-Morris, C. A., & Gough, N. (2020). Post-anthropocene imaginings: Speculative thought, diffractive play and women on the edge of time. In M. Krehl, E. Thomas,

& R. Bellingham (Eds.), *Post-qualitative research and innovative methodologies* (pp. 172–186). Bloomsbury Academic.

Allen, H. L., Estrada, K., Lettre, G., Berndt, S. I., Weedon, M. N., & et al. (2010). Hundreds of variants clustered in genomic loci and biological pathways affect human height. *Nature, 467*, 832–838.

Amigoni, D., & Wallace, J. (Eds.). (1995). *Charles Darwin's the origin of species: New interdisciplinary essays.* Manchester University Press.

Canadian Council on Learning. (2007). *Redefining how success is measured in first Nntions, Inuit and Métis learning, report on learning in Canada 2007.*

Comfort, N. (2018, September 25). Genetic determinism rides again. *Nature, 561*, 461–463. http://doi.org/10.1038/d41586-018-06784-5.

Cravens, H. (1985). History of the social sciences. *Osiris, 1*, 183–207.

Dean, C. (2007, October 26). James Watson quits post after remarks on races. *The New York Times.* https://www.nytimes.com/2007/10/26/science/26watson.html?rref=collection%2Ftimestopic%2FWatson%2C%20James%20D.&action=click&contentCollection=timestopics&region=stream&module=stream_unit&version=latest&contentPlacement=20&pgtype=collection.

Demos, T. J. (2018). To save a world: Geoengineering, conflictual futurisms, and the unthinkable. *eflux, 94*.

First International Eugenics Congress. (July, 1912). *Abstracts of papers read at the First International Eugenics Congress, University of London* (Vol. 1). Charles Knight & Co.

Galton, F. (1869). *Hereditary genius: An inquiry into its laws and consequences.* MacMillan & Co.

Galton, F. (1876). The history of twins, as a criterion of the relative powers of nature and nurture. *Journal of the Anthropological Institute of Great Britain and Ireland, 5*, 391–406.

Galton, F. (1889). *Natural inheritance.* MacMillan and Co.

Galton, F. (1901). The possible improvement of the human breed under the existing conditions of law and sentiment. *Nature, 64*(1670), 659–665.

Gearon, L. (2018). A very political philosophy of education: Science fiction, schooling and social engineering in the life and work of H. G. Wells. *Journal of Philosophy of Education, 52*(4), 762–777.

Gericke, N. M., & Smith, M. U. (2014). Twenty-first-century genetics and genomics: Contributions of HPS-informed research and pedagogy. In M. R. Matthews (Ed.), *International handbook of research in history, philosophy and science teaching* (pp. 423–467). Springer.

Gibbs, R. A. (2020). The human genome project changed everything. *Nature Reviews Genetics, 21*, 575–576.

Gottfredson, L. S. (1997). Mainstream science in intelligence: An editorial with 52 signatories, history, and bibliography. *Intelligence, 24*(1), 13–23.

Gould, S. J. (1981). *The mismeasure of man.* W. W. Norton & Company.

Hale, P. J. (2010). Of mice and men: Evolution and the socialist utopia. William Morris, H. G. Wells, and George Bernard Shaw. *Journal of the History of Biology, 43*(1), 17–66. http://doi.org/10.1007/s10739-009-9177-0.

Hamlin, K. A. (2014). *From eve to evolution: Darwin, science, and women's rights in gilded age America.* Chicago, IL: University of Chicago Press.

Haraway, D. J. (1975). The transformation of the left in science: Radical associations in Britain in the 30's and the U.S.A. in the 60's. *Soundings: An Interdisciplinary Journal, 58*(4), 441–462.

Haraway, D. J. (1979). The biological enterprise: Sex, mind, and profit from human engineering to sociobiology. *Radical History Review, 20*, 206–237.

Haraway, D. J. (2016). *Staying with the trouble: Making kin in the Chthulucene*. Duke University Press.

Haraway, D. J. (2017). Symbiogenesis, sympoiesis, and art science activisms for staying with the trouble. In A. L. Tsing, H. A. Swanson, E. Gan, & N. Bubandt (Eds.), *Arts of living on a damaged planet* (pp. M25–M50). University of Minnesota Press.

Haraway, D. J., Ishikawa, N., Gilbert, S. F., Olwig, K., Tsing, A. L., & Bubandt, N. (2016). Anthropologists are talking: About the Anthropocene. *Ethnos, 81*(3), 535–564. http://doi.org/10.1080/00141844.2015.1105838.

Herrnstein, R. J., & Murray, C. A. (1994). *The bell curve: Intelligence and class structure in American Life*. Free Press.

Huxley, T. H. (1895). *Evolution & ethics and other essays*. Macmillan and Co.

Igloliorte, H. (2017). Curating Inuit Qaujimajatuqangit: Inuit knowledge in the Qallunaat Art Museum. *Art Journal, 76*(2), 100–113. http://doi.org/10.1080/00043249.2017.1367196.

Jablonka, E., & Lamb, M. J. (2014). *Evolution in four dimensions: Genetic, epigenetic, behavioral, and symbolic variation in the history of life* (revised ed.). MIT Press.

Jablonka, E., & Lamb, M. J. (2020). *Inheritance systems and the extended evolutionary synthesis*. Cambridge, UK and New York, NY: Cambridge University Press.

Jackson, M. W. (2015). *The genealogy of a gene: Patents, HIV/AIDS, and race*. Cambridge, MA: The MIT Press.

Johnson, E. B. (2020, July 29). [Letter from Eddie Bernice Johnson to Dr. Marcia McNutt, re: STEM Racism].

Jones, A., Kim, A., & Reiss, M. J. (Eds.). (2010). *Ethics in the science and technology classroom: A new approach to teaching and learning*. Sense Publishers.

Joseph, J. (2017). *Schizophrenia and genetics: The end of an illusion*. BookBaby.

Joseph, J. (2018). *Twenty-two invalidating aspects of the Minnesota Study of Twins Reared Apart (MISTRA)*. https://www.madinamerica.com/wp-content/uploads/2018/11/Twenty-Two-Invalidating-Aspects-of-the-MISTRA-by-Jay-Joseph-Full-Version.pdf.

Keller, E. F. (2010). *The mirage of a space between nature and nurture*. Duke University Press.

Keller, E. F. (2015). The postgenomic genome. In S. S. Richardson & H. Stevens (Eds.), *Postgenomics: Perspectives on biology after the genome* (pp. 9–31). Duke University Press.

Kovas, Y., Malykh, S. B., & Gaysina, D. (Eds.). (2016). *Behavioural genetics for education*. Palgrave Macmillan.

Latour, B. (2017). *Facing Gaia: Eight lectures on the new climatic regime* (C. Porter, Trans.). Polity Press (Original work published 2015).

Lee, J. J., Wedow, R., Okbay, A., Kong, E., Maghzian, O., Zacher, M., Nguyen-Viet, T. A., Bowers, P., Sidorenko, J., Linnér, R. K., Fontana, M. A., Kundu, T., Lee, C., Li, H., Li, R., Royer, R., Timshel, P. N., Walters, R. K., Willoughby, E. A., . . . Cesarini, D. (2018). Gene discovery and polygenic prediction from a genome-wide association study of educational attainment in 1.1 million individuals. *Nature Genetics, 50*, 1112–1121. https://doi.org/10.1038/s41588-018-0147-3.

Malm, A. (2015). The Anthropocene myth. *Jacobin*. https://www.jacobinmag.com/2015/03/anthropocene-capitalism-climate-change/.

Marouli, E., Graff, M., Medina-Gomez, C., Lo, K. S., Wood, A. R., Kjaer, T. R., Fine, R. S., Lu, Y., Schurmann, C., Highland, H. M., Rüeger, S., Thorleifsson, G., Justice, A. E., Lamparter, D., Stirrups, K. E., Turcot, V., Young, K. L., Winkler, T. W., Esko, T., . . . Lettre, G. (2017). Rare and low-frequency coding variants alter human adult height. *Nature, 542*, 186–190.

Mills, M. C., & Rahal, C. (2019). A scientometric review of genome-wide association studies. *Communications Biology, 2*(9), 1–11.

Osborne, K. (2014). "One great epic unfolding": H. G. Wells and the interwar debate on the teaching of history. *Historical Studies in Education, 26*(2), 1–29.

Panofsky, A., & Bliss, C. (2017). Ambiguity and scientific authority: Population classification in genomic science. *American Sociological Review, 82*(1), 59–87. https://doi.org/10.1177/0003122416685812.

Partington, J. S. (2016). *Building cosmopolis: The political thought of H. G. Wells*. Routledge.

Pigliucci, M., & Müller, G. B. (Eds.). (2010). *Evolution, the extended synthesis*. MIT Press.

Plomin, R. (2018). *Blueprint: How DNA makes us who we are*. Penguin Random House.

Plomin, R., & Asbury, K. (2014). *G is for genes: The impact of genetics on education and achievement*. John Wiley & Sons.

Popejoy, A. B., & Fullerton, S. M. (2016). Genomics is failing on diversity. *Nature, 538*, 161–164.

Puar, J. K. (2017). *The right to maim: Debility, capacity, disability*. Duke University Press.

Reardon, J. (2005). *Race to the finish: Identity and governance in an age of genomics*. Princeton, NJ: Princeton University Press.

Reiss, M. J., & Harms, U. (2019). The present state of evolution education. In U. Harms & M. J. Reiss (Eds.), *Evolution education: Re-considered* (pp. 1–19). Springer.

Richardson, S. S. (2011). Race and IQ in the postgenomic age: The microcephaly case. *BioSocieties, 6*(4), 420–446. https://doi.org/10.1057/biosoc.2011.20.

Richardson, S. S., & Stevens, H. (Eds.). (2015). *Postgenomics: Perspectives on biology after the genome*. Durham, NC and London: Duke University Press.

Rimfeld, K., Ayorech, Z., Dale, P. S., Kovas, Y., & Plomin, R. (2016). Genetics affects choice of academic subjects as well as achievement. *Scientific Reports, 6*. https://doi.org/10.1038/srep26373.

Roberts, A. (2019). Education. In A. Roberts (Ed.), *H. G. Wells: A literary life* (pp. 295–302). Palgrave Macmillan.

The Science of Life. (1931). [Review of the book *The science of life*, by H. G. Wells, J. Huxley, & G. P. Wells]. *Nature, 127*, 477–479. http://doi.org/10.1038/127477a0.

Selleck, R. J. W. (1967). The scientific educationist, 1870–1914. *British Journal of Educational Studies, 15*(2), 148–165. http://doi.org/10.1080/00071005.1967.9973183.

Selzam, S., Krapohl, E., von Stumm, S., O' Reilly, P. F., Rimfeld, K., Kovas, Y., Dale, P. S., Lee, J. J., & Plomin, R. (2016). Predicting educational achievement from DNA. *Molecular Psychiatry*, *22*(2), 267–272. http://doi.org/10.1038/mp. 2016.107.

Smith-Woolley, E., Ayorech, Z., Dale, P. S., Von Stumm, S., & Plomin, R. (2018). The genetics of university success. *Scientific Reports*, *8*(14579), 1–9. http://doi. org/10.1038/s41598-018-32621-w.

Smith-Woolley, E., Selzam, S., & Plomin, R. (2019). Polygenic score for educational attainment captures DNA variants shared between personality traits and educational achievement. *Journal of Personality and Social Psychology*, *117*(6), 1145–1163. http://dx.doi.org/10.1037/pspp0000241.

Stalin, J. (1978). *Works Volume 14: 1934–1940*. Red Star Press Ltd.

Staub, M. E. (2018). *The mismeasure of minds: Debating race and intelligence between* Brown *and* The Bell Curve. The University of North Carolina Press.

Stern, F., & Kampourakis, K. (2017). Teaching for genetics literacy in the postgenomic era. *Studies in Science Education*, *53*(2), 192–225. http://doi.org/10. 1080/03057267.2017.1392731.

Tester, F. J., & Irniq, P.. (2008). *Inuit Qaujimajatuqangit*: Social history, politics and the practice of resistance. *Arctic*, *61*, 48–61.

Todd, Z. (2015). Indigenizing the Anthropocene. In H. Davis & E. Turpin (Eds.), *Art in the Anthropocene: Encounters among aesthetics, politics, environments and epistemologies* (pp. 242–254). Open Humanities Press.

Wagar, W. W. (2004). *H. G. Wells: Traversing time*. Wesleyan University Press.

Walkom, M. (Ed.) (1939). *Report of the twenty-fourth meeting of the Australian and New Zealand association for the advancement of science, Canberra meeting, January, 1939*. Sydney: Australasian Medical Publishing Company Limited. https://www.abc. net.au/radionational/programs/earshot/hg-wells/6452964.

Wells, H. G. (1893). *Text-book of biology*. University Correspondence College Press.

Wells, H. G. (1897). *The war of the worlds*. Edward Arnold.

Wells, H. G. (1903). *Mankind in the making*. Chapman & Hall.

Wells, H. G. (1920). *The outline of history: Being a plain history of life and mankind*. The Macmillan Co.

Wells, H. G. (1921). *The outline of history: Being a plain history of life and mankind* (Vol. II). The MacMillan Company.

Wells, H. G. (1938). *World brain*. Garden City, NY: Doubleday, Doran & Co.

Wells, H. G. (1945). *Mind at the end of its tether*. London, UK and Toronto, Canada: William Heinemann Ltd.

Wells, H. G. (1966). H. G. Wells on education. *Nature*, *211*(5053), 1061–1063.https://doi.org/10.1038/2111061a0.

Wells, H. G., Huxley, J. S., & Wells, G. P. (1931). *The science of life*. Cassell and Co.

Whitley, K. V., Tueller, J. A., & Weber, S. K. (2020). Genomics education in the era of personal genomics: Academic, professional, and public considerations. *International Journal of Molecular Sciences*, *21*(768), 1–19. http://doi.org/10.3390/ijm s21030768.

Williamson, B. (2018a, 26 July). *Genetics, big data science, and postgenomic education research*. https://codeactsineducation.wordpress.com/2018/07/26/postgenomic-education-research/.

Williamson, B. (2018b). *Personalized precision education and intimate data analytics*. https://codeactsineducation.wordpress.com/2018/04/16/personalized-precision-education/.

Winfield, A. G. (2007). *Eugenics and education in America: Institutionalized racism and the implications of history, ideology, and memory*. New York, NY: Peter Lang.

Wrong, G. M. (1921). The outline of history, being a plain history of life and mankind by H. G. Wells (review). *The Canadian Historical Review, 2*(2), 190–192.

Zudaire, I., & Napal Fraile, M. (2020). Exploring the conceptual challenges of integrating epigenetics in secondary-level science teaching. *Research in Science Education*. https://doi.org/10.1007/s11165-019-09899-5.

Creating Magical Research: Writing for a Felt Reality in a More-Than-Human World

Nicole Bowers

Riding the conceptual currents of the anarchive (E. Manning, October 30, 2018, personal communication) and non-representational methodologies (Vannini, 2015), I seek lively text in writing about nature, sustainability, and the environment and novel, playful ways of writing "research" of the same. Driven by a search for potentiated experiences for readers and writers in academic work, I reach toward magical realism to experiment with infusing my own writing practices with radioactive elements that exceed the static, informational text so often seen in normative science education articles. I propose to play with the writing of research through infusion of magical realism, engaging myself and readers in a tapestry of text that hopes to create "the real—the limit events that resist representation—as an immediate, felt reality" (Arva, 2008, p. 60).

It is no revelation that we live in precarious times. We, more-than-humans, have a new way of being in the world to grapple with; we face issues of climate change and environmental degradation as well as issues of poverty and displacement in a heightened and collective state. We have more forest fires, stronger hurricanes, more extreme floods, and longer droughts to look forward to (Allen et al., 2018). We can expect to deal with water shortages, food supply disruptions, land destruction, and mass migration of climate refugees (Miller, 2017; Rogelj et al., 2018). The International Panel on Climate Change (IPCC) report is quite clear on predicting what will happen,

N. Bowers (✉)
Arizona State University, Temp, AZ, USA
e-mail: nbowers1@asu.edu

M. F.G. Wallace et al. (eds.), *Reimagining Science Education in the Anthropocene*, Palgrave Studies in Education and the Environment, https://doi.org/10.1007/978-3-030-79622-8_5

and many scientists believe their predictions to be conservative (Butzer & Endfield, 2012).

Meanwhile, neoliberal politicians act to exacerbate the processes of unsustainability that brought us to this point in the first place. Late capitalism thrives on technologically enhanced modes of extraction and production to ensure continued economic growth with little regard for the natural and physical limits or consequences of that growth (MacGregor, 2014; Plumwood, 2002). Additionally, government programs have been gutted through deregulation and privatization, rendering them incapable of dealing with growing dire environmental and societal issues. The lack of market regulation has served to increase capital accumulation by the wealthy while decreasing opportunities and security for everyone else (Harvey, 2005). Over the past 40 years of neoliberalism, the growth rate of the economy remains the same as the rate prior to neoliberalism, but what has increased significantly is the percent of growth shared by wealthy elites (Harvey, 2005; Piketty, 2014). In addition to these economic and legislative mechanisms, neoliberalism acts as a "powerful public pedagogy" (Giroux, 2005, p. 14), so much so, that it may be easier to imagine the end of the world than the end of neoliberalism (Giroux, 2005).

These precarious times have been encapsulated by the contested term, the Anthropocene. Although the Anthropocene is seen by natural scientists as a problem of human activity and environmental degradation, many social scientists and researchers in the humanities also see it as a problem with entrenched ways of thought (Baskin, 2015; Johnson & Morehouse, 2014; Lövbrand et al., 2015). Analysis of the debates and discourse around the mainstream science narrative of the Anthropocene illuminate cracks in what, not long ago, was considered the impenetrable structure of Cartesian dualism (Lloro-Bidart, 2015; Lövbrand et al., 2015; Maggs & Robinson, 2016; Taylor, 2017). The Anthropocene's events have led to realizations around the interdependency of humans and non-humans, raised questions about politics and power, and indicate that a new, more transdisciplinary approach to education and decision making is needed (Castree, 2014; Davis & Todd, 2017; Latour, 2018; Swyngedouw, 2013). Momentum in many disciplines has built around addressing the issues of the Anthropocene, and it is important, particularly for those in educational research, to take this time to engage with the opportunity that the Anthropocene presents in rewriting our destructive narrative to be more sustainable (Lloro-Bidart, 2015; Lorimer, 2017).

In these turbulent times, science educators and researchers have been called to break with post-positivism, dualisms, and reductionism to settle on new onto-epistemological grounds (Bazzul & Kayumova, 2016; Haraway, 2016; Lather & St. Pierre, 2013). Theorists commonly acknowledge that traditional Western ideologies accompanied by their canonical toolkit that brought us the Anthropocene may be inadequate in ameliorating it (Haraway, 2016; Lloro-Bidart, 2015; Lövbrand et al., 2015; Maggs & Robinson, 2016; Tsing, 2015). One promising proposition (among many) lies in ontologies of process and epistemologies that expand to encompass affect with new combinations of knowing/experiencing/researching that honor the more-than-human world we need to navigate (Manning, 2013; Muraca, 2011). Although positing alternative onto-epistemologies propels researchers forward onto alternate and potentially more sustainable ground, possible movements on these new onto-epistemological grounds need expansive, plural exploration, and invention.

This chapter is one such exploration/invention and manifests as method-ological in nature. The approaches in this chapter could be considered post-qualitative (Lather, 2013; St. Pierre, 2013), post-human (Snaza & Weaver, 2014), non-representational (Vannini, 2015), and speculative (Shaviro, 2014), but following these burgeoning methodological orientations requires working with no guiding canon. Immanence provides the ontogenetic fabric for this chapter. I specifically work with the philosophies of Erin Manning, a philoso-pher whose work captured my imagination during an in-depth study of her writing, and who was kind enough to share her unpublished writing around the anarchive with all of the members of the study group exploring her work. I find resonance with Manning's work around Whitehead, who grounds my own commitments to an immanent and processual view of becoming. Addition-ally, Lather's and St. Pierre's (2013) call to push the boundaries of humanist qualitative research catalyzes this chapter's process of becoming.

Inspired by these theorists, I perform an alternative inquiry technique using elements of magical realism to explore tensions in the classroom. The genre of magical realism combines familiar realist techniques of thick description of everyday life with magical/fantastic elements to invite the reader to experience the mundane in different ways. This chapter invites the reader to consider alternative methodological approaches that may provide new openings, visions, experiences, and imagination amidst the problems of the Anthropocene and provides one example that I hope will inspire plural and divergent approaches to research. In this chapter, I will introduce anarchival writing, provide an example of magical realism-inspired research writing, and explain the elements of magical realism that may contribute to works that reverberate with the-more-than-human world of the Anthropocene (Faris, 2004; Manning, 2016; Richardson & St. Pierre, 2005).

Writing Beyond Findings

St. Pierre (2018) holds a view of connection between post-qualitative research and writing as a process of inquiry. Often, in traditional research, we are writing up our findings and reporting in a staid representational way that she refers to as the "linearity of the conventional qualitative research report" (St. Pierre, 2018, p. 605). She relates an instance where she did not intentionally write a sentence, but "it wrote itself" (St. Pierre, 2018, p. 605). I highlight this section of her text as it resonates with me; I have been involved in composing in dance, painting, writing, and even teaching where something occurs that I did not intend, but that I am not separate from either. This generative process that is not centered on the conscious intention of the researcher can be seen as part of research and may open up inquiry to the more-than-human. This feeling of only being a small part of the writing that "you" produce provides us with an entry point to talk about research that may be anarchival or beyond archival. It may imply that this new body created, in the case of St. Pierre (2018), her writing, is now a more active, living work resonant with ontologies of immanence. Not all writing seems to work this way; as Richardson and St. Pierre (2005) explain, traditional training in writing mires researchers in static representations or writing up findings.

Banks (2008) argues that the very notion of "findings... entail the idea that something preexists to be found" (p. 157), which indicates a specific view of reality, that of fully formed objects out there waiting to be known. This knower/known divide remains haunted with dualisms that have often been cited as fundamental problems of thought leading to our current unsustainable state, the Anthropocene (Haraway, 2016; Maggs & Robinson, 2016). Although, many researchers purposefully eschew positivist leanings, employing a variety of alternative theoretical frames, they often conform to conventional research format standards by reporting findings (Banks, 2008).

Manning (personal communication, October 30, 2018) would argue that the practice of reporting findings is more like archiving with its appearance of ordered, categorized, and authoritative truths. Reporting findings works well in ontologies of substance, but ontologies of immanence do not fit so squarely with the idea that findings pre-exist the reporting of them. Manning suggests anarchiving as a practice to move toward more active work as described by St. Pierre, but Manning understands that this task is not without obstacles.

A Case for Anarchival Writing

Manning (personal communication, October 30, 2018) begins her introduction of the anarchive by presenting a conundrum:

> The thing is, all accounting of experience travels through simplification—every conscious though, but all in a more minor sense, every tending toward capture of attention, every gesture subtracted from the infinity of potentions. And so, a double-bind presents itself for those of us moved by the force of potential, of processual, of the in-act. How to reconcile the freshness, as Whitehead might say, of process underway, with the weight of experience captured? (pp. 1–2)

The question that Manning (personal communication, October 30, 2018) poses is that if we capture an event in, "the ubiquitous model of description" (p. 2) how can we retain the activating potential that the event embodied in real time (the in-act of experience)? If we cannot retain this potential action, we are left with mere information to be transmitted and consumed in research reports with definite findings. An inability or unwillingness to include potential action leaves us with less likelihood of producing new movement or thought, as Manning (2013) cautions readers that "there are few starting points as lethal [to new thought] as the totalitarianism of Being" (p. 46).

This brings us to question not only the goal of research being finding preexisting somethings, but also to question representation. "Most contemporary social scientific research ends up... focusing on things that are stable, static, complete" (Vannini, 2015, p. 15), that totality of Being that Manning warns about. This focus results in representation, and "ordinarily, representation is bound to a specific form of repetition: the repetition of the same. Through representation, what has already been given will come to have been given again" (Doel, 2010, p. 117). The danger here is that traditional representational research, concerned with validly portraying what currently exists, may not lead us to any "politico-epistemic renewal" (Vannini, 2015, p. 3) needed to make the Anthropocene "as short/thin as possible" (Haraway, 2016, p. 160). In other words, the same systems of thought that makes it easier to imagine the end of the world than alternative politics may be, in part, supported by techniques of representation in that new thought is muted through continued reification. Additionally, many argue that attempting to represent an objective reality reinforces human exceptionalism, which is often cited as paving the road to the Anthropocene (Lövbrand et al., 2015; Muraca, 2011; Springgay & Truman, 2018; St. Pierre, 2018). This is not to say that conventional representational research has no place, but I join others in the call for more experimentation and space for multiple forms of research (Springgay & Truman, 2018; St. Pierre, 2018; Vannini, 2015).

Anarchival writing may be one type of writing that can bridge the conventional to the experimental, and Manning provides some guidance in the creation of such texts. She says that "anarchiving needs documentation—the archive—from which to depart and through which to pass. It is an excess energy of the archive" (Manning, personal communication, October 30, 2018, p. 21). Here the archive acts as a springboard. For example, thick description that comes with traditionally recognized forms of qualitative research may be one way of writing through/toward a more active creation as long as we go beyond just archiving and describing.

One last key to anarchiving needs to be explained, that of beauty. For Manning (personal communication, October 30, 2018) thinking with Whitehead, the strong concept of beauty is not the harmonious symmetry of "external aesthetic judgment" (p. 24). Instead, Whitehead (1933) calls the strong concept of beauty Intensity Proper, filled with difference and conflict, discordant and creative tension that intensifies toward novel becoming. Manning (personal communication, October 30, 2018) explains this beauty, saying "it is operative and felt more than it is seen. The beauty works on us more than we possess it, moving us with its more-than human intensity" (p. 24). Whitehead (1933) describes Intensity Proper as the strength of experience in the process of becoming, the excess that is so often excised when we aim to capture experience objectively through representation.

Manning suggests that when we try to convey this beauty in our experiences conscious intention may get in the way. Whitehead (1933) explains that "consciousness is the weapon which strengthens the artificiality of the occasion of experience. It raises the importance of the final Appearance relatively to that of the initial Reality" (p. 270). For Whitehead and Manning, actual objects exist, but they are artefacts of reality, or the process of becoming. Conscious intention and thought latches on to these objects, foregrounding them as actual and true (which they are), but placing all importance on their existence at the cost of excising the process of becoming. So, when St. Pierre (2018) claims that a sentence wrote itself, we might assume that she was part of a *process* that was not entirely conscious and subject to beauty in the form of Intensity Proper.

Including consciously thought description and moving toward beauty requires that we consider how events, entities, and occasions become. This paper takes Manning's view of immanence through process. For Manning (2013) and Whitehead (1933), definite bodies are ephemeral, phasing into individual existence and dephasing out in a constant process of becoming with/through a field of relation. Normal, conscious "knowing" is often conveyed in qualitative research through thick description. Now, imagine that there are other forces at play or at play differently as experience happens— shades of experience, what could have been, what is not, the intense beauty of the experience. These other forces are known "at a different register" (Manning, 2013, p. 24) that is not conscious thought. Manning collectively terms these forces as affect.

It is how to convey this affect, this excess of the static, captured events, entities, and occasions that this chapter is concerned with. This excess that is often excised in representation may be the site of novel creation, breaking open previously unimagined possibilities. Manning (personal communication, October 30, 2018) argues that techniques must be invented that allow for anarchiving, and hints that artfulness, intuition, changes in direction, and attunement to the minor gesture may aid in the invention of those techniques. Richardson (2003) suggests that writing can be a process of more-than also, but it depends on the forms and styles and involves approaches to writing research differently than traditionally "telling what you know." Manning herself engages in interactive art projects in a resonant way to that suggested by St. Pierre (2018) on writing as inquiry.

In response to this call for experimenting and inventing techniques for anarchiving, I explore the possibility of using magical realism in writing qualitative inquiry. This chapter marks the beginnings of my exploration, so next, I will share some magical elements in the following magical realism-inspired excerpt.

Live Science

Mrs. Foster sighed as she taped the last name tag on the red plastic crate. When Principal Trask told her three weeks ago that she would be the first teacher to receive new tables, she had begun planning immediately. The old desks were clunky, stocky numbers sporting rectangular laminated tops over a cavernous metal clad space where students kept their books and binders and

stuffed any number of odds and ends. Mrs. Foster shuddered thinking of some of the things she had pulled from those metal under-compartments just today.

"This will be better in so many ways," she murmured to herself.

Now the books and binders were stately arrayed, standing on end in the crates, each labeled with a student's name. No snot-filled tissue or exploded glitter pen would hide for months on end in the back of a desk compartment. Mrs. Foster looked out the window toward the teachers' parking lot, where only three cars remained. Although she had spent each day since learning of the impending arrival of the new tables ordering, organizing, clearing, and cleaning, this last day, the day of arrival, had left her at school several hours past the clamor and clang of students leaving for home.

Mr. Trask had selected Mrs. Foster as the new table pioneer for several reasons. Her classroom stayed meticulously arranged, she came an hour before and left an hour after students, and her test scores were among the top in the district.

When he informed her of her vanguard status with regard to the new tables, he said, "I have every confidence that you will work out the kinks without slowing down your class's pace, and I would very much appreciate it if you could present your strategies for classroom set-up and management around the new tables at the staff meeting on the Friday after you get your tables. Lots of pictures would be nice."

Mrs. Foster picked up her phone and started snapping pictures. She had arranged the twenty new desks into tables of four. Each desk had the name of a student printed in neat hand with dry erase marker. Directly on the desk! Upon first learning of this new feature, Mrs. Foster discussed it with some of the other teachers in the teachers' lounge.

She had admitted to the other teachers, "I am not sure if I like it or not."

The other teachers made some suggestions, but most agreed with Ms. Atkinson when she said, "It sounds like a nightmare, they will be distracted trying to write on their desks, they might try to write on other things inappropriately, and what if they use a permanent marker?".

Mrs. Foster made one last sweep of the room to make sure that all of the permanent markers were secured in her desk drawer and then headed out the door, flicking the lights off as she went.

The behavioral chart and some of the other anchor charts took an instant dislike to the new tables. They were a stretch to far out of their comfort zone. Instead of being rectangular or square, something they could easily relate to, they were triangular. The only straight lines apparent on their bodies were the two edges of the triangle making a rounded point, and those were obscured when joined to the other desks making a large round table. Even the outer curve of the tables was not a clean curved line; they were slightly wavy. Wavy, indeed! Who had ever thought of such a thing for children! The behavioral chart felt as if she would have to work extra hard now to add more structure and guidance to the classroom with the chaos of wavy edges added to the mix.

The tables, for their part, sat patiently, not too bothered by the whispers and tittering of their less solid classroom brethren hanging on the wall. They waited for the children, because of all of the other classroom objects, they and the chairs were most physically connected to the children, and it would be the children that shaped their lives most.

* * *

Tuesday morning burst through the windows in rays of sun, and Mrs. Foster surveyed the new set up with a more energetic eye. She smiled as she moved to the door. The rumble of conversation, barks of laughter, and squeals lessened as she appeared in the doorway. Facing her students down the hall and counting to ten in her head, she waited for her students to shuffle, squeeze, and bump into a fairly straight line.

With a smile she said, "We have a new seating chart, so after you enter, put your things away and find the desk with your name."

She turned her smile on the first student in line and asked, "So Analise, what are we feeling like today?".

Analise glanced to the right, and tapped the heart, indicating that her preferred greeting would be a hug. Mrs. Foster, being fairly short, only bent a little to embrace Analise before stepping aside and officially letting Analise into the classroom.

Mrs. Foster had been doing door greetings since the beginning of the year. She thought back to the day she sat at a long, molded plastic cafeteria table, taking precise notes during the district's professional development around student-centered teaching. In her spiral-bound notebook, she wrote:

Greet Students at Door

- creating a classroom community begins at the door
- greeting conveys to each expectation of engagement
- great way to build relationships (important for student-centered)

During the lunch break, she spoke with some of the other teachers, who suggested various greetings like a fist bump or a high five. Later that night, she surfed the web for more ideas and decided on fist bump, high five, hug, and dance (for students who did not want to touch). In the week leading to the first day of school, she printed out enlarged clip art of two fists bumping, a heart for a hug, the palm of a hand for a high five, and a musical note for dancing. After laminating them, she stuck them to the wall by her door with mounting putty in a neat vertical line. Most students liked to mix it up, one day with a hug another with a fist bump, but there were some students like Lupe, who always chose a hug, and Bella, who always chose a dance.

Some minutes later after a combination of hugs, dances, high fives, and fist bumps, Mrs. Foster stood by her desk at the far end of the room, counting in her head again as she waited for her students to settle down. Normally, by the

fifth month of school, her stillness and presence were enough to garner their attention, but today the students chatted excitedly past the count of fifteen. She held up her hand, high above her head in a peace sign. Slowly a wave of little peace signs rose in response as those students closest to her noticed her gesture for attention and stilled, thus signaling to those closest to them it was time to pay attention until the whole room was a forest of peace signs at different heights.

"I know the new desks are exciting, but we still have our rules and procedures," Mrs. Foster said as she pointed to the anchor chart of rules and the shiny, laminated behavioral chart adorned with clothes pins in a festive array of colors each with a student's name printed in black Sharpie.

"Remember," she added as she reached for the clothes pins, "everyone starts out on the 'ready to learn' block, but I did notice a few of you who were ready and organized when I first walked into the room. Great job Jasmyn, Aiden, and Marcus," she said as she moved each of their pins to the block labeled "great choice."

"Okay, before we start with math, I think maybe we should share what we think about the new tables. There is a dry erase marker in the small plastic tray under your desk, find it, and hold it up once you have it," Mrs. Foster said as she held up her own dry erase marker and walked slowly around the tables.

"You may write on your desk with this marker ONLY," she said, drawing out the word only

"What I would like you to do right now is write one word to describe what you think of your new desk."

As she walked around the room, she saw the word "awesome," several times, sometimes spelled without one of the e's. She also saw "weird," spelled "wird," "small," "shiny," "new," and a variety of other descriptors.

A bevy of hands shot up after Mrs. Foster announced, "Okay, time to share."

"Miguel, you wrote awesome, so why is the new desk awesome," she asked as she wrote the word awesome on the board to demonstrate the correct spelling.

"Well, um, I think it is awesome because it fits to make a table."

"Naomi, what is weird about your desk?"

"It is a weird shape, and it smells kind of weird."

Mrs. Foster paused while writing weird on the board, glancing back at Lupe, who had her eyes closed and was stroking the arm of Naomi who sat directly next to her. She finished the d with a sharp downward movement then glided over to stand behind Lupe and Naomi.

Leaning over, she pointed to Lupe's word, "happy," and asked, "Why happy, Lupe?".

"When I run my hands over the desk, it feels like it is laughing, like it is happy," Lupe responded as she took the hint to stop touching Naomi.

"Well these are some great descriptions, does anyone remember what kind of word describes a noun?"

After calling on various students giving a variety of answers, most examples of adjectives, Amanda said "adjective," and then the math block began.

Mrs. Forster gave a brief mini lesson as a review; after, the students sat quietly, working on their double-digit multiplication math worksheets. Mrs. Foster circulated around the tables stopping, checking students' work, occasionally exclaiming in praise or correcting some mistake. As she was telling Diego how nice his handwriting looked, she noticed Lupe stroking Naomi's arm again.

Crouching down by Lupe, Mrs. Foster said, "Lupe, what does number three of class rules say?".

Without looking up at the chart, Lupe murmured, "Keep hands, feet, and objects to yourself."

Mrs. Foster glanced at the desks standing off to the side of the room, the only old desk she had kept, and asked, "Lupe, would it help you to sit away from the other students?".

Lupe shook her head, never looking away from her math worksheet.

Mrs. Foster sighed, "Okay, well consider this a reminder then."

She straightened and walked over to the behavioral chart moving Lupe's clothes pin down from "ready to learn" to "reminder."

Mrs. Foster stared at the chart. She recalled speaking with a few of Lupe's other teachers in the teachers' lounge a few weeks ago.

"She just won't stop touching her classmates, and it does not seem to matter who she is sitting by, stroking arms, hair, sometimes just resting her hand on their shoulder. I don't know what to do about it. I was hoping that some of you may have come up with some strategies," Mrs. Foster had said, slightly deflated.

"I really haven't noticed that, but in art we are usually using our hands and Lupe has produced some amazing pieces."

"I haven't either, in robotics club, she is pretty talented—again we are using our hands a bunch there."

"Is she hurting other students? Have they complained?"

"No, but Lupe is sometimes so distracted with touching that she does not finish her work, and I worry that she is distracting other students. I think it might be a nervous habit, maybe if I gave her something touch, like a stress ball or something, it would help, but I also worry that something is wrong. She does it ALL the time."

Mrs. Foster glanced back at Lupe, who was stroking a small stuffed baby tiger with enormous eyes. At least she had stopped touching Naomi, but she still seemed more engaged in touching than doing her math worksheet. Despite having selected table mates for Lupe that usually did not liked to be touched, they did not seem to mind when Lupe touched them. Mrs. Foster had to remind her again to stop touching, this time, Adam, on the other side of her at the table. Lupe's pin on the behavioral chart was now on "stop and think," only one block up from "contact home."

When Lupe came back from recess, Mrs. Foster pulled her aside and said, "I think maybe we should move you to the desk over there," pointing to the lone, old desk at the side of the room.

"I would hate to have to call your parents again, and we still have one more lesson today—I think this will give you some space to focus."

"Can I write on my desk one more time?" Lupe asked.

Caught off guard by the request, Mrs. Foster stammered, "Um, sure, okay."

Lupe went to her new desk and wrote 'it's okay' then gathered her things including the bubble-eyed tiger and sat at the isolated desk off to the side as Mrs. Foster stood in the center of the classroom and announced amid glad whispers, "We have science today."

She instructed one table at a time to find their crates and get their science notebooks.

When all of the students' notebooks were sitting expectantly on their desks, she smiled and said, "I have a surprise for us today," as she held up a small circular clear plastic dish with something resting on it.

"Mr. Martin found a dead bee yesterday in his garden, and he heard that we were studying pollinators, so he brought it for us to look at."

Pointing to the board, she continued, "Here is a diagram of a flower with labelling boxes, underneath there are words with definitions. *If your table does not have the bee*, I want you to be working on the flower diagram, draw it in your science notebook, copy down the words and definitions, then try to label as many parts as you can. *If the bee is at your table*, I want you to draw the bee, *on a different page* and write down as many observations as you can. Remember, observations are done with the senses, in this case you are *not* going to touch or taste, but you *can* smell, listen, and look. Okay Shawn, what should you be doing if your table does not have the bee?".

"Working on the flower diagram in our notebooks."

"Great. Ali, what should you be doing if your table has the bee?"

"Um, draw it and write down observations."

"Yes, draw it and write down observations *on a different page*. Aiden, what senses are we *not* going to use during our bee observations?"

"Taste and touch."

"Great job paying attention," she said as she moved all students' pins that were low back to "ready to learn."

"Are there any questions before we get going? Yes, Adam?"

"Should we pass the bee to the next table when we are done?"

"Good question. I will give each table six minutes with the bee, then *I* will take it to the next table. Okay, let's get started."

In the last ten minutes of class, Mrs. Foster placed the bee on Lupe's desk. Lupe bent low, coming to eye level with the bee then raised up hovering her ear over the bee. Her head jerked up, and she glanced around, making sure that Mrs. Foster was occupied. Seeing her crouching at a table across the room pointing at Lily's notebook, Lupe reached out a finger and pressed it gently on the bee. Lupe closed her eyes and drew in a deep breath, letting it out slowly. Adam noticed Lupe touching the bee and stopped his drawing, staring at Lupe and the bee, waiting.

Lupe drew back her hand and stared intently at the bee. Adam raised up a bit in his chair to get a better look at the bee. It seemed to be wiggling, but he could not tell without standing up. Suddenly, the bee rose up, buzzing slightly, staggering a bit in flight, then zipped away. Adam could not control a delighted giggle, and Lupe glanced over at him and beamed.

Mrs. Foster looked up at the clock, and straightening hurriedly saying, "Sorry guys, I guess I lost track of time, go ahead and put up your notebooks, and we will start here next week during our science block."

In the rush to get everyone ready for the end of the day, Mrs. Foster forgot about the bee until after the students had gone. The plastic tray sat empty on her desk. After looking about the desk, she gave up with a shrug, thinking maybe one of the children took it. She remembered that Lupe had the bee last, so she went to Lupe's notebook to make sure that she had gotten a chance to make observations. There in her list of observations Mrs. Foster saw that Lupe had written buzzing and flying. She made a mental note to review inferences with the students at the beginning of the science block next week.

Magical Realism: A Guide

In magical realist text, authors weave impossible elements into realistic elements without explanation or world-building. The world in magical realism is our "real" world, recognizable through its thick description in the text. The magic in this world is treated just like everyday events. The characters and the classroom above are fictional; the details, events, and actions are all composites of real classrooms that I have observed or participated in during my years of teaching and researching—with the exception of the bee coming back to life. The excerpt demonstrates four of the main characteristics cited by Faris (2004) in her book *Ordinary Enchantments*, found in magical realism: the irreducible element of magic, defocalization, non-linear sequencing, and a realistic, everyday world.

The irreducible element of magic exists in the text defying universal laws of Western empiricism. Here the element of magic is irreducible because it is presented in text just as it is, presented with detail and certainty, with no extra explanation; it is presented just as factually and descriptively as everyday occurrences. Faris (2004) claims this element can evoke different responses from readers and embody central, political issues in the text.

The jarring appearance of the magical can interrupt the lulling realism, increasing a reader's participation in the text, creating an experience between the text and the reader that is novel. This contrast draws the reader into more attention and an active creation with the text. Magical elements disrupt normal reading habits and ask the reader to co-create/co-think with the text. For example, Lupe's resurrection of the bee may lead readers to reconsider her incessant touching of other students as something other than a nervous habit or further reconsider any student's odd behavior. Eliciting more consideration could be particularly valuable in creating anarchival texts that move forward from one author's writing to the reader's creation of novel ideas that are related to the text but not mimetic.

Additionally, the sensory detail that accompanies magical elements allows authors to textually embody forces at play that are not normally tangible in the moment of experience, like the effect of political or cultural realities on the shaping of a place and lives. Rushdie, the author of *Midnight's Children*, another magical realist text, writes that "realism can no longer express or account for the absurd reality of the world we live in—a world which has the capability of destroying itself at any moment" (as cited in Faris, 2004, p. 88). This suggests that if we include magical elements in realistic texts that we may be able to convey Manning's more than within the singularity of a specific experience. As Mrs. Foster does not know everything about Lupe's touching, this could indicate that, although she seems to be a diligent teacher, it is difficult to know everything about students' concerns and abilities and that students have their own reasons and worlds. The magical can bring in the threads of the rest of the world in both a temporal and geographical sense to the actual occasion of experience as magical realism has no allegiance to normal sequencing of events or static location. This is seen in the text above with the flashback sequences and Mrs. Foster's planning that seem to also be part of the present action, indicating that each action that we perform carries previous and anticipated actions.

This respects the breadth of the anarchive when Manning (2013) explains that "any occasion is at once the absoluteness of it-self in the moment of its concrescence and the will-have-become of its tendencies attunements, appetitions, both past and future" (p. 24). Our current experiences are heavy and pregnant with both the influx of the past and the pull of the future, and magical realism can provide a way to bring this together, often by collapsing time and bringing in specters of the past or with prophetic experiences.

Concerning the anarchive, this irreducible element may be key to writing in a different register so as to know in a different register as suggested by Manning. In other words, the contrast of the magical element allows the reader to reexperience the realist part of the text, inviting them to speculate on what else it might mean, opening the text to their own experiences through the disruption to the real that the magical element can cause. It may also be able to engage the reader in novel ways of creating/thinking by providing a text that offers glimpses of the whole world within the microcosm of magically enhanced mundane experience.

I also employed defocalization, a term coined by Faris (2004), referring to a narrative style that accentuates multivocal perspectives enhanced by the irreducible element of magic. In magical realist text, the perspective shifts fluidly from person to person and even to objects. Realism typically employs focalization that "originates in an individual's consciousness," Faris explains:

Because realism attempts to create an accurate picture of the world as it is experienced by ordinary human consciousness, readers of realistic fiction are most
familiar with focalizations that seem to be grounded in empirical evidence, the
quantity of sensory data enhancing the reader's confidence that this representation is accurate and causing the reader to invest in the narrative's picture of the
world with authority. (2004, p. 44)

Faris (2004) goes on to say that magical realism takes advantage of the realist
habit of deep sensory description but uses the irreducible element of magic
to decenter the perspective from the human to the more-than-human. The
presence of perceived magic, described sensually, creates a dissonance in the
reader. The reader knows that no human could perceive magic carpets, floating
priests, and bee resurrections, so they are left with a sense that the narrative
perspective comes from an other-than-human consciousness.

These shifts of narrative focus both between the magical and the real
and across different characters challenges representation's authority. Faris
(2004) explains that the realist description of the magical from an unfa-
miliar/unknown perspective challenges the norms of realistic representation
and undercuts the assumed authority of realism from within. Magical realism
questions realism's authority in the function of representation and creates
a space where realism is deconstructed while simultaneously employing its
representational power.

So, in addition to defocalized narration opening up a way to write the more-
than-human, it also provides a space to push representation past its inherent
stasis. Manning (2016) says that representations functions at the level of actual
bodies, archiving something static that can, at best, only weakly echo affect,
the live potential within the in-acted experience. It is this virtual potential that
the anarchive must create so as to move new, yet related thought forward from
one piece of work to another.

What makes magical realism so compelling in this sense is that it does not
discard realist, sensual representation but modifies it to take new directions.
The text above mostly describes the mundane and every day. This too is in
line with advice from Manning. As mentioned earlier, she contends that the
anarchive uses the archive as a springboard; they are not mutually exclusive.
Additionally, she sees movement of ideas within constraint to be a powerful
tool for creation of novel ways of thinking. She explains that habits, like that of
representational authority in qualitative research, structures our thinking and
doing, making our work predictable and comfortable, and although this seems
to be directly opposite of what she advocates in anarchiving, she sees a path
where habit is not discarded but pushed—asking "what else can habit do?"
(Manning, 2016, p. 87). She writes that:

Habit… is a mutable force. Habit directs our movements, constraining other
tendencies. These other tendencies, constrained as they are, can be said to still
be operative… The challenge is to make these minor tendencies operational,

> thereby opening habit to its subtle multiplicity and exposing the fact that habit was never quite as stable as it seemed. (Manning, 2016, p. 87)

This breaking open of habit is what we see Faris (2004) describe in magical realism; by employing traditional (habitual) realist representation but imbuing it with the irreducibly magical, representation is both challenged and transformed to encompass affect, the force of becoming. As writers, a technique that allows space for the more-than, is valuable if we wish our work to activate potential in other works not yet complete—or in short to anarchive.

Employing magical realism elements in qualitative writing as inquiry may allow us to decenter the human and evoke affect in our work. The added benefit remains, that the realist style of representational writing is not discarded, but instead coopted. It is transformed from its habitual mimicry, its static archiving into something more active, more live, possibly more able to drive change. Using realism as a springboard, magical realism presents the opportunity for the reader to be grounded in the concrete empirical observations that are so traditional in the field only to be invited to think more processually through magical realist elements.

I believe that moving with alternative onto-epistemological underpinnings provides a path away from entrenched ways of thought that have culminated in the crises of the Anthropocene. Of course, there exist no panaceas, and the process of reorientation requires adventurous and constant exploration and invention. Proliferation of new techniques and creative works in inquiry may be one way to challenge the status quo that has brought us the Anthropocene by allowing us to imagine, experience, and explore new politics. Although scholarly fiction is not a completely new proposal for qualitative inquiry, magical realism has not been a focus. I contend that anarchival writing of qualitative inquiry with magical elements may be a powerful technique that opens up new directions of thought needed as alternatives in the Anthropocene by challenging the habit of representation and the archive from within.

References

Allen, M. R., Dube, O. P., Solecki, W., Aragon-Duran, F., Cramer, W., Humphreys, S., Kainuma, M., Kala, J., Mahowald, N., Mulugetta, Y., Perez, R., Wairiu, M., & Zickfeld, K. (2018). Framing and context. In V. Masson-Delmontte, P. Zhai, H. O. Portner, D. Roberts, J. Skea, P. R. Shukla, A. Pirani, W. Moufouma-Okia, C. Péan, R. Pidcock, S. Connors, J. B. R. Matthews, Y. Chen, X. Zhou, M. I. Gomis, E. Lonnoy, T. Maycock, M. Tignor, & T. Waterfield (Eds.), *Global warming of 1.5°C: An IPCC special report on the impacts of global warming of 1.5° C above pre-industrial levels and related global greenhouse gas emission pathways, in the context of strengthening the global response to the threat of climate change, sustainable development, and efforts to eradicate poverty* (pp. 49–91). Intergovernmental Panel on Climate Change.
Arva, E. L. (2008). Writing the vanishing real: Hyperreality and magical realism. *Journal of Narrative Theory, 38*(1), 60–85.

Banks, S. (2008). Writing as theory: In defense of fiction. In J. G. Knowles & A. L. Cole (Eds.), *Handbook of the arts in qualitative research: Perspectives, methodologies, examples, and issues* (pp. 155–165). Sage.

Baskin, J. (2015). Paradigm dressed as an epoch: The ideology of the Anthropocene. *Environmental Values, 24*, 9–29.

Bazzul, J., & Kayumova, S. (2016). Toward a social ontology for science education: Introducing Deleuze and Guattari's assemblages. *Educational Philosophy and Theory, 48*(3), 284–299.

Butzer, K. W., & Endfield, G. H. (2012). Critical perspectives on historical collapse. *Proceedings of the National Academy of Science of the United States of America, 109*(10), 3628–3631.

Castree, N. (2014). The Anthropocene and geography I: The back story. *Geography Compass, 8*(7), 436–449.

Davis, H., & Todd, Z. (2017). On the importance of a date or decolonizing the Anthropocene. *An International Journal for Critical Geographies, 16*(4), 761–780.

Doel, M. (2010). Representation and difference. In B. Anderson & P. Harrison (Eds.), *Taking place: Non-representational theories and geography* (pp. 117–130). Ashgate

Faris, W. (2004). *Ordinary enchantments.* Vanderbilt University Press.

Giroux, H. A. (2005). The terror of neoliberalism: Rethinking the significance of cultural politics. *College Literature, 32*(1), 1–19.

Haraway, D. (2016). *Staying with the trouble: Making kin in the Chthulucene.* Duke University Press.

Harvey, D. (2005). *A brief history of neoliberalism.* Oxford University Press.

Johnson, E., & Morehouse, H. (2014). After the Anthropocene: Politics and geographic inquiry for a new epoch. *Progress in Human Geography, 38*(3), 439–456.

Lather, P. (2013). Methodology-21: What do we do in the afterward? *International Journal of Qualitative Studies in Education, 26*(6), 634–645.

Lather, P., & St. Pierre, E. A. (2013). Post-qualitative research. *International Journal of Qualitative Studies in Education, 26*(6), 629–633.

Latour, B. (2018). *Down to Earth: Politics in the new climatic regime.* Polity Press.

Lloro-Bidart, T. (2015). A political ecology of education in/for the Anthropocene. *Environment and Society: Advances in Research, 6*, 128–148.

Lorimer, J. (2017). The Anthropo-scene: A guide for the perplexed. *Social Studies of Science, 41*(1), 117–142.

Lövbrand, E., Beck, S., Chilvers, J., Forsyth, T., Hedren, J., Hulme, M., Lidskog, R., & Vasileiadou, E. (2015). Who speaks for the future Earth? How critical social science can extend the conversation on the Anthropocene. *Global Environmental Change, 32*, 211–218.

MacGregor, S. (2014). Only resist: Feminist ecological citizenship and post-politics of climate change. *Hypatia, 29*(3), 618–633.

Maggs, D., & Robinson, J. (2016). Recalibrating the Anthropocene: Sustainability in an imaginary world. *Environmental Philosophy, 13*(2), 175–194.

Manning, E. (2013). *Always more than one: Individuation's dance.* Duke University Press.

Manning, E. (2016). *The minor gesture.* Duke University Press.

Manning, E. (personal communication, October 30, 2018). Anarchive chapter: What things do when they shape each other.

Miller, T. (2017). *Storming the wall: Climate change, migration, and homeland security.* City Lights.

Muraca, B. (2011). The map of moral significance: A new axiological matrix for environmental ethics. *Environmental Values, 20*, 375–396.

Piketty, T. (2014). *Capital in the twenty-first century.* The Belknap Press of Harvard University Press.

Plumwood, V. (2002). *Environmental culture: The ecological crisis of reason.* Routledge.

Richardson, L. (2003). Writing: A method of inquiry. In Y. S. Lincoln & N. K. Denzin (Eds.), *Turning points in qualitative research: Tying knots in a handkerchief* (pp. 379–396). AltaMira Press.

Richardson, L., & St. Pierre, E. A. (2005). Writing: A method of inquiry. In Y. S. Lincoln & N. K. Denzin (Eds.), *The Sage handbook of qualitative research* (3rd ed.). Sage.

Rogelj, J., Shindell, D., Jiang, K., Fifita, S., Forster, P., Ginzburg, V., Handa, C., Kheshgi, H., Kobayashi, S., Kriegler, E., Mundaca, L., Séférian, R., & Vilarino, M. V. (2018). Mitigation pathways compatible with 1.5°C in the context of sustainable development. In V. Masson-Delmontte, P. Zhai, H. O. Portner, D. Roberts, J. Skea, P. R. Shukla, A. Pirani, W. Moufouma-Okia, C. Péan, R. Pidcock, S. Connors, J. B. R. Matthews, Y. Chen, X. Zhou, M. I. Gomis, E. Lonnoy, T. Maycock, M. Tignor, & T. Waterfield (Eds.), *Global warming of 1.5°C: An IPCC special report on the impacts of global warming of 1.5° C above pre-industrial levels and related global greenhouse gas emission pathways, in the context of strengthening the global response to the threat of climate change, sustainable development, and efforts to eradicate poverty* (pp. 93–174). Intergovernmental Panel on Climate Change.

Shaviro, S. (2014). *The universe of things: On speculative realism.* University of Minnesota Press.

Snaza, N., & Weaver, J. (2014). *Posthumanism and educational research.* Routledge.

Springgay, S., & Truman, S. E. (2018). On the need for methods beyond proceduralism: Speculative middles, (in)tentions, and response-ability in research. *Qualitative Inquiry, 24*(3), 203–214.

St. Pierre, E. A. (2013). The posts continue: Becoming. *International Journal of Qualitative Studies in Education, 26*(6), 634–645.

St. Pierre, E. A. (2018). Writing post qualitative inquiry. *Qualitative Inquiry, 24*(9), 603–608.

Swyngedouw, E. (2013). The non-political politics of climate change. *ACME: An International E-Journal for Critical Geographies, 12*12(1), 1–8.

Taylor, A. (2017). Beyond stewardship: Common world pedagogies for the Anthropocene. *Environmental Education Research, 23*(10), 1448–1461.

Tsing, A. L. (2015). *The mushroom at the end of the world: On the possibility of life in capitalist ruins.* Princeton University Press.

Vannini, P. (2015). Non-representational research methodologies: An introduction. In P. Vannini (Ed.), *Non-representational methodologies: Re-envisioning research* (pp. 1–18). Routledge

Whitehead, A. N. (1933). *Adventure of ideas.* The Free Press.

CHAPTER 6

Fire as Unruly Kin: Curriculum Silences and Human Responses

Annette Gough, Briony Towers, and Blanche Verlie

Introduction

When we started to write this chapter, large areas of Australia were on fire, including close to where we all live. The fire season in Australia in 2019 started earlier than previously, and continued into March 2020, a period that is now called the Black Summer. These fires have changed the way many Australians think of fire, their relationship with it, and the nature of society (Cave, 2020; Keneally, 2020; Marshall, 2020), and with climate change (Lippman et al., 2020; Flanagan, 2020; Marshall, 2020). The fires have opened up discussions about the relevance of traditional Indigenous burning practices to land management practices (Faa, 2019; Scolaro, 2018; Williamson et al., 2020) and strategies for reducing fire risk (Bonyhady, 2020). The fires have also opened up questions around bushfire education in schools, including bushfire prevention, mitigation, preparedness, and response in an Australian context (Towers et al., 2020).

The bushfires have impacted communities and their schools across most Australian states. Some schools were destroyed in the bushfires, with the New South Wales (NSW) government having 178 schools being bushfire affected (Chrysanthos, 2020). Even where the schools were saved, communities were affected. For example, in the small coastal town of Milton (NSW), a teacher at

A. Gough (✉) · B. Towers
RMIT University, Melbourne, VIC, Australia
e-mail: annette.gough@rmit.edu.au

B. Verlie
The University of Sydney, Sydney, NSW, Australia
e-mail: blanche.verlie@sydney.edu.au

© The Author(s) 2022
M. F.G. Wallace et al. (eds.), *Reimagining Science Education in the Anthropocene*, Palgrave Studies in Education and the Environment,
https://doi.org/10.1007/978-3-030-79622-8_6

91

one of the schools said he was glad the school had not been damaged so the children will have somewhere to come back to at the start of first term: "We have families at the school who've lost absolutely everything—and a school needs to be a safe, secure place" (in Fernandez & Lapham, 2020). Schools were also evacuated and closed during the pre-Christmas fires. School closures have significant effects on communities. Children are often left at home alone when schools are closed, and this creates significant risks for them and their families. They are also missing out on formal learning when not in school, so the fire becomes an anti-pedagogue (Rickards et al., 2019). Smoke from the bushfires also imposes on their learning as they are kept inside the school buildings rather than playing outside, and when outside they wear face masks because of the dirty air. In the coastal Victorian town of Mallacoota, 4,000 holidaymakers were stranded by bushfires cutting road access. They were then evacuated by naval ships and helicopters, often splitting families, with children going while parents remained (Millar & McMillan, 2020) or children leading their families to evacuate (Towers et al., 2020). The traumatic effects of such experiences on children and their learning cannot be underestimated, and it makes it very clear that fire is not just "in" the curriculum, but is inside, throughout, and outside the classroom, and it acts pedagogically.

Fire has a close relationship with human evolution, but humans continue to have an ambiguous relationship with fire, as it can be a friend and fiend, and more broadly, as we will argue, unruly kin. This uneasy relationship has become even more apparent in recent times in Australia. While debate about the relationship between the bushfires and climate change continues to rage, there is also much debate about how and what fire management practices should be applied in the Australian environment. Fire is very much part of the Australian landscape and the lives of Australian people. As Indigenous fire practitioner Victor Steffensen said, "We live in a country that needs fire" (SBS News, 2016), but not the supercharged blazes of recent months, which have been fuelled by climate change and severe drought, and that, since September 2019, have torched at least 46 million acres, killed at least 34 people, destroyed more than 6,500 buildings, and resulted in the death of more than a billion animals (reptiles, birds, and mammals) (RMIT ABC Fact Check, 2020). Capital cities have also been blanketed by heavy smoke from the bushfires for days, weeks, and months on end, with potential long-lasting health effects on the residents, especially children and the elderly, and an estimated 400 human deaths during the summer attributed to this. That the country (and its oceans) is getting hotter is not in doubt, except to the climate change sceptics, but even they cannot deny that we had the three hottest December days ever recorded in 2019 (Morton, 2019), and the drought continues.

In this chapter we discuss the relationship between fire and human evolution, which has led to the suggestion that recent times should be renamed the Anthropocene. We then look at how bushfires affect children and schools, and how fire is currently covered in the Australian Curriculum. This leads us to discussing different ways of approaching a fire pedagogy—one that is more respectful of fire as unruly kin. We acknowledge that calling fire "kin" may seem odd to some, but given that education has a role to socialise people to live in societies, and fire is now very much part of our societies, then it is not such a big leap to see fire as our unruly kin.

Unruly Kin: Fire, Human Evolution, and the Pyrocene

The discovery of fire was a pivotal moment in human evolution. Once humans had fire and could burn living and recently dead plants, they had light, had warmth, had a means of protection from other wild animals, could manage the landscape in novel ways, could make tools, and could cook their food (Glikson, 2013; Gowlett, 2016; MacDonald, 2017; Singh, 2017; Yin, 2016). Cooking also "detoxifies some foods that are poisonous when eaten raw, and it kills parasites and bacteria" (Adler, 2013). In addition, according to Andrew Glikson,

> the cooking of meat and therefore enhanced consumption of proteins allowed a major physiological development into tall hairless humans. . . . The utilization of fire has thus constituted an essential anthropological development, with consequences related to bipedalism, brain size and the utilization of stone tools. (2013, pp. 89–90)

Humans also have a genetic mutation that is not found in other primates that allows certain toxins, including those found in smoke, to be metabolised safely, thus inuring humans against some adverse effects from fire, when other species were not (Yin, 2016). Such evolutionary advantages afforded to humans by their relationship to fire leads Steve Pyne (2019) to call the human alliance with fire "a veritable symbiosis." According to Pyne (2019), the story of fire defines humanity's distinctive ecological agency: "humans are a uniquely fire creature, not only the keystone species for fire but a species monopolist over its manipulation." Fire has also become socially embedded, and its influence "has reached in some way into the human psyche, expressed in religion, in ritual, in ceremony and through ubiquitous myths about fire origins" (Gowlett, 2016, p. 7). Another aspect of human relationships with fire is its cultural consequences in relation to patriarchy: although historically some Australian Aboriginal women were involved in traditional burning practices (Steffensen, 2020), in general, "by allowing men to go out hunting while women stayed behind to cook by the fire, it spawned gender norms that still exist today" (Yin, 2016).

Fire is thus intertwined with human development, and fire can be seen as kin, in the sense used by Donna Haraway (2008, 2016). For Haraway, "kin" are those beings and bodies that we are entangled with, and who are central players in our processes of becoming—that is, our physical and social evolution. What she terms "companion species" (e.g., our gut bacteria, plantation crops, pets, etc.) can be our kin, but not all kin need to be living. Kin cannot be fully domesticated—and perhaps they domesticate us (Tsing, 2012)—but our relationships with kin do need keen attention and care. For Haraway, "the task is to make kin in lines of inventive connection as a practice of learning to live and die well with each other" (2016, p. 1). Complicated, messy, interwoven lives are not to be eschewed, but noticed and cultivated in ways that contribute to multispecies flourishing, and this includes the "complicated multispecies history" of fire, an agent that contributes to life, regeneration, and death (Haraway, 2016, p. 44).

In addition to the combustion of living and recently dead organic matter, there is evidence all around us that the combustion of ancient, decomposed, and compressed organic matter, i.e., fossil fuels, has also transformed human societies and systems. Indeed, our capitalist, globalised, consumer/market driven, urban systems, including transport, all stem from the burning of hydrocarbons. Fire is a process of combustion, and the products and outcomes of the combustion of fossil fuels are at the core of claims that we are now living in "the Anthropocene" (Crutzen & Stormer, 2000), the present period of geological time dominated by human impact on the Earth's processes. The starting date for the epoch is contentious, as is the term itself— often because of its human supremacy, and the way it hides troublesome differences between humans (including gender and cultural differences) and ignores specific humans' intimate relationships with technology and other animals (Åsberg, 2017; Gough, 2021). Indeed, Cecilia Åsberg (2017, p. 198) asserts that "nature is no longer separable from culture in this age of the Anthropocene." This has generated competing suggestions of names for the epoch, including Econocene, Capitalocene, Chthulucene, Plantationocene, and Symbiocene (Albrecht, 2016; Haraway, 2015, 2016). Given the central role of human practices of combustion in such debates, Pyne (2019) has argued that an appropriate name is the "Pyrocene": an era generated by multiple practices of combustion, and one that is characterised by increasing planetary flammability, the coming Fire Age.

The combustion of fossil fuels and its generation of the "Pyrocene" evidence our notion of fire as unruly kin. Although fossil fuels have provided great benefits to humans, their combustion has displaced—geographically but also temporally—heat energy, and this is now having unintended and increasingly unmanageable consequences, including climate change and more frequent and intense bushfires. Over Australia's Black Summer, firefighters found themselves in conditions with fires that could not be "fought," or even "contained," until heavy rains came. Who we are now is intimately entangled with our fiery histories, and who we will become is unavoidably tied to how the planet and our local ecologies burn in response. A keener attention to our past, present, and possible relations with fire, and acknowledgement of fire as unruly kin, is needed if we are to attempt the massive infrastructural, social, political, and economic transformations needed to avert widescale climate collapse. Yet colonial-industrial fire practices tend to aver such attunement and reflection. Thus, for a "pyro-pedagogy" (Reid, 2019, p. 771) responsive to the conditions we find ourselves in, we first turn to Indigenous fire stewardship, acknowledging that we are all white settlers and do not wish to appropriate Indigenous knowledge but rather let it be heard through drawing on Indigenous voices and sources.

Indigenous Fire Stewardship

As defined by Lake and Christianson, Indigenous fire stewardship is

> the use of fire by various Indigenous, Aboriginal, and tribal peoples to: (1) modify fire regimes, adapting and responding to climate and local environmental conditions to promote desired landscapes, habitats, and species; and (2) to increase the abundance of favored resources to sustain knowledge systems, ceremonial, and subsistence practices, economies, and livelihoods. (2019, p. 1)

Amy Christianson (2015) draws a fundamental distinction between wildfire and fire resulting from the burning practices of Indigenous people. As she explains, wildfires can be from natural processes or human activities and generally occur in the summer months or at other times when vegetation is dry and conducive to burning out of control. In contrast, Indigenous burning is highly controlled and generally takes place during low-risk conditions: decisions about where and when to burn are guided by a deep understanding of climatic cycles, ignition sources, fire behaviour, and landscape factors, such as how the topography and vegetation/fuels contributed to the natural fire regime and associated landscape fire effects (Huffman, 2013; Lake & Christianson, 2019; Steffensen, 2020). This knowledge, developed through multigenerational observations of fire-prone ecosystems, and the resultant Indigenous fire adaptations and cultural fire regimes, enabled Indigenous peoples on various continents to "live with fire" (Lake & Christianson, 2019).

In Australia, Aboriginal peoples applied fire to different ecologies, at different times, shaping their surroundings for more than 65,000 years (Neale et al., 2019; Pascoe, 2018). Bruce Pascoe (2018) cites evidence from palynologists, which suggests Aboriginal Australians began using fire as a tool as long as 120,000 years ago. Before colonisation, Indigenous fire stewardship on Country (a term commonly used to describe Aboriginal homelands, and which encompasses entangled relations between ancestors, spirits, species, and place), worked on five principles: (1) the majority of agricultural lands were burned on a rotating mosaic, which controlled intensity and allowed animals and plants to survive in refuges; (2) the time of burning depended on the type of land (e.g., soil type, dominant species of vegetation) and the condition of the bush at the time; (3) the prevailing weather was crucial to the timing of the burn; (4) neighbouring clans were advised of fire activity; and (5) the growing season of particular plants was avoided at all costs (Pascoe, 2018; see also Steffensen, 2020). Importantly, as Victor Steffensen (2020) describes, not every ecosystem was burned: he refers to "no-fire ecosystems" and cites wet rainforests and riverbanks as key examples. Engaging with Country in this way provided many benefits and resources, including food, water, medicine, access to key areas, fibres for basketry, and protection from dangerous bushfires, sustaining human life on the continent for millennia (Huffman, 2013; Pascoe, 2018; Russell-Smith et al., 2003; Steffensen, 2020).

The Aboriginal ontological concept of "the Dreaming" embeds fire within a metaphysical context (Eriksen & Hankins, 2015). Aboriginal people generally do not separate synergistic relationships between fire and other aspects of the physical and metaphysical: rather, those relationships are interdependent and interconnected. The scaling of those relationships extends from the individual to the universe, and is inclusive of the feedbacks within those levels (Eriksen & Hankins, 2015; see also Steffensen, 2020). Aboriginal fire knowledge is applied to the landscape to nurture and maintain those relationships, and the knowledge is passed from one generation to next through lore and cultural practice (Steffensen, 2020; Pascoe, 2018): "cultural fire helps prevent fire risks, rejuvenate local flora, [and] protect native animal habitat, all while restoring the kinship to the land" (Oliver Costello, cited in Higgins, 2020).

The process of colonisation severely limited Indigenous fire stewardship practices in countries around the world (Eriksen & Hankins, 2015; Lake & Christianson, 2019; Mistry, 2000), including Australia (Gammage, 2011; Neale, 2018; Neale et al., 2019; Pascoe, 2018; Weir et al., 2020). When the British invaded the Australian continent in 1788, colonial interests disrupted the Aboriginal use of fire through the forced removal of people from their sovereign lands and a policy of fire prohibition (Eriksen & Hankins, 2015; Gammage, 2011; Vale, 2002). The colonial worldview was that fires were destructive to the timber supply and dangerous to communities (Pyne, 2007). Interestingly, when the colonists invaded, they found what historian Bill Gammage has referred to as "the Greatest Estate on Earth"—vast park-like environments in which large trees were carefully situated within pampered grasslands, providing sustenance and shelter to an array of grazing animals. Ironically, these landscapes, so appreciated by the colonists, were created by the very fires they feared (Turner, 1999). Kohen (cited in Pascoe, 2018, p. 165) put it best:

> While Aboriginal people used fire as a tool for increasing the productivity of their environment, Europeans saw fire as a threat. Without low intensity burning, leaf litter accumulates, and crown fires can result, destroying everything in their path. European settlers feared fire, for it could destroy their houses, their crops and it could destroy them. Yet the environment that was so attractive to them was created by fire.

Fire and the Australian Curriculum

Such nuanced understandings of fire as part and parcel of life on this flammable continent, and something to be embraced, celebrated, and respected, is largely absent from Australian culture, which aligns with broader Western cultural relationships with fire. According to Pyne (2019),

> Fire disappeared as an integral subject about the time we hid fire into Franklin stoves and steam engines ... It lost standing as a topic in its own right. As with the fires of today, its use in history has been to illustrate other themes, not to track a narrative of its own.

An analysis of the Australian Curriculum (ACARA, 2019) supports this observation. Combustion is not covered as a topic in Science (or anywhere else) until Year 9. Here the Content Description statement is "chemical reactions, including combustion and the reactions of acids, are important in both non-living and living systems and involve energy transfer," and the two Content Elaborations are "recognising the role of oxygen in combustion reactions and comparing combustion with other oxidation reactions" and "describing how the products of combustion reactions affect the environment." Fire is also not discussed in Science until Year 9, and here it is only a "such as" in an Elaboration. The Content Description statement is: "Ecosystems consist of communities of interdependent organisms and abiotic components of the environment; matter and energy flow through these systems" and the associated Elaboration is "investigating how ecosystems change as a result of events such as bushfires, drought and flooding" (ACARA, 2019).

The most comprehensive location for the study of bushfires in the Australian Curriculum is in Year 5 Geography. Here the Content Description statement is: "The impact of bushfires or floods on environments and communities, and how people can respond," and this is elaborated as.

- mapping and explaining the location, frequency and severity of bushfires or flooding in Australia
- explaining the impacts of fire on Australian vegetation and the significance of fire damage on communities
- researching how the application of principles of prevention, mitigation and preparedness minimises the harmful effects of bushfires or flooding.

Climate change does not fare much better in the Science curriculum. It only appears in three Elaborations in Year 9 Science. The Content Description statement says, "Scientific understanding, including models and theories, is contestable and is refined over time through a process of review by the scientific community," and in the Elaboration there is "considering the role of science in identifying and explaining the causes of climate change." The Content Description "advances in scientific understanding often rely on technological advances and are often linked to scientific discoveries" includes "considering how computer modelling has improved knowledge and predictability of phenomena such as climate change and atmospheric pollution," and the Content Description "people can use scientific knowledge to evaluate whether they accept claims, explanations or predictions and advances in science can affect people's lives including generating new career opportunities" includes the Elaboration "considering the scientific knowledge used in discussions relating to climate change."

These silences around fire and climate change in the written curriculum are very concerning, as teachers increasingly only teach what is going to be assessed (Popham, 2001) and are less likely to teach that which is not explicitly written into the curriculum (Earp, 2019). This affects what is then taught and learned. The implementation and effectiveness of this curriculum has not been reviewed at a state, territory, or national level since it was developed. Given the curriculum isn't always taught in the same way as it is written, we should not assume bushfire education is being delivered as intended, or that it is being delivered at all.

One problem with the Australian Curriculum content statements is that they are relatively abstract and detached from children's lived experiences (Towers et al., 2020). Towers (2015) interviewed Australian children aged 8–12 to find out their knowledge of bushfire emergency responses and found many misconceptions about bushfire safety, which often came from a lack of knowledge about bushfire behaviour. For example, children often assume bushfires only travel through direct flame contact and think a nonflammable physical barrier (such as a river, a road, or a brick wall) will prevent a bushfire from reaching their property. But burning embers can travel many kilometres ahead of the fire front, and ember attack is a major cause of home ignitions. Such misconceptions are best addressed by making bushfire education more relevant to their own lives. Children need to explore and understand vulnerability to bushfire in their own communities as well as their capacity for reducing risk.

Bushfire education in schools is more effective when taught across the curriculum, rather than as isolated topics. One example is the bushfire education programme at Victoria's Strathewen Primary School for students in grades five and six. It incorporates science, art, civics and citizenship, design, English, and geography. An evaluation of this programme (Towers et al., 2018) showed it increased children's knowledge of local bushfire risks and the actions people can take to manage them. It also helped increase children's confidence for sharing their knowledge with others, gave them a sense of empowerment, and reduced bushfire-related anxieties. The programme's benefits extended to families, including increased bushfire planning at home with more participation from children in the process.

Traditionally, school-based disaster risk reduction and resilience education (DRRRE), such as bushfire education, has tended to adopt a transmission model of education, where a specific body of knowledge is transferred from adult to child. In this model, children are positioned as passive receivers of information as opposed to critical thinkers and problem solvers. This model also tends to be information driven rather than action oriented. Even when children are encouraged to take action, adults often prescribe the action, and there are limited opportunities for children to creatively address other issues that might concern them and for them to feel empowered to act at critical moments. What is needed is a more child-centred approach that is more holistic, place-based, and participatory. Such approaches not only provide children with essential knowledge and skills, but also empower them to actively participate in disaster risk reduction and resilience-building activities. We also need a pedagogy that embraces becoming-with fire: one where the boundaries between the students and fire "move, blur, dissolve and/or are (re)established due to their on-going intra-action" (Nakagawa et al., 2020, p. 11; see also Verlie, 2018, for an example of becoming-with climate, and with fossil fuels).

Pyro-Pedagogies of Becoming-With

"Our house is on fire" proclaimed Greta Thunberg (2019), and following her, thousands of young people around the world protested governmental inaction on climate change. These words reverberated ever louder as children and young people around Australia mobilised in public spaces with atmospheres thick with the airborne remains of native forests throughout the southern summer of 2019/2020 (Wahlquist, 2019; Regan & Yeung, 2020). Shouting from behind their P2 masks, we saw young people standing up for their futures, and becoming activists in the process. Exploring young people's lived experiences of bushfire within a colonial, carbon-intensive economy, pushes us to consider potentials for alternative pyro-pedagogies. The Black Summer demonstrates that fire operates pedagogically well beyond the confines of the classroom, and exerts agency exceeding the bounds of curriculum statements that dictate how to teach "about" fire. In multiple, intersecting, and reverberating ways, fire affects who we are and can be as humans. Our relationships with fire are changing, meaning that we are changing at a fundamental level of who we are, how we understand what it is to be human, and how we can live in this hot, dry, colonised country. Thus, we are calling for pyro-pedagogies founded on a notion of becoming-with fire, our unruly kin.

Reconsidering "what" "where" and "when" fire is, is a preliminary move towards this. As noted previously with respect to some children's recent lived experience of bushfire, fire is not just "in" the curriculum but is inside, throughout, and outside the classroom. Indeed, combustion co-composes what the classroom is: what it is made of, where it is, whether and how it is heated, how students and staff get to school, what they eat, and how flammable and at risk it is, etc. And classrooms co-compose fire, reciprocally: the school system more broadly is embedded in fossil fuel economies, and it positions children as future workers in capitalist systems. The material economies that build classrooms literally contribute to emissions which fuel fires. Classrooms and fire are much more closely and intimately related than we might typically think.

We are also advocating for understanding fire not just as something to be learned about, but as a pedagogue that operates through its role as unruly kin. Across Australia's Black Summer, children were kept home, sent home and evacuated from school, kept indoors, sent to hospitals, evacuated from holidays, evacuated with their families, participants in recovery efforts, and more. For young people who led their families to safety (Towers et al., 2020) this was surely a life-changing moment, and their sense of self will be affected in profound ways, even if we cannot anticipate that or even trace it with accuracy. That these experiences occur outside of classrooms does not mean they are not pedagogical or curricular. We don't know exactly what students are learning, or who or what they are becoming through these processes. The point is that fire is a pedagogue that exceeds the capacities of teachers, parents, and humans more broadly—it is unruly, and it changes us in significant, species- and society-changing ways. Further, even within "climate controlled" classrooms, and when mediated through imagery and narrative, fire acts pedagogically (Verlie, 2019a), especially in affective ways. All of this considered, we know that after this Black Summer things will not be the same: Australian culture will not be the same, and summer itself will not be the same. For example, people have gone from counting down the days until summer holidays start, to counting down the days until summer is over, indicating a significant affective change in our relationships with weather, seasonality, climate change, and our forests.

Part of fire's pedagogical capacities is its ability to shape who and how we become. Our focus on "kin" and "becoming" emphasises that human relationships with the non-human world are dynamic and open to the agency of the nonhuman world. Such a pedagogy does not provide clear-cut, practical suggestions of how and what to teach, but emphasises the need to continually be attuning to the agency of the world and how it interacts with and affects students' whole lives (Verlie, 2017; Nakagawa et al., 2020). However, we can become-with fire in different ways, and educators can play a role by "sparking" such changes and directing pyro-pedagogies in certain ways. For example, in traditional DRRRE, children might become hazard managers, replete with the reductive, top down, resource management approach to trying to keep nature in its place. Alternatively, they could become cultural burners and stewards of fire, if supported to learn from and engage with Aboriginal culture and ecological practice—such as through Marcia Langton's (n.d.) *Resources in Aboriginal and Torres Strait Islander Histories and Cultures for Teachers*. An Aboriginal-led pyro-pedagogy could be our moment for rejecting human exceptionalism, through attending to fire's unruly, massive, and terrifying but also beneficial and healing qualities.

A pyro-pedagogy of becoming-with fire would also be situated and place-based. As CSIRO bushfire expert Justin Leonard (in *SBS News*, 2016) said, "there's a way to live in every part of the landscape and it's this integrated way of understanding what, what fire is in that location, how to find the balance and manage the bush in the right way, and that easily unlocks how you build and live and behave and understand." Rejecting colonisers' logics that there is one right way to live in all places, and focusing on how particular combustive ecologies intersect with particular cultures and experiences will be fundamental to cultivating education that is relevant to children's whole lives. A situated and relevant pyro-pedagogy would also contribute to the transdisciplinary and participatory modes of education that Teresa Lloro-Bidart (2015) argues are necessary in the Anthropocene. This pedagogy also needs to adopt a whole-school approach, in line with Steffensen's (in *SBS News*, 2016) suggestion that we need to involve "communities in understanding fire. Not in its vicious form or its threatening form but understanding fire and its nurturing and... how beautiful it really is."

Briony Towers' (2019) research on bushfire education offers important insights for such pyro-pedagogies. For children to develop a coherent understanding of bushfire risk that is sufficient for identifying problems and solutions in their own local context, Towers argues that a holistic learning framework that incorporates the environmental and social dimensions is needed. "This requires teaching and learning activities that systematically build children's knowledge and awareness of the various dimensions of risk—the physical hazard, exposure, vulnerability and capacities—and how those dimensions interact to cause hazard impacts and disasters" (2019, p. 72). Towers also argues that there needs to be a.

> place-based pedagogy of bushfire risk, [where] the surrounding socio-environmental context serves as the learning ecosystem: abstracted environments are substituted with local landscapes; textbooks and worksheets are replaced by local experts and experiential activities in the field; and generic information about bushfire risk is augmented by local knowledge, data and predictions. (2019, p. 73)

Such an approach, that situates children in their specific worlds and sees them as active subjects who interact with and become-with fire, demonstrates that understanding fire as unruly kin that our children are unavoidably entangled with can generate pedagogies responsive to the conditions of the Pyrocene. Through acknowledging and teaching about fire's unruly capacities, we can steer students towards more ecologically attuned and responsive and resilient ways of being.

CONCLUSION

Robert Glasser (2019, p. 4) was quite prescient when he wrote in early 2019, "This emerging Era of Disasters will increasingly stretch emergency services, undermine community resilience and escalate economic costs and losses of life." What the bushfires of Australia's 2019/2020 Black Summer have shown us is that we need a different relationship with fire in the environment, one that recognises fire as our unruly kin and which foregrounds the agency of human and more-than-human materiality (i.e., "nature") and their entanglement. We see a pyro-pedagogy of becoming-with fire as building on each of our previous writings—Annette Gough's (2019) on more-than-human scientific inquiry in education, Blanche Verlie's (2018) on becoming-with climate change, and Briony Towers (2019) on child-centred DRRRE—and more. In doing this we have much to learn from Victor Steffensen and other First Nations peoples. We also hear young people's refrain, *our house is on fire*, calling ever more loudly for education to recognise that climate change is radically reconfiguring the fundamental conditions of children's lives today and in the future. Thus, we believe that thinking much more broadly about humanity's entanglement with fire—the histories, futures, geographies, and affective ecologies of various kinds of combustion—is necessary if we are to enable our young people to learn to live with climate change (Verlie, 2019b). This is very different from the current abstract, teacher-centred approaches to bushfire and climate change education in the Australian Curriculum, but one that is sorely needed.

Acknowledgements We acknowledge that this chapter was written by Annette and Briony on the land of the Wurundjeri people of the Kulin nation in which Melbourne now is located, and by Blanche on the land of the Gadigal people of the Eora nation in which Sydney is now located. We pay our respects to the Elders both past and present, and to those emerging.

REFERENCES

Adler, J. (2013, June). Why fire makes us human. *Smithsonian Magazine*. https://www.smithsonianmag.com/science-nature/why-fire-makes-us-human-72989884/.

Albrecht, G. A. (2016). Exiting the Anthropocene and entering the Symbiocene. *Minding Nature, 9*(2). https://www.humansandnature.org/filebin/pdf/minding_nature/may_2016/Albrecht_May2016.pdf.

Åsberg, C. (2017). Feminist posthumanities in the Anthropocene: Forays into the postnatural. *Journal of Posthuman Studies, 1*(2), 185–204. https://doi.org/10.5325/jpoststud.1.2.0185.

Australian Curriculum and Assessment Authority (ACARA). (2019). *Australian Curriculum*. http://www.australiancurriculum.gov.au.

Bonyhady, N. (2020, January 21). Prime minister says hazard reduction burns as important as emissions. *The Sydney Morning Herald*. https://www.smh.com.au/politics/federal/prime-minister-says-hazard-reduction-burns-are-climate-action-20200121-p53tha.html.

Cave, D. (2020, February 15). The end of Australia as we know it. *The New York Times*. https://www.nytimes.com/2020/02/15/world/australia/fires-climate-change.html.

Christianson, A. (2015). Social science research on indigenous wildfire management in the 21st century and future research needs. *International Journal of Wildland Fire, 24*, 190–200.

Chrysanthos, N. (2020, January 17). NSW to spend $20 million on bushfire recovery for 178 schools. *The Sydney Morning Herald*. https://www.smh.com.au/national/nsw/nsw-to-spend-20-million-on-bushfire-recovery-for-178-schools-20200116-p53s04.html.

Crutzen, P. J., & Stoermer, E. F. (2000). The "Anthropocene." *Global Change Newsletter, 41*, 17–18.

Earp, J. (2019, November 13). Education reform: Curriculum content and deep learning. *Teacher Magazine*. https://www.teachermagazine.com.au/articles/education-reform-curriculum-content-and-deep-learning.

Eriksen, C., & Hankins, D. (2015). Colonisation and fire: Gendered dimensions of indigenous fire knowledge retention and revival. In A. Coles, L. Gray, & J. Momsen (Eds.), *The Routledge handbook of gender and development* (pp. 129–137). Routledge.

Faa, M. (2019, November 14). Indigenous leaders say Australia's bushfire crisis shows approach to land management failing. *ABC News*. https://www.abc.net.au/news/2019-11-14/traditional-owners-predicted-bushfire-disaster/11700320.

Fernandez, T., & Lapham, J. (2020, January 16). Back to school—NSW Education Minister says 140 bushfire-hit sites will be safe for students. *ABC News*. https://www.abc.net.au/news/2020-01-16/nsw-fire-affected-schools-will-be-safe-for-returning-students/11870494.

Flanagan, R. (2020, January 25). How does a nation adapt to its own murder? *The New York Times*. https://www.nytimes.com/2020/01/25/opinion/sunday/australia-fires-climate-change.html.

Gammage, B. (2011). *The biggest estate on Earth: How aborigines made Australia*. Allen & Unwin.

Glasser, R. (2019, March). *Preparing for the era of disasters*. Special report. Australian Strategic Policy Institute.

Glikson, A. (2013). Fire and human evolution: The deep-time blueprints of the Anthropocene. *Anthropocene, 3*, 89–92. https://doi.org/10.1016/j.ancene.2014.02.002.

Gough, A. (2019). Symbiopolitics, sustainability, and science studies: How to engage with alien oceans. *Cultural Studies ↔ Critical Methodologies*. https://doi.org/10.1177/1532708619883314.

Gough, A. (2021, February 23). Education and the anthropocene. In C. Mayo (Ed.), *Oxford encyclopedia of gender and sexuality in education*. Oxford University Press. https://doi.org/10.1093/acrefore/9780190264093.013.1391.

Gowlett, J. A. J. (2016). The discovery of fire by humans: A long and convoluted process. *Philosophical Transactions Royal Society B, 371*, 20150164. https://doi.org/10.1098/rstb.2015.0164.

Haraway, D. (2008). *When species meet*. University of Minnesota Press.

Haraway, D. (2015). Anthropocene, Capitalocene, Plantationocene, Chthulucene: Making kin. *Environmental Humanities, 6*, 159–165. https://doi.org/10.1215/22011919-361593.

Haraway, D. (2016). *Staying with the trouble: Making kin in the Chthulucene*. Duke University Press.

Higgins, I. (2020, February 9). Indigenous fire practices have been used to quell bushfires for thousands of years, experts say. *ABC News*. https://www.abc.net.au/news/2020-01-09/indigenous-cultural-fire-burning-method-has-benefits-experts-say/11853096.

Huffman, M. (2013). The many elements of traditional fire knowledge: Synthesis, classification, and aids to cross-cultural problem solving in fire-dependent systems around the world. *Ecology and Society, 18*(4). https://doi.org/10.5751/ES-05843-180403.

Keneally, T. (2020, February 1). Thomas Keneally: 'These fires have changed us.' *The Guardian.* https://www.theguardian.com/australia-news/2020/feb/01/thomas-keneally-these-fires-have-changed-us?CMP=share_btn_link.

Lake, F., & Christianson, A. (2019). Indigenous fire stewardship. In S. L. Manzello (Ed.), *Encyclopedia of wildfires and wildland-urban interface (WUI) fires.* https://doi.org/10.1007/978-3-319-51727-8_225-1.

Langton, M. (n.d.). *Resources in Aboriginal and Torres Strait Islander histories and cultures for teachers.* https://indigenousknowledge.research.unimelb.edu.au/.

Lippmann, T., Abram, N., Sharples, J., Clarke, H., Sen Gupta, A., Meissner, K., Boer, M., Henley, B., Rohling, E., Tapper, N., Alexander, L., Sawyer, R., Mcgowan, H., Yebra, M., Russell-Smith, J., Murphy, B., Grierson, P., Nolan, R., Penman, T., ... Bell, T. (2020, February 3). *There is no strong, resilient Australia without deep cuts to greenhouse gas emissions: An open letter on the scientific basis for the links between climate change and bushfires in Australia.* https://australianbushfiresandclimatechange.com/.

Lloro-Bidart, T. (2015). A political ecology of education in/for the Anthropocene. *Environment and Society: Advances in Research, 6,* 128–148. https://doi.org/10.3167/ares.2015.060108.

MacDonald, K. (2017). The use of fire and human distribution. *Temperature, 4*(2), 153–165.

Marshall, K. (2020, January 25). The 'forever fires' and Australia's new reality. *The Age Good Weekend.* https://www.theage.com.au/national/the-forever-fires-and-australia-s-new-reality-20200122-p53tk0.html.

Millar, R., & McMillan, A. (2020, January 5). Victoria bushfires: Exhausted families farewell fire-racked Mallacoota. *The Age.* https://www.theage.com.au/national/victoria/victoria-bushfires-exhausted-families-farewell-fire-racked-mallacoota-20200105-p53p1d.html.

Mistry, J. (2000). *World savannas: Ecology and human use.* Prentice-Hall.

Morton, A. (2019, December 21). Climate of chaos: The suffocating firestorm engulfing Australia. *The Guardian.* https://www.theguardian.com/australia-news/2019/dec/20/climate-of-chaos-the-suffocating-firestorm-engulfing-australia.

Nakagawa, Y., Verlie, B., & Kim, M. (2020). Collectively engaging with theory in environmental education research. *Australian Journal of Environmental Education.* https://doi.org/10.1017/aee.2020.6.

Neale, T. (2018). Digging for fire: Finding control on the Australian continent. *Journal of Contemporary Archaeology, 5*(1), 79–90. https://doi.org/10.1558/jca.33208.

Neale, T., Carter, R., Nelson, T., & Bourke, M. (2019). Walking together: A decolonising experiment in bushfire management on Dja Dja Warrung country. *Cultural Geographies, 26*(3), 341–359.

Pascoe, B. (2018). *Dark emu: Aboriginal Australia and the birth of agriculture.* Magabala Books.

Popham, W. J. (2001). Teaching to the test. *Educational Leadership, 58*(6), 16–20.

Pyne, S. J. (2007). *Awful splendour: A fire history of Canada.* UBC Press.

Pyne, S. (2019, August 25). Winter isn't coming. Prepare for the Pyrocene. *History News Network*. https://historynewsnetwork.org/article/172842.

Regan, H., & Yeung, J. (2020, January 10). Tens of thousands protest Australian PM's climate policies amid bushfire crisis. *CNN*. https://edition.cnn.com/2020/01/10/australia/australia-fires-climate-protest-morrison-intl-hnk/index.html.

Reid, A. (2019). Climate change education and research: Possibilities and potentials versus problems and perils? *Environmental Education Research, 25*(6), 767–790.

Rickards, L., Verlie, B., Towers, B., & Lay, B. (2019, November 19). Climate is disrupting children's education. *Eureka Street*. https://www.eurekastreet.com.au/article/climate-is-disrupting-children-s-education.

RMIT ABC Fact Check. (2020, February 4). Have more than a billion animals perished nationwide this bushfire season? Here are the facts. https://www.abc.net.au/news/2020-01-31/fact-check-have-bushfires-killed-more-than-a-billion-animals/11912538.

Russell-Smith, J., Yates C., Edwards A., Allan, G., Cook, G., Cooke, P., Craig, R., Heath, B., & Smith, R. (2003). Contemporary fire regimes of northern Australia, 1997–2001: Change since Aboriginal occupancy, challenges for sustainable management. *International Journal of Wildland Fire, 12*(4), 283–297.

SBS News. (2016, February 16). Insight S2016 Ep2—Line of Fire. https://www.sbs.com.au/programs/video/614446147894/Insight-S2016-Ep2-Line-of-Fire.

Scolaro, N. (2018, August 20). Victor Steffensen listens to the land. *Dumbo Feather*. https://www.dumbofeather.com/conversations/victor-steffensen-listens-to-the-land/.

Singh, P. (2017, April 24). Importance of discovery in human brain development. *Medium*. https://medium.com/@Prakhar__Singh/importance-of-fire-for-human-brain-development-24b78638fdbc.

Steffensen, V. (2020). *Fire country: How indigenous fire management could help save Australia*. Explore Australia.

Thunberg, G. (2019, January 29). 'Our house is on fire': Greta Thunberg, 16, urges leaders to act on climate. *The Guardian*. https://www.theguardian.com/environment/2019/jan/25/our-house-is-on-fire-greta-thunberg16-urges-leaders-to-act-on-climate.

Towers, B. (2015). Children's knowledge of bushfire emergency response. *International Journal of Wildland Fire, 24*(2), 179–189. https://doi.org/10.1071/WF13153.

Towers, B. (2019). School-based bushfire education: Advancing teaching and learning for risk reduction and resilience. *Australian Journal of Emergency Management, Monograph No. 5*, 71–74.

Towers, B., Gough, A., & Verlie, B. (2020, January 22). Bushfire education is too abstract. We need to get children into the real world. *The Conversation*. https://theconversation.com/bushfire-education-is-too-abstract-we-need-to-get-children-into-the-real-world-129789.

Towers B., Perillo, S., & Ronan, K. (2018). *Evaluation of survive and thrive: Final report to the Country Fire Authority*. Bushfire and Natural Hazards CRC.

Tsing, A. (2012). Unruly edges: Mushrooms as companion species. *Environmental Humanities, 1*, 141–154.

Turner, N. J. (1999). Time to burn. In R. Boyd (Ed.), *Indians, fire and the land in the Pacific Northwest* (pp. 185–218). Oregon State University Press.

Vale, T. (Ed.). (2002). *Fire, native peoples, and the natural landscape*. Island Press.

Verlie, B. (2017). Rethinking climate education: Climate as entanglement. *Educational Studies, 53*(6), 560–572. https://doi.org/10.1080/00131946.2017.1357555.

Verlie, B. (2018). From action to intra-action? Agency, identity and "goals" in a relational approach to climate change education. *Environmental Education Research,* 1–15. https://doi.org/10.1080/13504622.2018.1497147.

Verlie, B. (2019a). "Climatic-affective atmospheres": A conceptual tool for affective scholarship in a changing climate. *Emotion, Space and Society, 33,* 100623. https://doi.org/10.1016/j.emospa.2019.100623.

Verlie, B. (2019b). Bearing worlds: Learning to live-with climate change. *Environmental Education Research, 25*(5), 751–766.

Wahlquist, C. (2019, November 29). Climate change strike: Thousands of school students protest over bushfires. *The Guardian.* https://www.theguardian.com/world/2019/nov/29/climate-change-strike-thousands-of-school-students-protest-over-bushfires.

Weir, J., Sutton, S., & Catt, G. (2020). The theory/practice of disaster justice: Learning from indigenous peoples' fire management. In A. Lukasiewicz & C. Baldwin (Eds.), *Natural hazards and disaster justice: Challenges for Australia and its neighbours* (pp. 299–317). Palgrave Macmillan.

Williamson, B., Weir, J., & Cavanagh, V. (2020, January 10). Strength from perpetual grief: How Aboriginal people experience the bushfire crisis. *The Conversation.* https://theconversation.com/strength-from-perpetual-grief-how-aboriginal-people-experience-the-bushfire-crisis-129448.

Yin, S. (2016, August 5). Smoke, fire and human evolution. *The New York Times.* https://www.nytimes.com/2016/08/09/science/fire-smoke-evolution-tuberculosis.html.

Part II

Decolonizing Anthropocene(s)

Redrawing Relationalities at the Anthropocene(s): Disrupting and Dismantling the Colonial Logics of Shared Identity Through Thinking with Kim Tallbear

Priyanka Dutt, Anastasya Fateyeva, Michelle Gabereau, and Marc Higgins

What does it mean to respond to the *Anthropocenes*, plural, when doing science education? As Davis and Todd (2016) query, "if the Anthropocene is already here, the question then becomes, what can we do with it as a conceptual apparatus that may serve to undermine the conditions that it names" (p. 763)? What is gained and what is lost through the rallying cry of the naming of this epoch after *humankind* (i.e., the Anthropos), particularly if and when such a naming of a shared catastrophe masks the uneven responsibility for this contemporary moment, as well as the unequal ways in

---

The original version of this chapter was revised: ESM has been updated. The correction to this chapter is available at https://doi.org/10.1007/978-3-030-79622-8_24

Supplementary Information The online version contains supplementary material available at (https://doi.org/10.1007/978-3-030-79622-8_7).

P. Dutt (✉)
University of Lethbridge, Lethbridge, AB, Canada
e-mail: pdutt@ualberta.ca

A. Fateyeva
Edmonton Public Schools, Edmonton, AB, Canada
e-mail: fateyeva@ualberta.ca

M. Gabereau
Ehpewapahk Alternate School, Maskwacis, AB, Canada
e-mail: gabereau@ualberta.ca

M. Higgins
Department of Secondary Education, University of Alberta, Edmonton, AB, Canada
e-mail: marc1@ualberta.ca

© The Author(s) 2022, corrected publication 2022
M. F.G. Wallace et al. (eds.), *Reimagining Science Education in the Anthropocene*, Palgrave Studies in Education and the Environment, https://doi.org/10.1007/978-3-030-79622-8_7

which it is felt? More pointedly, can we critically engage with *the* Anthropocene, singular, without acknowledging the multiplicity of moments in which Indigenous Land and its ecology of humans, other-than-humans, and more-than-humans were at risk of extinction from "Man"? Within this chapter, we contend that if the ways of thinking and practicing both science and science education continue to be rooted in the same *settler*[1] colonial, capitalist, and toxic ways-of-knowing and -being that have and continue to produce Indigenous erasure and support the acquisition of Indigenous Land, that responses to *the* Anthropocene, singular, will be fraught (e.g., Bang & Marin, 2015; Liboiron et al., 2018). As Davis and Todd (2016) state:

> without recognizing that from the beginning, the Anthropocene is a universalizing project, it serves to re-invisibilize the power of Eurocentric narratives, again re-placing them as the neutral and global perspective. By linking the Anthropocene with colonization, it draws attention to the violence at its core, and calls for the consideration of Indigenous philosophies and processes of Indigenous self-governance as a necessary political corrective, alongside the self-determination of other communities and societies violently impacted by the white supremacist, colonial, and capitalist logics instantiated in the origins of the Anthropocene. (2016, p. 763)

Beginning from the recognition that both science and science education are most often premised on *othering* Indigenous peoples (Bang & Marin, 2015; Tallbear, 2013), we leverage this unusual turn towards *a* shared identity as a means of provoking new ways for the Anthropocene(s) to be felt. In this work, we take seriously Davis and Todd's (2016) call for the "consideration of Indigenous philosophies and processes … as necessary political corrective" by engaging in a citational politics: we privilege and think with Indigenous thought-practices that have been working within, against, and beyond Western humanism since the beginnings of colonization. Notably, we turn to Dakota scholar Kim Tallbear (2013), drawing on her work illuminating and troubling the relations between identity, science, settler colonialism, and Indigeneity. Following this, we offer a series of aesthetic provocations as a means of making the Anthropocene(s) felt otherwise.

Troubling Shared Identity as a Settler Move to Innocence

Western modern science, from its very beginnings, operationalizes processes through which some are deemed "non-scientific" as a means of legitimizing and lifting those that are "scientific." Such logics extend more broadly to Western modernity and through colonial logics: dichotomously framing some as having culture, civilization, knowledge, and rationality, amidst other things,

---

[1] Settler colonialism is a structure, and not only an event, through which settlers continue an ongoing project of Indigenous erasure and Land acquisition (see Tuck & Yang, 2012).

because others have not. While these culturally shape(d) who can and could participate in science, as well as how, it also affects the kinds of sciences practiced.

While the genetic science of race is largely discounted today, Tallbear (2013) reminds us that "gene discourses and scientific practices are entangled in ongoing colonialisms. What 'they' [settlers] think and do have always determined how much trouble 'we' [Native Americans] have" (p. 9). Despite science's distancing from race, we continue to see contemporary examples of how science is entangled in practices of defining, categorizing, policing, and perpetuating identities through practices such as direct-to-consumer DNA tests. In turn, science continues to inherit the responsibility to face the ways in which it has had and continues to have a part in defining *identity*, as well as how DNA continues to be systematically used to (re)direct power and privilege towards some (e.g., white settlers) at the expense of others (e.g., Indigenous peoples). For example, Tallbear (2013) invites us to consider the ways in which genetics is leveraged in the U.S. political system through property rights:

> Property rights accorded to whiteness are protected by the U.S. legal system. One of those rights is control of the legal meanings of group identities. Whites have legally defined who counts as black or Indian. This is an important right, for the racialization and subordination of those black and red "others" has been necessary to solidify the exclusive parameters of whiteness. (TallBear, 2013, p. 136)

This difference has become so normalized and normative that white settlers can unironically approach the question of difference vis-à-vis a stance of "helping" Indigenous peoples without recognizing their own complicity: such as the colonial impulse to improve or introduce new technologies, framing traditional ways-of-knowing and -being as backwards. One such dubiously helpful techno-scientific extension is the genetic theory of the mitochondrial Eve, which would have us all originate from Africa, and the notion that if we could come to realize that we were all connected that racism would end: racism is incompatible with knowledge of genetics.

Challenging the notion that scientific knowledge is enough for change to occur, Tallbear (2013) reminds us that "racism does not need to be scientifically 'correct' to thrive. ... [It is an] ahistorical ... hope that scientific knowledge can make the crucial intervention of halting centuries of race oppression" (p. 149). As Tallbear (2013) suggests, it may not even be enough to halt racist and oppressive myths proffered by the same scientists who offer "we are all African" as corrective: genetic research *on* Indigenous peoples continues under the guise of "preserving" the DNA of "vanishing" Indigenous peoples. While it is beyond the scope of this chapter to fully unpack *why* this is problematic (e.g., not acknowledging the colonial complicities that accompany the production of the image of Indigenous peoples as vanishing, either culturally or physically), we are wary of what calls to shared identities make visible and what they mask in terms of colonial logics and practices.

In turn, while we agree that we are *all* affected by and need to respond to the Anthropocene, we wish to attend to what the illumination of shared culpability serves to conceal. Since the 1950s, the period in which *the* Anthropocene is often stated to have begun, "carbon dioxide levels, mass extinctions, and the widespread use of petrochemicals, ... and radioactivity left from the detonation of atomic bombs" (Davis & Todd, pp. 762–763) have been (re)shaping the globe. However, while we are all are impacted by and inherit the Anthropocene, it is irresponsible to frame the issue as one that impels and affects us all equally, as this serves to "mask power with innocence" (McKinley, 2001). *The* Anthropocene does not and cannot account for the ways that the responsibility for this current moment are unevenly distributed or how it unevenly impacts diverse groups: the Global South, endangered animals and species, Indigenous peoples, and marginalized urban and rural folk of colour, amidst others. Rather, not unlike Tallbear's (2013) lines of questioning of colonial complicities of genetic science, Tuck and Yang (2012) invite us to consider such actions as *settler moves to innocence*:

> Settler moves to innocence are those strategies or positionings that attempt to relieve the settler of feelings of guilt or responsibility without giving up land or power or privilege, without having to change much at all. (p. 10)

Here particularly, the settler move to innocence is one of equivocation. Through equivocation, "or calling everything by the same name" (Tuck & Yang, 2012, p. 17), the intent is to signal the ways in which we are all affected and must respond to this shared catastrophe. While not wholly untrue, equivocation serves to mask the ways in which settlers have conveniently ignored and have actively participated in *Anthropocenes* which precede and come to constitute the present one. For example, the "Orbis spike" of 1610 (Lewis & Maslin, 2015) in which atmospheric CO_2 levels drastically dropped as a result of the genocide of Indigenous peoples, having planetary consequences.

Further, Tallbear (2013) helps us to think about the ways in which colonial logics fetishize origins (e.g., *the* Anthropocene). Complicating the notion that Indigenous identity *originates* in DNA, Tallbear (2013) offers an Indigenous conception of being that is relational. Not only is such a means of honouring longstanding and ongoing Indigenous philosophies, but it is also a means of calling into question and understanding the ecology of settler colonial thought-practices which constitute problematic extractivist genetic practices through which Indigenous peoples continue to be the *object* of science. In turn, we believe that choosing to observe *this* Anthropocene (as *the* Anthropocene) has much to do with how the world's settlers have displaced and destroyed through extractivism (i.e., the production of "value" through ever accelerating extraction of resources).[2]

---

[2] It is also worth noting that even *positive* identity construction can also be in the service of settler colonial logics within. For example, Indigenous peoples who take up the

After Davis and Todd (2016), we want to take seriously the notion that "the Anthropocene betrays itself in its name: in its reassertion of universality, it implicitly aligns itself with the colonial era" (Davis & Todd, 2016, p. 763). However, rather than offer a corrective identifying (beyond the plural form: Anthropocenes), we turn to more relational forms of meaning-making in response to the ways in which *a* meaning, like *an* identity, often works to reassert and reproduce settler colonial ways-of-knowing and -being. Specifically, we take seriously Tallbear's (2013) call to not make Indigenous peoples the object of study when discussing the manifestations of settler colonialism in science but rather invert the gaze back onto science itself. Further, an important move in not offering a corrective is that, as Plains Cree scholar Cash Ahenakew (2017) states, "the work of decolonization is not about what we do not imagine, but what we cannot imagine from our Western ways of knowing" (p. 88). We, as science educators, may need to sit with and in the difficult question of *why* we cannot or have not been able to respond to the Anthropocene(s). In turn, we recognize the need for new ways to (re)open what we can even imagine within science education as we respond to Anthropocenes. As Ahenakew states, "using metaphor and poetry to disrupt sense-making and prompt sense-sensing in the experience of readers" (2016, p. 337) because "modern academic literacies and technologies can make what has been made invisible by colonialism visibly absent, but they cannot make it present" (2017, p. 89). Accordingly, not only do we avoid offering *the* meaning because it risks reasserting and reproducing settler colonial ways-of-knowing and -being, but also because it may not be heard: science education has had and continues to have an active part in rendering unintelligible Indigenous ways-of-knowing and -being. Nonetheless, in endnotes, we offer *a* meaning as a means of engaging the possibilities that the aesthetic provocations below make possible.

Below, you will find a series of image-texts[3] that are intended to invite an unpacking and undermining of the problematic ways in which Indigeneity, settler colonialism, science education, and the Anthropocene(s) intersect and coalesce (e.g., producing non-Western epistemologies and ontologies as lesser-than). Inspired by diverse influences such as Indigenous storywork (e.g., Archibald, 2008), political cartoons, and meme culture we wish to produce a field of meaning-making that implicates the reader within the question of the Anthropocene(s) in the every day as well as its settler colonial and neo-colonial implications.

---

roles of Land protectors, while perpetually positioned as enemies of economic progress, are simultaneously positioned as stewards of the Land in ways that can and do let settlers *off-the-hook* in terms of their ecological responsibilities (El-Sherif, 2020).

[3] All of the following images have been illustrated by Anastasya Fateyeva.

TURTLE ISLAND: A HAUDENOSAUNEE CREATION STORY[4]

Long before the world was created, there was an island in the sky inhabited by sky people. One day, a pregnant sky woman drops through a hole created by an uprooted tree and begins to fall for what seems like eternity.

Coming out of darkness, she eventually sees oceans. The animals from this world congregate, trying to understand what they see in the sky. A flock of birds is sent to help her. The birds catch her and gently guide her down onto the back of Great Turtle. The water animals like otter and beaver have prepared a place for her on turtle's back. They bring mud from the bottom of the ocean and place it on turtle's back until solid earth begins to form and increase in size.

Turtle's back becomes Sky Woman's home and the plants she brought down with her from Skyworld, including tobacco and strawberries, are her medicine. She makes a life for herself and becomes the mother of Haudenosaunee life as we know it today.

UNPACKING EVERYDAY (NEO-)COLONIALISMS[5]

Let's talk about unpacking your recent purchase from Whiteazon. We can see that you're either in the pursuit of achieving a level of whiteness as defined and dictated by present-day, settler colonial society because you may not be white passing or perhaps you're educating yourself on the privileges that are associated with the notion of white passing and what that means. In the off chance that you're making this purchase as a white person, please leave a review when you can!

About:

Colonialism Illustrated *by Wyatt Mehn*

Avg. Rating ☆☆☆☆☆

---

[4] The first provocation is a depiction of Turtle Island, the name used by many Indigenous peoples when addressing what is colonially known as North America, followed by a Haudenosaunee creation story (Niro et al., 1999).

[5] The second provocation is a satirical approach to the understandings of "belonging" in Western contemporary society. At once, it speaks to the requisite participation in consumeristic materialism: acquiring and holding valuable assets such as property, goods, and other forms of capital (which are then read as social capital). Simultaneously, it expresses a relation to whiteness and settler colonial logics: belonging often requires conformity of those not presenting in the image of MAN (e.g., changing or altering one's behaviour through speaking one language over another, changing the dress or traditional adornments one chooses to wear, etc.). Further, this provocation also helps to exemplify the ways in which colonial science excludes contributions to the field of STEM from the East or from different origins. Finally, there is the notion of identity and what that means, how it is defined, and how we use said definitions to create laws. Of particular note here is the tendency to dictate identity based on direct-to-consumer DNA tests, which may be used to govern bodies, but may also be used as ways to validate (or invalidate) certain identities over others, or invalidate them altogether (TallBear, 2013). It also begs the question: does one need a quantitative measure to know their identity or can identity be rooted in something else altogether?

Wyatt Mehn knows just how confusing colonialism can be. In his illustrative piece, Mehn attempts to depict the ways in which one can attain a colonial body short of bleaching one's skin. This can be done by "acting white." In this piece, Mehn shows examples of the colonized person, i.e., straightening otherwise naturally curly hair when it is on a brown body, reducing the number of cultural or traditional adornments worn by a brown body, or negating accents or heritage-related tonal changes in order to establish a "white" voice. Mehn also shows the converse for each example, highlighting the spectrum of which these dichotomies exist.

Important White Men by *Ano Therwy Atman*

Avg. Rating ☆☆☆☆☆

Perhaps you need to establish your place in a predominantly Eurocentric Western society. To do so, Atman has created a comprehensible guide to the Important White Men whose names replace the contributions from Eastern societies or from non-male bodies. From Watson and Crick's discovery of DNA and not Rosalind Franklin to the erasure of Ibn al-Haythem's contributions to the understanding of light and optics, this book will create space to understand and internalize the ways in which Eurocentrism and Western ideologies actively work against non-white, non-male bodies.

Property Management by *Sam Guy*

Avg. Rating ☆☆☆☆☆

Owning property is a tough endeavour but Sam Guy has created a simple, multi-step approach that any person of colour can follow. In it, he speaks to the Five Fundamentals of Owning Property: Having means and money, Nepotism, Blankets laced with smallpox, Treaties that may or may not contain loopholes in which Indigenous peoples are tricked into giving up their livelihood for the sake of their nations, and general Whiteness. Guy's work also emphasizes that the fifth fundamental, whiteness, is perhaps the most precious and valuable property to have and hold in Western society.

Principles of Debate, Vol. 1—10 by *Al SoSam Guy*

Avg. Rating ☆☆☆☆☆

While the connections and sense of community we make on a day-to-day basis are valuable, Al Guy believes that the one true way to create identity is through the definitions posed by raw data or otherwise cultivated by science. In his inspiring collection of historical data and journal articles, Guy creates a map from genetic markers to the definitions and depictions of identity. No longer is a total of 7% Japanese DNA markers irrelevant to your identity. Further, Guy provides explanations into the world of defining and labelling identity while also providing insights into what is or isn't part of those leading definitions, complete with an interview from presidential candidate Melizabeth Twarren and her embracing of her Cherokee DNA.

Indigenous Erasures: Supersessionism and Scientific Origin Stories[6]

~~Long before the world was created, there was an island in the sky inhabited by sky people. One day, a pregnant sky woman drops through a hole created by an uprooted tree and begins to fall for what seems like eternity.~~

~~Coming out of darkness, she eventually sees oceans. The animals from this world congregate, trying to understand what they see in the sky. A flock of birds is sent to help her. The birds catch her and gently guide her down onto the back of Great Turtle. The water animals like otter and beaver have prepared a place for her on turtle's back. They bring mud from the bottom of the ocean and place it on turtle's back until solid earth begins to form and increase in size.~~

~~Turtle's back becomes Sky Woman's home and the plants she brought down with her from Skyworld, including tobacco and strawberries, are her medicine. She makes a life for herself and becomes the mother of Haudenosaunee life as we know it today.~~ (Niro et al., 1999).

When the solar system settled into its current layout about 4.5 billion years ago, Earth formed when gravity pulled swirling gas and dust in to become the third planet from the Sun. Like its fellow terrestrial planets, Earth has a central core, a rocky mantle, and a solid crust (NASA, 2020).

[6] The third provocation is a revisiting of the first, a newly rendered version of Turtle Island, but now with red pen and altered definitions. Following the previous "unpacking" of whiteness in settler colonial Western societies, we suggest that these settler colonial, Eurocentric, and strict views of science begin to produce particular ways-of-relating within and across places and communities. Specifically, we understand the project of "identity" and "origins" as being entangled in the production of hard and fast definitions which shape how the North and the rest of the global population is "mapped out" and understood. For example, the need to define a people: in the ability to quantify, by per cent, their DNA, or by the presence of haplogroups, or by the defined borders that we draw between societies, or by the hypotheses we create regarding a nation's potential migratory patterns (e.g., Bering land bridge theory). These ways-of-defining are inseparable from and entangled with settler colonial ideas of science and the larger project of Western modernity which desires the creation of lines and boundaries rather than bridges and connections between the self and the other, the observer and the observed, as a means of privileging settler colonial lifeways (e.g., lifting MAN, the subject of modernity, be they a scientist, educator, or other).

Within this provocation, the stark contrast between the Haudenosaunee creation story and NASA's explanation of the formation of the universe is included in order to truly show not only the dichotomy that exists within our present-day teachings of science and the stories by which Indigenous elders and other communities disseminate knowledge. It is also meant to invite a consideration of the way in which Western modern science supersedes other-ways-of-knowing and -being, erasing or devaluing them in the process. The question lingers: why does or must science education value one (Western modern science) so highly above all others?

Indigenous Erasures: The Genographic Project[7]

The Genographic Project cannot, for example, tell me how I am related to my various Dakota tribal kin, the ultimate set of relations in tribal life. ... The question of how we as Dakota got to where we are has already been answered, and the answer does not lie in genetics. I could reference Dakota creation stories that give us values for living, narrate our common history, cohere us as a people with a common moral framework, and tie us to a sacred land base. ... "Who we really are" is not a question that most, if any, Dakota think can be answered by finding out that they have mtDNA markers that "originated" in Mongolia. (TallBear, 2013, p. 152)

Conclusion

The Anthropocene is the epoch under which "humanity"—but more accurately, petrochemical companies and those invested in and profiting from petrocapitalism and colonialism—have had such a large impact on the planet that radionuclides, coal, plutonium, plastic, concrete, genocide, and other markers are now visible in the geologic strata. (Davis & Todd, 2016, p. 765)

The Anthropocene(s) are a time in which our present geological age is significantly defined by the impact of *Man* (e.g., the white, masculine subject of Western modernity). While the notion and framing of *the* Anthropocene is not wrong—increasing human influence has led to greater impacts on climate change and the surrounding ecosystems—we suggest that this admission of culpability masks more than it reveals: *the* Anthropocene cannot be separated from the scientific production of identity which upholds and reproduces settler colonial thought practices. Importantly, when 100 companies produce 71% of the world's pollution, the creation of a shared identity may divert attention away from addressing the "petrochemical companies and those invested in and profiting from petrocapitalism and colonialism" (Davis & Todd, 2016, p. 765) who hold primary responsibility for the contemporary moment.

It is our desire that the above aesthetic provocations will generously generate different configurations of pedagogy which will allow for new ways to attend to the interconnected matrices of power and privilege which shape this contemporary Anthropocenic epoch that is predicated on so many

---

[7] The fourth provocation illustrates the Manhattan plot: data that is extracted from a sample of one's DNA. Within a single sample, there may be dozens of genetic markers depicting dozens of differing identities within a single individual. However, as TallBear (2013) puts it, these tests cannot define who she is because her personhood emanates from her relations, rich culture, long-standing traditions, and passed down stories. Further, rather than a shared identity, TallBear (2013) speaks to how her Dakota kin are connected through their experiences, not by a direct-to-consumer DNA test that spits out an array of percentages based on genetic markers tied to Mongolia.

other Anthropocenes. With this, we hope that these might allow for open, vulnerable, self-reflexive dialogue around the problematic ways in which the Anthropocene(s), Indigeneity, and science education intersect.

In refusing the ways in which we are discursively called to leave behind our relationalities in the universalizing project of the Anthropocene, we suggest turning to Indigenous thought-practices to break the cyclical storm of anthropogenic violence: to remember that the same land we begin to separate, colonize, commodify, and pollute is also the land in which we may find coalition and community.

References

Ahenakew, C. (2016). Grafting Indigenous ways of knowing onto non-Indigenous ways of being. *International Review of Qualitative Research*, *9*(3), 323–340.

Ahenakew, C. R. (2017). Mapping and complicating conversations about Indigenous education. *Diaspora, Indigenous, and Minority Education*, *11*(2), 80–91.

Archibald, J. (2008). *Indigenous storywork*. UBC Press.

Bang, M., & Marin, A. (2015). Nature–culture constructs in science learning: Human/non-human agency and intentionality. *Journal of Research in Science Teaching*, *52*(4), 530–544. https://doi.org/10.1002/tea.21204

Davis, H., & Todd, Z. (2016). On the importance of a date, or decolonizing the Anthropocenes. *ACME: An International Journal for Critical Geographies*, *16*(4), 761–780.

El-Sherif, L. (2020). Six Nations Land Defenders in Caledonia reveal hypocrisy of Canada's land acknowledgements. *The Conversation*. https://theconversation.com/six-nations-land-defenders-in-caledonia-reveal-hypocrisy-of-canadas-land-acknowledgements-145158.

Lewis, S. L., & Maslin, M. A. (2015). Defining the Anthropocene. *Nature*, *519*(7542), 171. https://doi.org/10.1038/nature14258

Liboiron, M., Tironi, M., & Calvillo, N. (2018). Toxic politics: Acting in a permanently polluted world. *Social studies of science*, *48*(3), 331–349. https://doi.org/10.1177/0306312718783087

McKinley, E. (2001). Cultural diversity: Masking power with innocence. *Science Education*, *85*(1), 74–76.

NASA Science. (2020). *Earth*. https://solarsystem.nasa.gov/planets/earth/in-depth/

Niro, S., Oneida, K. G., & Brant, A. (1999). Origin stories - Sky woman. *Canadian Museum of History*. https://www.historymuseum.ca/cmc/exhibitions/aborig/fp/fpz2f22e.html

TallBear, K. (2013). *Native American DNA: Tribal belonging and the false promise of genetic belonging*. University of Minnesota Press.

Tuck, E., & Yang, W. (2012). Decolonization is not a metaphor. *Decolonization: Indigeneity, Education & Society*, *1*(1), 1–40.

CHAPTER 8

Decolonizing Healing Through Indigenous Ways of Knowing

Miranda Field

The field of psychology is embarking on a process of interrupting the historical colonial cycle of harm and beginning to work with and alongside Indigenous communities to understand the healing journey (Drees, 2013). Understanding the healing journey provides a foundation for both those who seek and guide healing. As a society, disconnection with the land influenced how psychology teaches those who guide healing, the practices the field establishes as evidenced-based, and what elements contribute to positive outcomes. Exploring healing from a decolonizing perspective provides insight into the essential understanding where learning and healing coexist along this journey. Researchers, mental health service providers, and educators alike move towards acknowledging that challenging and displacing hegemonic Western knowledge systems is required to engage and enable multiple forms of literacy and knowledge to coexist (Kermoal & Altamirano-Jimenez, 2016). This work is not possible without the guidance of Indigenous communities, ensuring that healing is engaged in a strength-based, ethically responsible manner and for those guiding healing, an individual process of decolonizing Western psychology practices can begin.

To better support trajectories of decolonizing practices in psychology, this chapter aligns with the Truth and Reconciliation of Canada (TRCC) Calls to Action (2012); specifically, Call to Action (Legacy, Health) 22, which calls upon "those who can effect change within the Canadian health-care system to

M. Field (✉)
University of Regina, Regina, SK, Canada
e-mail: mirandafield@uregina.ca

© The Author(s) 2022
M. F.G. Wallace et al. (eds.), *Reimagining Science Education in the Anthropocene*, Palgrave Studies in Education and the Environment, https://doi.org/10.1007/978-3-030-79622-8_8

recognize the value of Aboriginal healing practices and use them in the treatment of Aboriginal patients in collaboration with Aboriginal healers and Elders where requested by Aboriginal patients" (TRCC, 2012). Acknowledging the notion that mental health healing may not occur without learning, and vice versa, shifts the paradigm to one where healing and learning coexist through meaningful relationships.

This chapter seeks to answer the call put forward by McGinnis et al. (2019), which calls on mental health practitioners and educators to recognize, support, and incorporate Indigenous knowledge systems in to practice and to consider the ethical implications of integrating and supporting Indigenous ways of knowing into current psychology practice. Through decolonizing land-based healing experiences and drawing on my personal journey—as an educator trained in psychology—I seek to integrate lived experience with supporting academic literature to provide an understanding of the historical, relational disconnection from land. Through Indigenous ways of knowing, the field of psychology can begin to decolonize individual practice and begin to integrate traditional ways of healing into current practice and embark in the process of interrupting the historical colonial cycle of harm. A foundational understanding within decolonizing psychology is understanding relationships through Indigenous ways of knowing and being. The personal journey is the beginning of reconnecting relationships through learning and healing.

I acknowledge the ways in which my upbringing as a European-descent Christian from rural Canada has a foundational influence on how I see myself, my community, and my world. I have dedicated friendships, education, and travel to allow myself the opportunity to broaden my understanding of the world and those around me. I understand and acknowledge that I do not know everything there is to know about any topic and seek support and further understanding in many areas. My network of personal and professional relationships continues to grow as I seek ongoing and continued support, which establishes a foundational network which I can draw upon when professional guidance and additional knowledge is required.

I acknowledge my minimal lived experience with oppression, racism, discrimination, and stereotyping and understand that my foundation is built around my world (growing up and now) being situated among the dominant ethnicity, culture, and belief systems of my surroundings. I continue to seek support, guidance, understanding, and patience from those knowledgeable in education and health care practices. I seek to engage in decolonized Indigenous understandings in areas from academia to land-based healing.

My training in both education and psychology has been greatly influenced by legally defensible writing and not that of nuance or publicized self-reflection. To decolonize my practices and understanding in psychology, I must disrupt the colonial influences of my foundation and embrace Indigenous ways of knowing and connecting to these lands. By acknowledging my settler influences and teachings, exploring Indigenous worldviews and the interconnectedness to wellness, and engaging in strength-based approaches guided by Indigenous communities, I have initiated personal and professional steps towards reconciliation.

A Path of Decolonizing Healing Through Learning from the Land

Indigenous Peoples have lived for thousands of years in harmony with the land, where deep spiritual connections to the land impact overall health and wellbeing (Alt, 2017; Kingsley et al., 2013). Through colonial disruptions, Indigenous Peoples experienced and continue to experience a disconnect from traditional ceremonies, traditional medicines, and traditional land uses. To connect further, McGinnis et al. (2019) bring attention to the impact of colonialization through the fragmenting of relationships among humans and more-than-humans. Colonization has influenced the detachment of traditional connections and relationships with humans, plants, animals, and the land (Elder B. McKenna, personal communication, May 4, 2019). The deep connections include both learning and healing from and with the land. The process of decolonizing healing involves reconnecting and engaging with the land to situate healing and learning. This chapter will operate with the following definitions offered by Indigenous scholars and teachers:

Styres (2019) defines decolonizing as "an unsettling process of shifting and unraveling the tangled colonial relations of power and privilege" (p. 30). Furthermore, Linda Tuhiwai Smith (2012) provides guidance on the careful consideration and critical thinking of each act of reading, writing, and research one partakes in when engaging with decolonization. Contrary to many thoughts on decolonization, Smith explains that:

> Decolonization, however, does not mean and has not meant a total rejection of all theory or research or Western knowledge. Rather, it is about centering [Indigenous] concerns and world views and then coming to know and understand theory and research from [Indigenous] perspectives and for [Indigenous] purposes. (p. 41)

From an Indigenous perspective, this healing journey has no definite beginning—it is not linear. It is based in meaning-making and relationships along the journey. The field of psychology may begin to move away from pathologizing healing to a strength-based healing process with this understanding when the focus shifts to relationships; relationships with self, relationships with community, relationships with the more-than-human, and relationships with the land. Elder Alma Poitras shares that from an Indigenous perspective, healing incorporates more than the physical recovery (personal communication, May 4, 2019). Indigenous healing includes the physical, spiritual, emotional, and mental aspects of one's being, which includes the use of ceremony and traditional medicines. Western medical practices have incorporated physical modalities, and Western psychology practices have incorporated emotional, mental, and spiritual healing modalities. From an Indigenous perspective, physical, emotional, mental, and spiritual healing exists through relationships which occur alongside the healing journey.

Learning is often understood as the acquisition of knowledge or skills through experience, study, or by being taught. But as Indigenous teacher Dustin Brass shares:

> If [Indigenous people] go back to [their] roots and we look at where education started for us as Indigenous people, it interacted with the land. . . . The basis of the education is delivered through the land. It's a way to interact with learning from the land, learning from the environment, the things that surround us. (Blue Sky, 2020)

Centering Indigenous voices and definitions of healing work to decolonize the Western understanding of the healing journey. The definitions reinforce that the land, relationships, and spiritual connections are essential to understanding the healing and learning journeys for Indigenous peoples.

The following chapter is informed by my personal journey of decolonized land-based healing through learning experiences and supporting literature. This integration intends to illuminate the historical influence of human's disconnection with the land through articulating the experiential relationships. Through my personal learning journey via decolonizing practices, it is apparent that healing and learning coexist through engagement with the land.

LEARNING AND HEALING FROM THE LAND

The journey of healing begins with acknowledging a crisis—a crisis in the current practice of psychology and Indigenous Peoples (Johnston Research, 2010). Throughout colonial history, Indigenous peoples have not been given the opportunity to heal with dignity through their Indigenous ways of knowing and being (Raphael, 2018). To begin the journey of decolonization is to acknowledge our collective history. If it is called upon to integrate Indigenous ways of knowing into the field of psychology, an overall understanding of and critical engagement with current psychology practices must be present. Many current psychology practices are grounded in the Western psychology which refers to psychology influences of Europe and were established as the normative behaviours all cultures were to be judged and examined against. Western psychology integrates ideals of colonial cultures—explorers, conquerors, and settlers—where the goals of the perceived dominant culture enforced their policies and processes over the people (Raphael, 2018). Behaviours and views towards self, community, more-than-human, and land varied greatly and continue to contribute to a historical colonial cycle of harm.

Jordan (2015) shares the historical work of Levy-Bruhl and Jung, which describes how "Indigenous peoples do not distinguish themselves sharply from the environment, believing that what went on outside also went on inside of self" (p. 17). This exemplifies how the natural environment and Indigenous peoples are interwoven. "It is a culture so intrinsically linked to the land that what the [Indigenous peoples] saw (and still see) was not an environment with different geographical aspects, but a profoundly metaphysical landscape capable of expressing their deepest spiritual yearnings" (Jordan, 2015, p. 70). This interwovenness includes physical, emotional, mental, and spiritual wellness and provides affirmation that the natural environment is imperative for Indigenous wellness and healing.

Each human has experiences with the land, whether they recognize or acknowledge them as valid (Ball, 2013). The beginning of awareness is to explore and acknowledge each experience with, on, and within the natural environment—the land. These interactions with nature are important in our relationship to nature and to our healing (Jordan, 2015). Jordan continues that "it is through the presence of the natural world that the therapeutic process is facilitated; nature in this sense acts as another presence which both guides and provokes the therapeutic work" (p. 50). This awareness is not only for those receiving treatment, but for those providing support throughout the healing process.

BUILDING ON STRENGTHS OF RELATIONSHIPS

McGinnis et al. (2019) build upon previous literature which promotes the requirements of strength-based healing approached when working with Indigenous peoples. With the understanding that postmodern approaches move away from pathologizing people and move towards discovering an individual's strengths and connections, solutions are created in the present and in the future (Corey, 2016). Indigenous approaches are distinctly different from postmodern approaches but share strength-based foundations of individuals creating meaning in their lives through relationships. Strength-based approaches include connections to self, community, more-than-human, and the land and promote a balanced care approach which includes physical, emotional, spiritual, and mental healing. Strength-based approaches are client-centered and include client choice, user-center environments for the client and families, and remained focused on the relationship (Alt, 2017; Elder A. Poitras, personal communications, May 4, 2019).

The continued integration of strength-based approaches requires an understanding of the social and cultural contexts of behaviour. Viewing behaviour and individual actions through this lens, assists in moving towards the present and future without situating solely in the past. Clients are not to blame for their problems; rather they are involved in creating solutions in moving forward in a positive direction (Corey, 2016). Stories assist the client with honouring their unique journey, both socially and culturally, and anchor this understanding in the physical, emotional, mental, and spiritual experiences within their story (Archibald et al., 2019).

Within healing processes, reciprocal mutual relationships focus on the meanings given to experiences rather than focusing primarily on the biological influences of disorders. This lens provides an important understanding of the experiences of the client, which impacts the therapeutic process (Jordan, 2015).

Relationship with Self

To further articulate my journey of awareness, I share a story of my journey learning alongside Indigenous healers and how, through experiences such as this, I embarked on my journey of awareness and understanding of Indigenous ways of knowing. My personal journey of reconnecting with the land was solidified in a retreat which served as a foundational period where I actively engaged in addressing my own worldview and opened myself to exploring Indigenous connections to the land:

I had the opportunity to participate in an autumn retreat at Wanuskewin (the name *wanuskewin* means gathering place), located in Saskatchewan, Canada. Wanuskewin is known to be a deeply spiritual place where, for more than six thousand years, Indigenous peoples of the Northern Plains gathered to hunt buffalo, gather medicine and food, and escape the winter winds. Archaeological dig sites located at Wanuskewin have been dated to before the time of the Egyptian pyramids, and artifacts such as tipi rings, pottery fragments, plant seeds, and projectile points give insight into the ancient societies (Wanuskewain Heritage Park Authority, 2020).

This retreat was my first immersive experience which joined Elders, Indigenous students, non-Indigenous students, and researchers through workshops, lectures, and speakers, which were held in conjunction with hands-on land activities. Traditional foods, nehiyawak (Plains Cree) language lessons, snaring, bow shooting, beading, plant and medicine walks, campfires, sleeping in tipis, and archeology hikes were activities I took part in during this immersion retreat.

An archeology student gifted her time to the retreat participants to share her current work located at Wanuskewin. We began our walk into the small valley on the warm autumn evening. The leaves had begun to change colours from green to golden brown, orange, and red; the grass was damp at our feet. As we descended into the valley nearing sunset, we followed the winding path in and among the bushes and trees, over the creek, and came upon the open clearing. In the center of this clearing sat a blue tarp and wooden pit cover. As the archeology student opened the wooden cover, she shared stories of the discoveries of ancient tools, toys, and cooking ware. What I was not aware of in the stories she shared, was the ethical application of archeology digs. To learn of the ethics surrounding excavation, to leave items along the edges untouched, illuminated the great respect that occurs within this practice. There is great respect for sites that are explored and are specific rules in place to ensure voyeuristic exploration does not occur. Even when these items may benefit or further our understanding of the ancient peoples, these rules apply.

As I took part in the gathering of medicines, grasses, and berries alongside Elders I listened to them; the teachings continue to only take a little and leave a lot. I take these teachings into practice, and with respect to the teachings about the animals and their need for food, these support sustained growth of the plants and animals. The notion of not taking all of a plant is understandable in the literal sense. If you take it all, it will be gone (Kimmerer, 2013). But, to observe and practice these understandings offers a profound respect to the teachings offered. This immersion experience was my first dedicated step in actively engaging with the land.

As researchers, scholars, and teachers, the want to know more is always present. My personal journey is no different. To be ok with not knowing everything is a teaching in and of itself. To know it is not your place, or your story must be understood and accepted.

Gaining awareness of how our inner processes can be represented in relation to our outer reality can be facilitated through focusing on the land (Jordan, 2015). Gaining awareness of self is essential and being in the natural environment facilitates healing for both the practitioner and the client. Elder Vee Whitehorse shares that "wisdom can not be given, it has to be experienced on your own" (personal communication, May 11, 2019). Through this, individuals are honoured as the experts of their own lives. The awareness that is gained along this journey is based on the understanding that the individual is not the problem; the problem is the problem (Corey, 2016), and knowing this works towards building healing capacity, which is facilitated through learning about one's self.

Relationships with the Community

Healing in isolation is difficult; clients heal with the support of their community (Alt, 2017). Relationships with other communities and people who enjoy wellness and health and who have connections to others, nature, and the spiritual world have been indicators of individual healing (Canadian Collaborative Mental Health Initiative, 2006). These relationships create collaborative dialogues where solutions are co-created (Corey, 2016). Collaborative dialogues establish clear and observable goals which work towards creating a self-identity based in culture, individual strengths, and community resources to address present and future concerns. Incorporating Indigenous ways of knowing and being into the healing journey may move forward through Marsh et al.'s (2016) recommendation for future studies to include and focus on the importance of collaboration between mainstream and Indigenous healers.

Relationships with the More-Than-Human

As a white settler, I have begun to establish and reflect on my personal connections to healing and learning from the land. I acknowledge I will never be able to comprehend the extent of Indigenous peoples' interconnection to the land, but I seek to ethically engage my personal and professional practice.

The idea of a neutral space exists when working in the natural world; it operates as a space where neither the client nor therapist own or control the space (Berger, 2006; Jordan, 2015). A neutral space facilitates the concept of a transitional space which "exist[s] within and between geographical and relational 'spaces,' and between mind and nature" (Jordan, 2015, p. 54). This transitional space integrates Indigenous ways of knowing and being—the transition space represents the spiritual connection. Taking time to observe, listen, and engage in the natural environment is the natural way of healing (Elder A. Poitras, personal communication, May 4, 2019) and provides space to connect with one's self, one's community, the more-than-human, and the land.

My personal experience with transitional space occurred while observing the join up of a Curly Horse and a human. Curly Horses are considered Indigenous to North America and in this experience, the Curly Horses are participants in Equine Assisted Learning and Animal Assisted Therapy. This experiential learning opportunity is grounded in non-violent, trust-based partnerships between human and horses. A join up refers to a bonding method during the first formal time a horse interacts with a human and through this process is where I was able to observe this transitional space. A space where the human and more-than-human create a connection, a bond, that is true and real. This bond between horse and human occurs through body language, body positioning, and eye contact and this connection invites the horse to approach.

To observe the respect and trust that is required for any form of connection to occur between human and horse is difficult to describe. To see this in person is a piece of indescribable beauty. The most significant observation from my experience with Curly Horses is the human allowing and providing the opportunity to the animal the option to retreat (to move away physically and emotionally) when overcome. Not only offering the retreat but honouring the retreat. The relationship is built upon trust and respect that allows space when it is needed, but to always welcome them back when they are ready. This is a teaching which honours the relationship and the spirit of both humans and more-than-humans.

In contrast to many Western notions of land, Indigenous understandings situate spirit as a central component of knowing and recognizes the spirit's innate existence in all life forms (Dell et al., 2011). To understand the connection between human and horse is to understand the spirit works in relation with the natural physical environment—animate and inanimate—and is central to Indigenous ways of knowing. Where spiritual and physical realities meet is described as a space where spiritual exchange occurs (Dell et al., 2011). This space is described within First Nations, Inuit, and other Indigenous languages throughout the world. In Dell et al. (2011), Te Pou references a concept of the Samoan peoples of the Pacific Islands. The concept of "Va" refers to the "space between", which maintains respect to scared space, harmony, balance, and the importance of relationships between our physical, spiritual, emotional, and mental dimensions. Te Pou elucidates that "relationships are not unidirectional, but mutually linked and reciprocal, and the space between, is not space that separates, but space that relates" (Dell et al., 2011, p.9). The foundation to individual and communal healing is formed upon the understanding that space recognizes the culture and creates meaning which is informed by spirit (Dell et al., 2011).

RELATIONSHIPS WITH THE LAND

Many Indigenous worldviews express that all must live in harmony with nature, which entails taking care of the earth—and not controlling it (CCMHI, 2006). Jordan (2015) believes "an understanding of the human problem of need and dependency can help us more fully understand why we... have developed an insecure, avoidant and ambivalent relationship to nature and the planet" (p. 52). Within Indigenous cultures the natural environment, including plants, animals, and the land, has played a primordial role within traditional knowledge, ceremony, and healing (Gendron et al., 2013). Generations of Elders, adults, youth, and children have and continue to spend time learning with and from the land.

Experiential approaches to learning and healing build upon strengths of Indigenous ways of knowing and being in relation with the land (Greenwood et al., 2018). Having opportunities where one can authentically interact with the land begin to initiate healing effects. Elder Alma Poitras (personal communications, May 4, 2019) shares that that natural way of healing includes seeing, hearing, touching, smelling, and tasting the natural environment. Through these acts, Elder Alma guides those to ask: What I am supposed to learn from this? My understanding in this is where meaning-making interconnects with learning and healing.

Exploring current mental health practice, design, and experiences creates discussions around how the land is or is not integrated into healing practices. In psychology, what do we value as important along the healing journey and how does this impact our clients? Along my journey, the interconnection between healing and learning became evident when reflecting on how my own children see and experience nature. Here I will share a personal story of observations during my children's play.

Endless piles of rocks and sticks. Wherever we travel, the park, the school, the driveway, the beach, or across Canada, little pockets return full of sticks and stones. These objects caught my children's eyes in a way that my own eyes could not. The colours, the sparkles, the flecks, and patterns along with the bends, twists, and bumps become an irresistible desire. These objects become items of barter and negotiation to determine how many of these treasures will be allowed to make the journey to our home. It is easy to dismiss these perceived treasures as such. With the dirt, sand, and bugs, it is easy to ask for them to be tossed aside and left where they are. But to the children, these are the finest of treasures that have been found on the most enduring of archeological digs. These sticks and stones are the tokens, the glorious prizes of the adventure that was had. To the children, sticks and stones represent their accomplishments, a sense of pride and a sense of wonder to the events of the day. When we, as adults, quickly dismiss these treasures as dirty and not worthy, we begin to tell our children that what they see as valuable is not. What they work hard for is not important. And if it is not pleasing to an adult, it does not matter. These sticks and stones, as small as they might seem, teach us lessons of value and difference, and as adults, we very quickly exchange what is valuable, what is worthy, and what is to be treasured.

BUILDING HEALING THROUGH LEARNING

The healing journey is not intended to be linear. Decolonizing personal and professional practice involves acknowledging where the crisis is, moving from awareness to understanding, and organizing these understandings into action. Similar to the healing journey, the decolonizing journey is not linear and involved a process of learning. To clearly identify my personal learning and healing experiences, Ritchhart et al. (2011) present a starting point to make thinking visible and to articulate the learning process. A numbered list highlights my journey of identifying the steps which I have taken to describe how my understanding is developed (Ritchhart et al., 2011).

HEALING THROUGH LEARNING ABOUT SELF

If healing occurred with my relationship to self, then I speak to that healing as being assisted by decolonizing learning from the land. Along a private walk during the retreat, I came upon the most beautiful old rocks covered in various kinds of moss. I wanted to know their story, their purpose. This experience allows me to (1) observe closely and describe what is there (Ritchhart et al., 2011). Sitting, listening, feeling, and smelling—those were the stories I was given. Through this, I begin (2) building explanations and interpretations (Ritchhart et al., 2011) of what I have been given. Although, being OK with not knowing has to occur. Reflection time allows for the individual to be present in the experiences and the space to learn about and from the land. Unrestricted reflective time fosters additional focus for connections, meaning-making, and clarity. Knowing there may not be answers, or connections may not be evident, allowed me to reduce my feelings of nervousness and insecurities towards the need for answers. The acceptance of the not knowing brings comfort in the healing process.

HEALING THROUGH LEARNING ABOUT COMMUNITY

If healing occurred to my relationship with community, then I speak to that healing as being assisted by decolonizing learning from the land. Within the Indigenous communities (students, scholars, peers, Elders) I have been working alongside, many have granted me the gift of acceptance within the community. To be welcomed within a community has reignited my interest in many topics where I felt as though I could not continue. Community members have aided in my understanding and acceptance of my own physical, emotional, mental, and spiritual limits—both professionally and personally. Through (3) reasoning with evidence (Ritchhart et al., 2011), the community has assisted me in strengthening my connections to the community. I am more confident in who I am and where I belong in this community. These strengthened relationships have supported and guided me through the process of decolonizing my personal and professional practice. I am more confident in the work I partake in and more willing to accept the challenges life offers.

Healing Through Learning About the More-Than-Human

If healing occurred through my relationship with the more-than-human, then I speak of this healing as being assisted by decolonized learning from the land. While observing Curly Horses, it was evident trust was an essential component of the human and more-than-human relationship. This relationship requires respecting the spirit and journey while establishing safe forms of communication. (4) Making connections (Ritchhart et al., 2011) between how the Curly Horses and humans form relationships, Nadasdy (2007) draws on literature from anthropology which chronicles Indigenous world views regarding human and more-than-human relationships as reciprocal. This reciprocation involves both learning and healing existing and moving through the space between. Understanding that the spirit is a central component of knowing, having practitioners acknowledge and recognize the spirit's innate existence in all life forms is essential to integrating Indigenous ways of knowing and being into current practice.

Healing Through Learning About the Land

If healing occurred with my relationship with the land, then I speak of this healing as being assisted by decolonized learning and active engagement with the land. The experience of observing my own children and their interactions with the land encourages the (5) consider[ation of] different viewpoints and perspectives (Ritchhart et al., 2011) and allows all to analyze why these items, topics, and relationships are of importance to the individual. This time allows space to accept the unknown or understood and through trust and respect, honour and value the significance of these relations. Fostering our relationship with the land is the act of healing in relation (A. McGinnis, personal communication, May 11, 2019). A sense of pride is felt when I observe the unadulterated wonder and joy my children have when interacting with the land. These relationships have created an ongoing reciprocation involving both learning and healing existing and moving through the space between.

Honouring the Journey

The final step to making my thinking visible is step (6), capturing the heart and forming conclusions (Ritchhart et al., 2011). To decolonize healing is to embrace the notion that healing and learning coexist through relationships with the land along the journey. Within the journey lies a reciprocal, interwoven relationship between learning about self, learning about the community, learning about the more-than-human, and learning about the land. Through acknowledging and understanding our colonial past and taking action to begin decolonizing ourselves, actively engaging in this process will facilitate an open space to promote and integrate Indigenous ways of healing and learning into ethical psychology practices. Through these relational experiences, the space between facilitates healing effects through spiritual connections for both the practitioner and the client and begins to address the historical relational disconnect with the land.

It remains an ethical responsibility for the field of psychology to include, incorporate, and provide opportunities for the integration of the natural environment to support Indigenous ways of knowing and being in current practice (Elder A. Poitras, personal communication, May 4, 2019). The journey for practitioners begins with acknowledging the crisis in our relationship with the natural environment. The journey continues through bringing awareness to our understandings and teachings about the world, and through the ownership of our ways, practitioners can view the natural environment as actively supporting Indigenous wellness.

Through this perspective shift it is possible to release colonial influences. Utilizing a strength-based understanding the field of psychology can practice in collaboration with Indigenous communities to incorporate Indigenous ways of knowing, working with and within the natural environment, into our practice. When practitioners begin on this journey of meaning-making, it may contribute to the reconciliation process of decolonizing individual practice, understanding, and move towards authentically supporting, respecting, and integrating Indigenous ways of knowing and being into the therapeutic process.

Building healing capacity through learning elucidates the understanding of the past, the needs of the present, and lays foundations for the future to work towards restoring integrity and prompting ethical balanced care. In confidence, healing with dignity may be facilitated through relationships with self, community, more-than-human, and the land. McGinnis et al. (2019) eloquently concludes that "reconciliation across Western and Indigenous contexts requires learning to work together with the more-than-human world and developing ethical spaces for health research in which holistic wellness is appreciated and understood in the context of all our relations" (p. 1).

REFERENCES

Alt, P. (2017). Sacred space and the healing environment. *Annuals of Palliative Medicine, 6*(3), 284–296.

Archibald, L., Lee-Morgan, J., & De Santolo, J. (2019). *Decolonizing research: Indigenous storywork as methodology*. Zed Books.

Ball, N. (2013). Just do it: Anishinaabe culture-based education. *Canadian Journal of Native Education, 36*(1), 36–58. https://login.libproxy.uregina.ca:8443/login?url=https://search-proquest-com.libproxy.uregina.ca/docview/1523947533?accountid=13480.

Berger, R. (2006). Beyond words: Nature-therapy in action. *Journal of Critical Psychology, Counseling and Psychotherapy, 6*(4), 195–199.

Blue Sky. (2020, January 9). Former Balfour Collegiate teacher lauds benefits of land-based education. CBC News. https://www.cbc.ca/news/canada/saskatchewan/former-balfour-collegiate-teacher-lauds-land-based-learning-1.5420882.

Canadian Collaborative Mental Health Initiative. (2006). *Pathways to healing: A mental health guide for First Nations people*. Canadian Collaborative Mental Health Initiative.

Corey, G. (2016). *Theory and practice of counseling and psychology*. Cengage Learning.

Dell, C., Chalmers, D., Bresette, N., Swain, S., Rankin, D., & Hopkins, C. (2011). A healing space: The experiences of First Nations and Inuit youth with equine-assisted learning. *Child Youth Care Forum, 40*, 319–336. https://doi.org/10.1007/s10566-011-9140-z.

Drees, L. (2013). *Healing histories: Stories from Canada's Indian hospitals*. The University of Alberta Press.

Gendron, F., Bourassa, C., Cyr, D., McKenna, B., & McKim, L. (2013). The medicine room: A teaching tool for elders and educational opportunity for youth. *First Nations Perspectives, 5*(1), 83–97.

Greenwood, M., de Leeuw, S., Lindsay, N., & Reading, C. (2018). *Determinants of Indigenous peoples' health in Canada: Beyond the social*. Canadian Scholars' Press.

Johnston Research. (2010). *Waawiyeyaa evaluation tool*. https://www.johnstonresearch.ca/the-waawiyeyaa-evaluation-tool/.

Jordan, M. (2015). *Nature and therapy: Understanding counselling and psychotherapy in outdoor spaces*. Routledge.

Kermoal, N., & Altamirano-Jimenez, I. (2016). *Living on the land: Indigenous women's understanding of place*. AU Press.

Kimmerer, R. W. (2013). *Braiding sweetgrass: Indigenous wisdom, scientific knowledge, and the teachings of plants*. Milkweed Editions.

Kingsley, J., Towsend, M., Henderson-Wilson, C., & Bolam, B. (2013). Developing an exploration framework linking Australian aboriginal peoples' connection to country and concepts of well-being. *International Journal of Environmental Public Health, 10*, 678–698. https://doi.org/10.3390/ijerph10020678.

Marsh, T., Cote-Meek, S., Najavits, L., & Toulouse, P. (2016). Indigenous healing and seeking safety: A blended implementation project for intergenerational trauma and substance use disorders. *The International Indigenous Policy Journal, 7*(2). https://doi.org/10.18584/iipj.2016.7.2.3.

McGinnis, A., Kincaid, A., Barett, M., Ham, C., & Community Elder Research Advisory Group. (2019). Strengthening animal-human relationships as a doorway to

Indigenous holistic wellness. *Ecopsychology*, *11*(3), 162–173. https://doi.org/10.1089/eco.2019.0003.

Nadasdy, P. (2007). The gift in the animal: The ontology of hunting and human-animal sociality. *American Ethnologist*, *34*(1), 25–43. https://www.jstor.org/stable/4496783.

Raphael, D. (2018). The social determinants of health of under-served populations in Canada. In A. Arya & T. Piggott (Eds.), *Under-served: Health determinants of Indigenous, inner-city, and migrant populations in Canada* (pp. 23–38). Canadian Scholars.

Ritchhart, R., Church, M., & Morrison, K. (2011). *Making thinking visible: How to promote engagement, understanding, and independence for all learners.* Jossey-Bass.

Smith, L. T. (2012). *Decolonizing methodologies: Research and indigenous peoples.* Zed Books.

Styres, S. (2019). Literacies of land: Decolonizing narratives, storying, and literature. In L. Smith, E. Tuck, & W. Yang (Eds.), *Indigenous and decolonizing studies in education: Mapping the long view* (pp. 24–37). Routledge.

Truth and Reconciliation Commission of Canada (TRCC). (2012). Call to Action (Legacy, Health).

Wanuskewain Heritage Park Authority. (2020). *Wanuskewain.* https://wanuskewin.com/.

CHAPTER 9

Still Joy: A Call for Wonder(ing) in Science Education as Anti-racist Vibrant Life-Living

Christie C. Byers

In this chapter I share a meta-assemblage research-creation (Manning, 2016): a researcher-created experimental exhibit of found poetic assemblages about wonder, joy, Black life, neurodiversity, love, science, and science education. The intention of this meta-assemblage research-creation is to explore the affective flows of the phenomenon of wonder, while also inviting consideration of how the multiple forces and co-components of the body(ies) assembled here move together in an uneasy and historically traceable tension. These co-movements suggest how "traditional" science and school science education are not only complicit with, but also may be directly implicated as primary protagonists in the violent anti-Black racism and planet-wide suffering happening in our world today.

As an emerging science education researcher and long-time elementary science teacher in the United States, I have been thinking about and dwelling with the phenomenon of wonder for many years. I will share here an overview of some of my theorizing about the concept thus far as a means of providing

---

[1] I attach the (ing) to the base word "wonder" to denote one of the many nuanced dimensions of this complex phenomenon/affect: it assumes the form of both a noun (e.g., an object of wonder, a state of wonder, or the concept of wonder itself), and a verb—a dynamic process of wonder*ing*—movement that includes thinking-feeling through encounters, becoming aware of the extraordinary in-of the ordinary, connecting/integrating, exploring, imagining new possibilities, and creative improvisation.

C. C. Byers (✉)
George Mason University, Fairfax, VA, USA

© The Author(s) 2022
M. F.G. Wallace et al. (eds.), *Reimagining Science Education
in the Anthropocene*, Palgrave Studies in Education and the Environment,
https://doi.org/10.1007/978-3-030-79622-8_9

context for engaging with the meta-assemblage that follows. Through a slow and transdisciplinary study, I have come to think-feel wonder(ing)[1] as a dynamic, multidimensional, multimodal, and somewhat sensitive and slippery *affect* (Byers, 2021); a common, ubiquitous, and catalytic force that works to open up possibilities and make felt potential in and through everyday events and encounters. Further, I have been wondering about the relationship of wonder with what psychologist Daniel Stern (2004, 2010) has termed "vitality affects"—affects we feel and sense in others, that permeate our everyday lives, and "are felt experiences of force, in movement – that have a temporal contour, and a sense of aliveness, of going somewhere" (Stern, 2010, p. 8). In a similar vein of thinking with movement and felt *aliveness*, biologist, philosopher, and biosemiotician Andreas Weber has developed the concept of *enlivenment*, a feeling of aliveness experienced in reciprocal relationships with other feeling beings in his books *Enlivenment: A Poetics for the Anthropocene* (2019), *The Biology of Wonder: Aliveness, Feeling and the Metamorphosis of Science* (2016), and *Matter and Desire: An Erotic Ecology* (2017). Rosi Braidotti, in her intellectually rigorous work mapping the proliferation of posthuman knowledges and scholarship, discusses an *ethics of joy* embedded in feminist posthuman thinking, related to vitality and increased capacities of bodies to affect in her many works, including *Posthuman Knowledge* (2019) and the co-edited volume *Posthuman Glossary* (2018, pp. 221–224). The concept of vitality affects has also been taken up by Felix Guattari as central to his ethico-aesthetic paradigm in his text, *Chaosmosis* (1995).

In coming to think of wonder(ing) as an opening-up and enlivening vitality affect, I have noticed that the phenomenon, as experienced, is never quite *containable* within human bodies or singular categories or disciplines. It resists enclosure, capture, and narrow or flat representation as it emerges in and through encounters (or perhaps it is always already there) and moves around and across "things" or "bodies" in a trans- (and transformational) manner. Wonder does move through and can be felt by human bodies, and even linger there for a little while, but this affect, like all affects, tends to behave more like a fleshy ghost (Stewart, 2007); sometimes appearing, moving through, between, and across bodies (Brennan, 2004), and then quickly vanishing again. Wonder(ing) has a habit of entering into and fleeting across encounters, popping up, or perhaps mediating the popping up, of little shocks of semi-awareness (Byers, 2021; Massumi, 2015). Wonder is also sometimes experienced as an intensity of feeling that slides across bodies more slowly, like a warm washing-over, registering as moments of clarity or awareness, deep presence, and a comforting sense of connectedness; *the felt reality of relation* (Massumi, 2002, p. 17, emphasis in original), with and across other bodies in the world (Di Paolantonio, 2018; Washington, 2019; Weber, 2016). The three main points that I am interested in focusing on here are that as an affect or intensity that is felt emerging in and through encounters, wonder(ing) (1) is always available (like an open invitation), (2) is always on the move (across), and (3) carries with it the powerful potential *to move bodies* (Ahmed,

2015; Braidotti, 2019; Irigaray, 1993; Massumi, 2002). This capacity to move bodies might be expressed as increasing bodies' capacities to affect (Braidotti, 2019; Massumi, 2002, 2015; Spinoza, 2009). Wonder has indeed been linked with joyful exploration, imagination, creativity, and becoming inspired to take action (Glăveanu, 2020), and is increasingly being thought in association with the concept of human flourishing in education (Schinkel, 2020). Importantly, this increase in capacity to affect and association with human flourishing is always relational, and not individual; it involves co-created and co-moving intensities enabled through reciprocal relations with other bodies—regardless of whether these bodies are human or more-than-human matter (e.g., ideas, objects, animals, institutions, forces, organizations, plants, water, books, etc.) (Bazzul & Kayumova, 2016; Braidotti, 2019; Haraway, 2016; Manning, 2016; Massumi, 2015; Weber, 2016).

The increase in capacity activated through wonder(ing) is quite palpable—and is likely even measurable for those inclined toward measuring things and valuing things that are measurable. A few empirical studies have initiated efforts in this area (Hadzigeorgiou, 2012; Gilbert & Byers, 2017; Girod et al., 2010; Girod & Wong, 2002). In an empirical study I am currently grappling with writing up (Byers, 2021), for example, preservice elementary teachers who explicitly engaged with wonder(ing) in a science methods course through wonder journaling and digging into and sharing their wonders with one another science-fair style experienced the phenomenon as related to:

1. enthusiastic, motivated entry into learning through their interests and experiences;
2. feeling more "in tune" with self, "science," and one another;
3. joyful engagement and a marked increase in "aliveness" or lively energy;
4. broadening views about knowledge/knowing and more comfort with uncertainty;
5. imagining and enacting possibilities of doing things differently; rethinking "structures" as "structured";
6. perceiving the world as more open, more wonder-filled;
7. increased feelings of confidence and competence, or the capacity to affect;
8. a sense of wellbeing extending beyond the course into other areas of their lives.

The future teachers in the study exclaimed: "I am much more creative than I ever thought I was!" and "I learned I have so many ideas and they are not stupid!" and "I want to wonder more in my life!" in addition to dreaming up ways to promote and welcome wonder(ing) in their future teaching practices with elementary children.

Some of the data shared by these preservice teachers is included in the poetic meta-assemblage research-creation that follows. I juxtapose these with other forms of data that move with similar rhythms from scholarly texts, conversations, video recordings, photos, memos, etc. The exhibit, titled *Still Joy*, is an attempt to illustrate how wonder(ing), as an enlivening/enabling and life-affirming force/affect, is present, and has been moving, and continues to move in and through all areas of life-living, including troubled and racialized spaces. With respect to the wicked and ongoing problem of racism and anti-Black violence in the US, this assemblage attempts to provoke thought about how wonder(ing) carries with it the possibility, if intentionally attuned to, foregrounded, and allowed, of both calling out and troubling this trouble, while also (always) opening up new possibilities for vibrant and richly variegated life-living. Reflecting on this particular exhibit as it emerged through my experimental writing (Massumi, 2002) has me wonder(ing) whether foregrounding wonder in science education contexts may be a critical component of an anti-oppressive, life-affirming, plurality-celebrating mode of living science education otherwise.

Though I might have chosen to create this experimental, rhythmic collage of process philosophy and wonder thinking with any of the many rich areas of critical scholarship that work to interrogate and critique hegemonic violence wielded against marginalized "others" as my entry point, for this particular assemblage, I chose to focus on Black thought, highlighting especially the exact words of Black feminist, Black pedagogical, and Black philosophical and critical scholars. I make this intentional choice as a move toward taking on more responsible action in my scholarship—given the urgent need to directly address the current (and lengthy historical) circumstances of racial unrest and ongoing violence against Black people in the United States (Hattery & Smith, 2018). My intended audience is mainly US teachers who teach science, elementary teachers especially, the majority of whom are white women like me. It is urgent that they/we recognize that doing things differently, by making decisive, ethical, and responsible moves in our classrooms is both possible and immediately necessary.

It is important to point out that this crying out for doing differently in science education is not an urging for yet another way of "helping" Black, Indigenous, Latinx, neurodiverse students, etc. through some kind of an assimilation project, where these "others" are deemed broken, behind, or less than and in need of being fixed, saved, and/or helped along by/through initiation into white Western structures of knowing and being. Rather, this work represents a call for caring for and valuing (and thus making space for) the rich and varied experiences, perspectives, ideas, modes of being/becoming, and diverse sense-making repertoires (Bang et al., 2017) that all children enter learning spaces with; an expansion of what it means to be "scientific"—and a pointing out that much of Western (white) science and traditional school science education is getting it wrong in the life-affirming department. Much (science) education today works to actively still, snuff out, and police any "out of bounds" movements. Attempts are made to tame the vibrant, reciprocal, relational life-living and joyfully active learning that is constantly emerging and trying to move and thrive in education spaces. A great deal of harm and suffering in the form of closing off capacity happens as a result of this stilling and policing. The argument here is that science learning, as a part of life itself in all of its ongoing emergence, diversity, playfulness, feeling, and vibrancy involves bursts of exuberance, difference-in-relation, creative and supernormal improvisations (Massumi, 2012), movements toward and through joy, and the forging of often unexpected and caring/reciprocal relations. If allowed, learning science could be and become so much more than the forced fragmentation, siloing, neutrality, stillness, hierarchical, distancing, and inert/static "rational" objectivity promoted today (Visvanathan, 2002; Weber, 2016, 2019; Whitehead, 1929). There is no view from nowhere (Haraway, 1988), no evolutionary hierarchy (Marks, 2017). I see this project as resonating with the marginal and minor yet ubiquitous transdisciplinary idea that vibrant, diverse, plurality of learning-as-life-living, and improvisational becoming-with others is always already moving and happening in and through encounters, through ongoing reciprocal processes of influx and efflux (Bennett, 2020), within locally situated ecological webs of relations (Braidotti, 2019; Cordova, 2007; Escobar, 2017; Haraway, 2016; Weber 2016, 2017, 2019). What I am attempting to amplify here is that wonder(ing) is a catalyst or primary force involved in a non-erasable life process: the ongoing ontogenesis of lively, movement-moving and becoming; movements that are always working with and gesturing toward what else is possible (Glăveanu, 2020; Manning, 2016; Massumi, 2002). What emerged for me (and continually moved me and moved through me) while creating this meta-assemblage is that policing and regulating Black life is intimately related to policing and regulating wonder. Both of these acts involve the violent disruption and attempted regulation and control of vibrant and precious life processes. What would happen if we valued and cared for these ongoing, relational emergences of difference and attempts at mutual flourishing? What if we moved with and followed these indeterminate *felt elsewheres* and *otherwise possibilities* (Vossoughi, 2021), instead of persistently enclosing, policing, and snuffing them out?

What I aim to make felt through this work is that making the time and space for wonder could be a relatively simple move to allow for these vital, deeply relational and emergent/divergent processes to do their work of (re)energizing bodies toward new and healthier/livelier modes of living, becoming, and creating together. Thriving. Not just surviving (Love, 2019). Notably, wonder(ing) is particularly powerful in that it registers as both critical *and* creative (Braidotti, 2019; Byers, 2021; Glăveanu, 2020). Welcoming wonder(ing) always carries with it the potential for a beginning again, for new beginnings; for a re-viewing (Glăveanu, 2020) and a re-newing. An opening-things-up that as Sara Ahmed (2015) points out "allows us to see the world *as made*," a world "that does not have to be" (p. 180), and "can be unmade as well as made. Wonder energizes the hope for transformation" (p. 181). Following this thinking, recognizing and welcoming wonder(ing) might allow educators and students to move together with (instead of constantly fighting against) the already-happening, relational, inclusive, affirmative, generative, joyful, and socially-just (science) education potential squirming around, putting out tendrils, and trying to thrive in the everyday; right there in plain sight.

Wonder already has one foot wedged into the proverbial science education door (well, perhaps just a winding tendril or a toe) through a couple of brief mentions in the Framework document underpinning the Next Generation Science Standards (NGSS), and appendix H of the NGSS document itself (see p. 28 of NRC, 2012 and p. 1 of appendix H, NGSS Lead States, 2013). Additionally, it gains support through scientists themselves who name wonder as central to their work and one of the greatest rewards they experience (Cantore, 1977; Carson, 1965; Hadzigeorgiou, 2014). The key for science educators is to allow for the time and space, and to be willing to create the necessary conditions (Byers, 2021) that would enable a full, multi-dimensional, and multimodal version of wonder(ing) to be welcomed and supported. As far as I can tell thus far in this wonder-tracking project, it is only a very narrow, known-facts, standards-based, and anemic version of wonder(ing) that is currently allowed in our current, impoverished version of (science) education (diAgnese, 2020; Visvanathan, 2002); something more akin to curiosity. Although curiosity is important, it represents a single, rather limited and superficial (Kingwell, 2000; Opdal, 2001) dimension within a pool of related concepts and phenomena that fall under the umbrella of wonder (Glăveanu, 2020). Even then, when curiosity, or a single-dimensional aspect of wonder(ing) is deemed allowable for children in (science) education contexts, it is too often with the intention that this phenomenon might be "captured" and/or (re)directed for exploitive, and/or predetermined, instrumental purposes, rather than for the opening up, exploratory, and trans-connective and enlivening possibilities that a fuller version of wonder(ing) might activate.

Importantly, these imaginative and connective possibilities catalyzed through wonder(ing) can never be fully anticipated in advance. This is perhaps one of the most challenging obstacles to foregrounding wonder in education today. Though the simple acts of slowing down, allowing and/or curating encounters, valuing diverse movements, and hesitating to regulate and police thought might promote wonder(ing), how do these kinds of moves align with the goals of education as they are conceptualized and enacted today? The common cultural approach in modern US schools is to speed up, standardize, fragment, and regulate knowledge for predetermined outcomes (diAgnese, 2020). In order to accomplish this, it is imperative to maintain authority and control, to "civilize" the perceived feral nature of children, perhaps most seriously affecting non-white and/or neurodivergent children (Shalaby, 2017). This regulating process results in the dampening of interest, (en)forced homogenization, and the bounding and swaddling of creative thought (Cant, 2017; Gilbert & Gray, 2019). Notably, this control is often maintained through policing and punishing children's lively, moving bodies (Leafgren, 2016; Shalaby, 2017).

And now, finally, hoping the phenomenon of wonder(ing) has begun to take on a shape for readers through this overview, I turn to the meta-assemblage research-creation itself. Although some portions of the work shared here might look and sometimes read like poetry, I am not quite a poet, and the work does not carry the trained artfulness and musicality of the kind of beautiful poetry those with poetic gifts are able to create. I am a (becoming) science education researcher who has been following wonder around and reading/thinking/tinkering with concepts that appear to resonate and relate with it. The forms presented here are researcher-data co-created assemblages; an experimental exhibit of my thinking-in-progress put in motion along with the thinking of many others. This research-creation (Manning, 2016) is composed of and with snippets of found data, found texts, conversations, research memos, intuitive juxtapositions, photographs, and the following of creative impulses (Ulmer, 2018). The exact words I cut and pasted from other works into this moving collage are those that I felt resonating with the flow of the project. These works are cited in the footnotes. For copyright reasons, brief analysis notes of how I retroactively made sense of the inclusion of these words and works is included, though I do worry about the potential of these notes to foreclose thought and "capture" meaning—a move that does not align with spirit and movement of wonder(ing). During the actual event of the creation of this exhibit, the bits and pieces flowed together in ways that I, as the researcher-composer, did not feel entirely in control of; or as Erin Manning (2016) might describe it: The material intuited its relational movement, and edged its way into form. These snippets of thought, or data scribbles, came together in an exhibit that (hopefully) works to make felt some of wonder(ing)'s meandering affects and effects, connections to contemporary social justice concerns, and movements toward more generative and affirmative modes of living out (science) education.

I also see this project as aligning with and gathering inspiration from Jasmine Ulmer's poetic call for critical qualitative inquiry is/as love (2017). Although this assemblage ends on a mournful note, this is the tragic reality of where we are in the United States today. Working for wonder, however, is, at its core, about making space for hope, for dreams, for play, for affirmative movement, for the joyful, vibrant, and deeply connected and mutually-flourishing existence that at some level we all know (feel) is possible.

Still Joy

Insurgent Black life exceeds the bounds;
it moves too much.[2]

[2] Manning (2016, p. 5). *There are policed boundaries and norms for movement in US culture (including schools); Black life is perceived to exceed these, and thus is actively regulated/policed.*

My son likes to move and talk a lot,
but his school required stillness, silence[3]

(it's like the monkeys are running the zoo)
A white teacher I know actually spoke these words,
referring to a new school initiative
to not punish or expel Black children
quite as much.

I mean, it steals joy.
I think school steals joy—
it crushes wonder,
stifles imagination[4]

It happened with my son—
he's so curious about everything,
and yeah, he loves to move and talk.[5]

He's at home with me now, it's
been a gift for both of us—creating and
feeding curiosity together. We've really been co-learners,
and it's been joyful, absolutely full of wonder.[6]

I'm sorry, did you say steals joy, or.
stills joy?

My Southern/midwestern Twang
can be confusing—
but I was musing …
I said steals, but stills works too.[7]

---

[3] Wendi Manuel-Scott, professor and scholar of race, gender, the African American experience, and the history of Black women in the Atlantic world, personal conversation. *There are policed boundaries and norms for movement in US culture (including schools); Black life is perceived to exceed these, and thus is actively regulated/policed.*

[4] Ibid. *Wonder is thought here as a life process at the heart of ontogenesis/emergence of becoming: ongoing relational movements toward joy, imagination, creativity, and difference. There are policed boundaries and norms for movement in US culture (including schools); wonder exceeds these, and thus is actively regulated/policed.*

[5] Ibid. *There are policed boundaries and norms for movement in US culture (including schools); Black life is perceived to exceed these, and thus is actively regulated/policed.*

[6] Ibid. *Outside of the (highly regulated/policed) school environment, wonder has been allowed to move around and flourish, with positive outcomes for both mother and child.*

[7] Wendi Manuel-Scott, personal conversation. *There are policed boundaries and norms for movement in US culture (including schools); wonder(ing), as movement, exceeds these, and thus is actively regulated. I/we connect school's stilling/stealing of wonder with stilling her (Black) son's movements back when he was in school.*

Steals. Stills.

"A rich science education has the potential to *capture* students' sense of wonder about the world and to spark their desire to continue learning about science throughout their lives."[8]

Capture?

"If a child is to *keep alive* his inborn sense of wonder, he needs the companionship of at least one adult who can share it, rediscovering with him the joy, excitement, and mystery of the world we live in."[9]

Keep Alive?

"I don't even know where to start. At what point did wondering become so difficult? It's almost as if that part of my brain is just *locked away* very deep, and I'm so structured and I'm so used to, 'no, this is how it's gonna be, so this is how you should do it.'"[10]

Locked Away?

****

So Intimidated about Teaching Science[11]

Science is a source of a lot of anxiety

---

[8] National Research Council (2012, p. 28). *Science education is poised to capture (and thus potentially exploit and regulate) the life process of wonder(ing) as expressed and experienced by children.*

[9] Carson (1965, p. 55). *Adults can help prevent the stilling/stealing of wonder (and thus keep wonder "alive") by attending to, encouraging, and sharing in children's wonder.*

[10] Preservice elementary teacher, science methods course, 2017. *Policing/regulating wonder might lead to wonder, as a life process, being forced to become "locked away" or hidden from view and replaced with the performed repetition of the normative order of things (receiving knowledge, not creating it), the traditional, acceptable mode of living/learning in schools.*

[11] Byers (2021) Found Poetic Data Assemblage, 20 preservice elementary teachers, science methods course. All phrases from original data and (re)arranged by researcher, no new words added. *Preservice teachers describe experiences with wonder(ing) being shut down over time (though recognizable in children—both the children they once were, and the children they are now observing as part of their teacher education placements in schools), as well as the regulating, intimidating nature of the kind of science education that they came to know through their past schooling experiences. Though these past experiences and memories have affected their views/confidence about teaching science, engaging with wonder (again) is opening up their thinking.*

for a lot of people,
So no, science has never been my strength.

Like back in school, I didn't get it super quickly
so it wasn't something I was ever really confident in.

Because to me it's always been a very structured,
like there's a black and white answer
to everything that you do.

It wasn't even a thing really,
it was like—you just filled out a packet.
Like, 'this is the answer'—take it or leave it.

I remember being little, being in school—
I would get in trouble for getting off task
and wondering about something.
My mind was always racing,
I always had a million questions,
a million things running through my head.

Things would just like really grab my attention,
like black holes, blood, life, God, death—
just like big things in general.

I would wonder about them, and ask questions,
but no, that was *not* what we were talking about.
Like "you're off task, you're off topic."
"That is definitely *not* in the curriculum."

We like shut down the "off topic" conversation.
We just shut it down.
We squash the idea of wondering.

Kids are taught to just keep it to themselves.

I love watching children wonder about things
they're curious little beings,
constantly wondering little creatures—
like their brains are so cool!
It's like their whole body—and their whole mind—
I mean, they have wonders about everything!
Everything is new to them!

But, they kind of learn to shut off that wonder.
We just shut it down. We squish it.
It's like we shut down all these

like interests and fascinations,
and it's like *why*?

I know that I am more passionate about things
and I'm more interested in things
when I like wonder about them
and when I'm fascinated by them—

Why shouldn't we like, make *that* the standard?

How can we expect them to do well in school
if they're not fascinated by anything?
If we don't *allow them* to be fascinated by anything?

And if we stifle wonder,
they're not gonna even *want* to wonder any more.

Kids are like,
all right, let's get through this,
let's get to recess, or like
let's get to the end of the day.

I think that's why I was so discouraged—
my wonders were put on hold.

And yeah, I was really worried,
so intimidated
about teaching science.

****

Science: A Fractured Pilgrimage[12]

My favourite metaphor is of science as a journey -
a fractured pilgrimage that began as a search
for man, nature and God;
an attempt to be at home in the cosmos
that became a homelessness.[13]

[12] Byers (2021) Found Poetic Data Assemblage in Visvanathan (2002). Phrases (re)arranged by researcher. Minimal wording changes for flow and emphasis. *This assemblage emphasizes Visvanathan's analysis from outside of Western science—detailing the "progress" of science as it became entangled with neoliberal capitalism and Cartesian dualisms, and the harm this collaboration has caused. As a solution, he suggests science might return to where it began: as embedded in (not separate from) life/nature; with wonder.*

[13] Visvanathan (2002). *Science as a discipline started out differently than what it later became. It began in a similar way to how a preservice teacher in the previous section described her own wonder(ing) as a child: as a search for answers to big questions; for finding a home/sense of belonging in the universe. It began with wonder, but later became a fragmenting process of separating self/other.*

The scientific self, an invention,
perfected self as spectator;
separated from the world,
the object becomes a spectacle,
a specimen.[14]

The constitution of self/object,
self/"other" begins through
this constitution of distance.[15]

Science, a hegemonized form of knowledge,
reduced a whole series of "others"
to lesser orders of being
subject perpetually to the
scientific gaze.[16]

Science as "neutral" is a lie;
science creates its own 'microphysics of power'
by determining discourses,
by pre-empting the ways one thinks.[17]

And the link between science and market is essential;
nature is seen as dead, a resource;
a mountain becomes a repository of ore.[18]

We must move from a glossary of distanced objectivity—
of "other" - to a language of relation and celebration -
to a dance of possibilities.[19]

But first science needs to cry.[20]

---

[14] Visvanathan (2002). *Science became a discipline that separates self/other, subject/object.*

[15] Ibid. *Science became a discipline that separates self/other, subject/object.*

[16] Ibid. *Science has separated self/other in a hierarchical manner, and claimed power/hegemony.*

[17] Ibid. *Science polices/regulates culture/society by claiming power and bounding discourses and thought through the guise of "neutrality."*

[18] Ibid. *Science, as a discipline that separated self/other, subject/object also separated humans from nature, hierarchically, thus enabling the exploitation of ("dead, inert") nature for human purposes. Science is entangled with capitalism, applying market values to living beings and life processes.*

[19] Ibid. *A solution to the suffering caused by the hegemony of science may be through embracing relations and other possibilities beyond those currently narrowly defined, regulated, and policed by Western science.*

[20] Ibid. *Science might open up to the fullness of human experience, including the affective/emotive, beginning with a time of mourning for what it has become and the pain it has caused.*

Crying is essential—
for the earth and what science has become.[21]

When you cry, you care;
caring and crying may inaugurate
new social practices.[22]

There is more truth in crying
Than in any Cartesian meditation.[23]

And then science also needs laughter -
to not forget to be playful.[24]

Science should mimic nature in all its playfulness;
it is through play that ecology restores
by multiplying alternatives, lifestyles, life forms.[25]

And with this crying and laughing and playing,
science could begin again in wonder.[26]

****

Black people have always needed
the gift of wonder,
to be able to imagine, to dream
to believe in the impossible—
to imagine what might be[27]
and move toward that.

Movement-moving[28]

---

[21] Ibid. *Science might open up to the fullness of human experience, including the affective/emotive, beginning with a time of mourning for the pain it has caused.*

[22] Visvanathan (2002). *Science might open up to the fullness of human experience: affect, emotion, care, empathy, etc. may be included as allowable scientific/social practices.*

[23] Ibid. *Crying (and other human modes of being) represent reality more accurately than Cartesian dualities that separate and hierarchize reason/emotion, dismissing the latter.*

[24] Ibid. *Opening up to the fullness of human experience would include laughter and play.*

[25] Ibid. *Play is an ecological principle; a life process that includes exuberance and diversity. This process might be embraced and celebrated as a regular, allowable aspect of what life does; not regulated and policed.*

[26] Ibid. *It is through wonder that science began. Wondering is a primary process of life-living. Science might begin again in wonder in order to produce more life-affirming modes of living.*

[27] Manuel-Scott, personal conversation. *Black life, as regulated and policed life, has always needed wonder(ing) for its movement toward imagining what else is possible.*

[28] Manning (2014). *Life, thought here with the phenomenon of wonder catalyzing ontogenesis at the center, is always moving. Movement-moving is central to process philosophy.*

dream-dreaming

*Our ancestors dreamed us up
and then bent reality to create us.*[29]

Black women gardeners[30]
Black mothers tending gardens
*deep, thoughtful women
where music was always on*[31]
*She's got to sing,
she can't help but sing!*[32]
I know why the caged bird sings.[33]

*If they ask you, tell them we were flying.
Freedom is (in) the invention of escape.*[34]

*The birds they sang
at the break of day
start again, I heard them say.*[35]

Black women mother gardeners
dreaming what might grow
in wide open fields,
and small brown plots—
dark rich soil full of
dream-drenched seeds.

Virtual possible/impossible
speculative fabulations[36]

---

[29] Imarisha (2015). *Black life, as regulated and policed life, needs wonder(ing) for its relation to dreaming, for activation of movement, for feeling and imagining the "what else is possible."*

[30] Manuel-Scott, personal conversation. *The recognition and affirmation of Black women for their lively, joyful wondering, life-affirming dreaming, hoping, and for planting seeds of possibility.*

[31] Fred Moten interview (2018). *The recognition and affirmation of Black women for their lively, joyful wondering, life-affirming dreaming, hoping, and for planting seeds of possibility.*

[32] Ibid. *The recognition and affirmation of Black women for their lively, joyful wondering, life-affirming dreaming, hoping, and for planting seeds of possibility.*

[33] Angelou (2009). *Black women, despite facing regulation/policing/captivity are recognized and affirmed for their lively, joyful wondering, generative dreaming, singing, hoping, and planting seeds of possibility.*

[34] Harney and Moten (2013). *Connecting to birds/flight of the previous passage, wondering is viewed as related to/resonating with freedom/dreaming up escape from regulating structures.*

[35] Cohen (1984). *Following the affective flow of bird metaphors from the lines before; here resilience/persistence is added and woven in.*

[36] Haraway (2016). *Connecting wonder with dreaming, to joyful speculation of what else is possible.*

of dark/bright futures
spectral shimmers, *Beloved*[37] *fleshy ghosts*[38]
sowing seeds, dispersing,
remembering,
intuiting.

The edgings into form of the material's intuition[39]
soil sweat blood Hughes poems dreams laughter
grandma *Fitzgerald musical notes*
playing hide and seek[40]
wings seeds land dance
all *vibrant matter*[41]

Matter intuits its relational movement,
the capacity to think the more-than
as a memory of the future,
a time that makes its own way—
a time schism.[42]

A crack, a break
There is a crack, a crack in everything.[43]
Freedom is (in) the invention
of escape, stealing away in the confines,
in the form, of a break.[44]

These cracks, *time schisms*
are slower places, fugitive spaces
the there-here, then-now
of everyday encounters

---

[37] Morrison (1987). *Wonder evokes spectral, the ineffable.*

[38] Stewart (2007). *Wonder evokes spectral affects, the ineffable.*

[39] Manning (2016). *Manning's process philosophy includes the ontogenesis, or in-forming of material relations; wonder is thought here as being at the heart of this process.*

[40] Fred Moten interview (2018). *These words and affects contribute to the various entities/bodies/materials co-moving and becoming in-through the relational field of ontogenesis.*

[41] Bennett (2010). *Bennett's contribution here is her theorizing of the agency of matter.*

[42] Manning (2016). *In the midst of events and encounters (where ontogenesis is occurring), there is felt a more-than, an excess, beyond linear or regulated/policed conceptions of time and space; this might be thought of as a crack, or a schism—a place for escape, or where new modes of life-living might emerge and thrive.*

[43] Cohen (1984). *In the midst of events and encounters (where ontogenesis is occurring), there is always felt a more-than, an excess, beyond linear or regulated/policed conceptions of time and space; this might be thought of as a crack, or schism—a place for escape, or where new modes of life-living might emerge and thrive.*

[44] Harney and Moten (2013). *In the midst of events and encounters (where ontogenesis is occurring), there is always a more-than, an excess, beyond linear or regulated/policed conceptions of time and space; this might be thought of as a crack, or schism—a place for escape, or where new modes of life-living might emerge and thrive.*

the in-between here-where
time-space opens to the
more-than of what it seems.
The extraordinary in-of the ordinary.[45]

That nothing is quite what it seems
in the movement-moving of an event
suggests a kind of wonder—
a wondering in movement,
a wondering at the world directly.[46]

Wonder(ing) here is thought-felt
as liminal—a liminal experience,[47]
the extraordinary in-of the ordinary.[48]
An ordinary affect.[49]
A minor gesture[50]*-gesturing,*
disorienting-orienting-reorienting,[51]
unsticking, sticking,[52]
disassembling, assembling.[53]
A sort of shimmering apprehension
on the threshold in between knowing and not knowing.[54]

A daydreamy schism in place-time

[45] Glăveanu (2020). *These schisms, or thresholds/openings toward other modes of being, are available in the ordinary/everyday.*

[46] Manning (2014, pp. 165, 168). *Manning connects movement and perception of the more-than in-through events with wondering at-with the world, unmediated. Wonder as a life-process, as catalyzing ontogenesis. This thought is developed from the thinking of many process-based philosophers (Massumi, Bergson, Simondon, Deleuze and Guattari, Whitehead, James, etc.).*

[47] Pearce and MacLure (2009, p. 254). *Wonder(ing) is thought here as taking place at the edges of experience, in the in-between, the threshold spaces, or the previously mentioned cracks or time-schisms.*

[48] Glăveanu (2020). *These schisms, or thresholds/openings gesturing toward other modes of being, are available in the ordinary/everyday.*

[49] Stewart (2007). *Wonder is common, ordinary; an everyday affect.*

[50] Manning (2016). *Analysis: these schisms, or thresholds/openings, are felt as (minor) gestures, in otherwise regulated/policed spaces (the major). They gesture toward other modes of being.*

[51] Ahmed (2006). *Ahmed discusses how wonder can help reorient what was previously oriented a different way: according to hegemony, power, structures that structure, regulate, police.*

[52] Ahmed (2010). *Ahmed theorizes regulated structures, norms, and affects as "sticky"— thus wonder may help to open things back up, to help them get unstuck.*

[53] Deleuze and Guattari (1987). *There is a plurality/multiplicity of potential assemblages in a field of potentiality, or the plane of immanence making up the milieu of every event.*

[54] Pearce and MacLure (2009, p. 254). *Wonder takes place at the edges of experience, in the in-between, threshold spaces, or time-schisms, troubling certainty and knowledge.*

other here-there
slowed down deep presence
feeling-thinking-moving
where the event is still welling[55]
where aesthetic cognitive and spiritual modes
are simultaneously mobilized,[56]
where force has not yet turned to form,
here, *in the midst—*
there is potential for new diagrams
of life-living to be drawn.[57]

Space for *rhizomatic*[58] sprouting,
vining tendrils branching winding bursting forth
writhing, thriving

(not just surviving)

See Figure 9.1.

We want to do more than survive!
We want our full humanity recognized with dignity![59]

dancing pulsing laughing
fullness of life living
the flux of liveliness coursing
through existence unlimited[60]

It's about Black joy—
and always putting love

---

[55] Manning (2016, p. 15). *In the midst of events and encounters (where ontogenesis is occurring), there is always felt a more-than, an excess, beyond linear or regulated/policed conceptions of time and space; this might be thought of as a crack, or schism—a place for escape, or where new modes of life-living might emerge and thrive.*

[56] Pearce and MacLure (2009, p. 254). *In the "midst" referred to in the previous line, there are primes mobilized across areas more traditionally thought of as separate (or that science has forced to be separate): spiritual, cognitive, aesthetic, realms, etc. Wonder is thought here as a connecting/synthesizing force, rather than one of fragmenting and separating these.*

[57] Manning (2016, p. 15). *In the midst of events and encounters (where ontogenesis is occurring), there is always felt a more-than, an excess, beyond linear or regulated/policed conceptions of time and space; this might be thought of as a crack, or schism—a place for escape, or where new modes of life-living might emerge and thrive.*

[58] Deleuze and Guattari (1987). *Rhizomatic movement is a non-hierarchical, non-linear form of movement.*

[59] Love (2019). *Black life is human life, and life itself (ontogenesis, desiring, thriving), and deserves/demands being treated as such; schooling for Black children needs to be about thriving, not just surviving.*

[60] Manning (2016, citing Deleuze, 2005). *The movement-moving (ontogenesis, emergence, liveliness) courses through existence eternally; it is always there and can be felt/attuned to (perhaps through wonder).*

Fig. 9.1   Tendril-twisting, vine-vining: Wondering and thriving plant articulations of solidarity (*Source* Byers [2021], Takoma Park, MD)

> at the center.[61]
>
> bright buildings cracks, openings
> that look like someone is ready
> to love us in that space.[62]
>
> Black joy, life-living
> vibrating at edges-edging
> *Movement-moving,*
> *edging into form.*[63]
>
> *I don't need to disavow the notion*
> *that black people have rhythm.*[64]

(Look at them go! Look at them grow!)

[61] Love (2019). *Life, wonder, joy, love are part of the same "alternative" or minor life processes (forced into minority; minoritized) described throughout the assemblage that need to be attuned to and centered. This is especially imperative for marginalized/racialized life. There are policed boundaries and norms for movement in US culture (including schools); Black life is perceived to exceed these, and thus is actively regulated.*

[62] Love (2019). *Life, wonder, joy, love are part of the same "alternative" or minor life processes described throughout the assemblage that need to be attuned to and centered. This is especially imperative for marginalized/racialized life. There are policed boundaries and norms for movement in US culture (including schools); Black life is perceived to exceed these, and thus is actively regulated.*

[63] Manning (2014, 2016). *Ontogenesis is an aspect of a process philosophy, creating diversity, new forms. Wonder is thought here to be at the heart of this process.*

[64] Moten, as cited in Manning (2016, p. 5). *There are policed boundaries and norms for movement in US culture (including schools); Black life is perceived to exceed these, and thus is actively regulated.*

Unpredictable, loud, sudden, bright, booming
(and often also) softly threading, sliding,
silvery traces, windings crossings
shimmery snail trails[65] of
looping relational *enlivenment.*[66]
A lingering-tingling *feeling*
of aliveness-in-connection.[67]

Yes, *this is this and that is that,*[68] and yet
what about and *what if* [69] this, and this, and this, and, and, and ...
(opening up to the trans-than, more-than, we teachers could ever have
imagined, or ever planned, on our own)
And with such joy!

An ethics of joy![70]
Such *dynamic supernormal improvisations.*[71]

★★★★★★★★★★★★★★★★★★★★★★★★★★★★★★★★★★

Something Useful, Something Beautiful[72]

You see something, you wonder about it,
It ignites a fascination
and it sparks you to then
                    do something ...

To, to, to, to
          reach for something
          connect to something

---

[65] Wallace and Byers (2018). *There is no "essence" of life/subjectivities, including Black life.*

[66] Weber (2019). *Life, including Black life, moves toward joy, liveliness, aliveness-in-connection.*

[67] Ibid. *Life, including Black life, moves toward relation, joy, liveliness, aliveness-in-connection.*

[68] Byers (2021). *Representations of stasis exist; but these are not all that are possible.*

[69] Glăveanu (2020). *Wonder is about the "what else is possible" always happening and available in plain sight.*

[70] Braidotti (2019). *Life moves toward joy; it is an ethical imperative to highlight and support this movement.*

[71] Massumi (2012). *Improvisation beyond the "norm"—movements toward difference is a life principle in a process-based philosophy.*

[72] Byers (2021) Found Poetic Data Assemblage, 20 preservice teachers, science methods course. Words and phrases from original data and (re)arranged by researcher, minimal words added for flow and emphasis. *preservice teachers describe experiences with wonder(ing) as opening things up, involving movement/ fluidity, forging relationships, and inspiring joy, liveliness and new ideas/actions.*

make something
change something

I-we want to make something useful or beautiful!

With you and you and you—
(we all left feeling so good about each other)

it sparks you (us!)
into a kind of sudden awareness, like
there is so much more to this world
that we don't even know (yet)!

A spark! A pop! A flow! A flowing!
It all just flowed out.
(Like a stream of consciousness on the side of the page.)
It's fluid, beautifully fluid, and I love that!

I was pumped, to say the least!

It sparked me to come up with
a question experiment design idea
like to do actual science with a why
and a reason behind it

For-with us, for-with people, for-with the planet

To, to, to, to

talk about
write about
create build say do
something

something useful
something beautiful

****

Did you feel that?
Like a *supernormal*[73] supernova.
A more expansive, inclusive, and ethical science.

---

[73] Massumi (2012). *Improvisation beyond the "norm"—movements toward difference as a life principle.*

A broader sense of everything.[74]

A whole-ness of diverse part-ness.
A part-ness of diverse whole-ness.
A simplicity-complexity of wonder-wondering.
movement-moving[75]
simplicity multiplicity plurality vibrancy
variegating vibrating aliveness.

Life-worth-living! A thriving even.
We want to do more than just survive![76]

An embodied and embedded ecological ethics.[77]
An ethico-aesthetic paradigm.[78]
A mutual flourishing.

Black neurodiverse joyful insurgence,
is *movement-moving.*[79]
Insurgent Black life is neurodiverse
through and through.[80]

Somewhere along the line
Black flesh held the responsibility
of protecting generativity,[81]
neurodiversity.[82]

When you say Black life matters,
you are saying that life matters, and when

---

[74] Byers (2021). *A direct quotation from a preservice teacher; wonder as the opposite of narrowing.*

[75] Manning (2014). *This theorizing of wonder aligns with process philosophy.*

[76] Love (2019). *Black life is human life, and life itself (ontogenesis, desiring, thriving), and demands being treated as such; not just regulated and marked for (barely) surviving.*

[77] Braidotti (2019) and Weber (2016), mash-up. *This theorizing of wonder(ing) aligns with posthumanist feminist theory and ecological life processes.*

[78] Guattari (1995). *This theorizing of wonder aligns with Guattari's thinking about an alternative paradigm based on ethics, ecology, and aesthetics. An anti-fascist paradigm.*

[79] Manning (2014). *This theorizing of wonder aligns with process philosophy.*

[80] Manning (2016, p. 5). *Manning connects the neurodivergence movement (honoring, celebrating a plurality of ways to be human, highlighting, especially, autistic perception attunement to a broad field of as-yet determined potential) with Black life, perceived as moving/being outside of the policed/regulated boundaries of life-living.*

[81] Moten as cited in Manning (2016, p. 5). *All of life is generative, improvisational, exuberant, always moving and moving toward difference, etc. Moten and Manning theorize how Black life came to represent this generativity and is thus is most violently policed.*

[82] Manning (2016, p. 5). *Manning makes the connection between neurodiversity and Black life.*

you say life matters, you are saying
Black life matters.[83]

Pressed tendril Black leg twitching
under taut white *normopathy*[84]
scared stiff reflex muscle
A threat to the order he represents
and that he is sworn to protect.[85]

Please man[86]
Stills. Joy, movement-moving
stilled.
Momma!
Momma I'm through![87]

This is the threat:
Insurgent life[88] as
movement-moving, bloom-blooming
tendrils-twisting, voice-booming
edge-edging, dance-dancing
wonder-winding, walk-walking,
path-pathing, dream-dreaming,
cry-crying, laugh-laughing,
wild-wilding
unfettered *feral*[89]
life-living.

All of this.
This *insurgent life cannot be properly regulated*[90]

---

[83] Moten, as cited in Manning (2016, p. 5). *All of life is generative, improvisational, exuberant, always moving and moving toward difference, etc. Somehow Black life came to represent this generativity and thus is most violently policed.*

[84] Massumi (2014). *The policing/regulating of norms is pathological (turning psychological/medicalizing phrasing on its head).*

[85] Moten, as cited in Manning (2016). *There are policed boundaries and norms for movement in US culture (including schools); Black life is perceived to exceed these, and thus is actively regulated by the police, who are the border control. They are sworn to uphold these norms.*

[86] George Floyd, May 25, 2020, words uttered while dying, his neck pinned under the knee of a white police officer in Minneapolis, Minnesota.

[87] Ibid.

[88] Manning (2016, p. 5). *Insurgent life is life moving outside of the regulated boundaries (Though insurgency, ontogenesis, improvisation, etc. is a regular, ecological life process).*

[89] Gilbert and Gray (2019). *Life is exuberant, moving, unpredictable, ever-changing.*

[90] Manning (2016, p. 5). *The policing of norms and boundaries becomes difficult, as these are regular, ongoing life processes. It is how life works.*

It is *a profound threat to the already existing order of things.*[91]

Capture wonder, still joy.

Insurgent life must be stilled.
Dispossessed, *disposed of;*
the only way to respond is to of get rid of them.[92]

Blackness, life-living, is life at the limit.[93]

Life is Black life.[94]

Anti-blackness is anti-life.[95]
Anti-life is anti-wonder.
Anti-wonder is anti-difference.
Anti-difference is anti-neurodiversity.
Anti-neurodiversity is anti-capacity.
Anti-capacity is anti-generativity.
Anti-dreaming
anti-movement
anti-joy
stilled joy

Anti-wonder is anti-blackness.
Anti-blackness is anti-flourishing.

Neurotypicality as founding identity politics
discounts discards, disposes of, stills all life
all generative force,
all unbounded, unpredictable,
rhythmic insurgent life[96]
steals joy

[91] Moten, as cited in Manning (2016, p. 5). *Life outside of the norms/boundaries is a threat and needs to be policed/regulated.*

[92] Thom (2016). *The threat to hegemony (excessive movement beyond the bounds) must be eliminated.*

[93] Manning (2016, p. 5). *Black life, holding the responsibility for generativity, for living life at the limit, is profoundly threatening to the perceived order of things—to white supremacy, to normopathy.*

[94] Moten, as cited in Manning (2016, p. 5). *Black life is all life. Black life represents here the ontogenesis, generativity, joy, affirmation, connection, and movement processes at the heart of life itself.*

[95] Ibid. *Black life is all life. Black life represents here the ontogenesis, generativity, joy, affirmation, connection, and movement processes at the heart of life itself.*

[96] Manning (2016, p. 5). *Analysis: Generative, unpredictably moving life (non-neurotypical life) is discounted; it must be snuffed out. This resonates with wonder(ing) being stilled, snuffed out.*

stills joy
stills.

We need to ask:
Can there still be joy?
Is there joy still?

In the stilling of Black life,
in the stilling of insurgent life,
in the stilling of generativity,
in the stilling of neurodiversity,
in the stilling of feeling,
in the stilling (and dis-stilling)
and the placing of a leg on
the neck of
a breathing,
(and now nearly breathless)
planet—

Is there still joy?

References

Ahmed, S. (2006). *Queer phenomenology: Orientations, objects, others*. Duke University Press.

Ahmed, S. (2010). Happy objects. In M. Gregg & G. J. Seigworth (Eds.), *The affect theory reader* (pp. 29–51). Duke University Press.

Ahmed, S. (2015). *The cultural politics of emotion*. Routledge.

Angelou, M. (2009). *I know why the caged bird sings*. Ballantine Books. (Original book published 1969.)

Bang. M., Brown, B., Barton, A. C., Rosebery, A. & Warren, B. (2017). Toward more equitable learning in science: Expanding relationships among students, teachers and science practices. In C. V. Scharz, C. Passmore, & B. J. Reiser (Eds.), *Helping students make sense of the world using next generation science and engineering practices* (pp. 33–58). NSTA Press.

Bazzul, J., & Kayumova, S. (2016). Toward a social ontology for science education: Introducing Deleuze and Guattari's assemblages. *Educational Philosophy and Theory, 48*(3), 284–299.

Bennett, J. (2010). *Vibrant matter: A political ecology of things*. Duke University Press.

Bennett, J. (2020). *Influx and efflux: Writing up with Walt Whitman*. Duke University Press.

Braidotti, R. (2019). *Posthuman knowledge*. Polity Press.

Braidotti, R., & Hlavajova, M. (Eds.). (2018) *Posthuman glossary*. Bloomsbury Academic.

Brennan, T. (2004). *The transmission of affect*. Cornell University Press.

Byers, C. C. (2021). *"A broader sense of everything": The affects/effects of wonder(ing) in a science methods course* (Unpublished manuscript). College of Education and Human Development, George Mason University.

Cant, A. (2017). *Unswaddling pedagogy*. Canadian ISBN Service.

Cantore, E. (1977). *Scientific man: The humanist significance of science*. ISH Publications.

Carson, R. (1965). *The sense of wonder*. Harper and Row.

Cohen. L. (1984). Anthem [Song]. On *The future* [Album]. Columbia Records.

Cordova, V. (2007). *How it is: The Native American philosophy of V. F. Cordova*. University of Arizona Press.

Critical Resistance. (n.d.). *Do black lives matter? Robin D. G. Kelley and Fred Moten in conversation*. [Video file]. https://vimeo.com/116111740.

diAgnese, V. (2020). Contrasting the neoliberal educational agenda: Wonder reconsidered. In A. Schinkel (Ed.), *Wonder, education, and human flourishing: Theoretical, empirical, and practical perspectives* (pp. 23–39). VU University Press.

Deleuze, G. (2005). *Pure immanence: Essays on life*. Zone Books.

Deleuze, G., & Guattari, F. (1987). *A thousand plateaus: Capitalism and schizophrenia* (B. Massumi, Trans.) University of Minnesota Press. (Original work published 1980).

Di Paolantonio, M. (2018). Wonder, guarding against thoughtlessness in education. *Studies in Philosophy and Education, 38*, 213–228. https://doi.org/10.1007/s11 217-018-9626-3.

Escobar, A. (2017). *Designs for the pluriverse: Radical interdependence, autonomy, and the making of worlds*. Duke University Press.

Gilbert, A., & Byers, C. C. (2017). Wonder as a tool to engage preservice elementary teachers in science learning and teaching. *Science Education, 101*(6), 907–928.

Gilbert, A., & Gray, E. (2019). Wonder in the science classroom In S Fifield & W. Letts (Eds.), *STEM of desire: Queer theories in science education* (pp. 109–123). Sense Publishing.

Girod, M., Twyman, T., & Wojcikiewicz, S. (2010). Teaching and learning science for transformative, aesthetic experience. *Journal of Science Teacher Education, 21*, 801–824.

Girod, M., & Wong, D. (2002). An aesthetic (Deweyian) perspective on science learning: Case studies of three fourth graders. *The Elementary School Journal, 102*(3), 199–224.

Glăveanu, V. P. (2020). *Wonder: The extraordinary power of an ordinary experience*. Bloomsbury Academic.

Guattari, F. (1995). *Chaosmosis: An ethico-aesthetic paradigm*. Indiana University Press.

Hadzigeorgiou, Y. (2012). Fostering a sense of wonder in the science classroom. *Research in Science Education, 42*(5), 985–1005.

Hadzigeorgiou, Y. (2014). Reclaiming the value of wonder in science education. In K. Egan, A. Cant, & G. Judson (Eds.), *Wonder-full education: The centrality of wonder in teaching, and learning across the curriculum* (pp. 40–65). Routledge.

Haraway, D. (1988). Situated knowledges: The science question in feminism and the privilege of partial perspective. *Feminist Studies, 14*(3), 575–599.

Haraway, D. (2016). *Staying with the trouble: Making kin in the Chthulucene*. Duke University Press.

Harney, S., & Moten, F. (2013). *The undercommons: Fugitive planning and black study*. Minor Compositions.

Hattery, A., & Smith, E. (2018). *Policing black bodies: How black lives are surveilled and how to work for change*. Rowman & Littlefield.

Imarisha, W. (2015). Introduction. In W. Imarisha & a. m. brown (Eds.), *Octavia's brood: Science fiction stories from social justice movements* (pp. 3–6). AK Press.

Irigaray, L. (1993). *An ethics of sexual difference* (C. Burke and G. C. Gill, Trans.). The Anthlone Press. (Original work published 1984).

John Hope Franklin Center at Duke University. (2018, November 5). *Left of black with Fred Moten* [Video file]. https://www.youtube.com/watch?v=fFTkoZTFd1k.

Kingwell, M. (2000). Husserl's sense of wonder. *The Philosophical Forum, 31*(1), 85–107. https://doi.org/10.1111/0031-806X.00029.

Leafgren, S. (2016). *Reuben's fall: A rhizomatic analysis of disobedience in kindergarten.* Routledge.

Love, B. (2019). *We want to do more than just survive: Abolitionist teaching and the pursuit of educational freedom.* Beacon Press.

Manning, E. (2014). Wondering the world directly—or how movement outruns the subject. *Body & Society, 20*(3&4), 162–188.

Manning, E. (2016). *The minor gesture.* Duke University Press.

Marks, J. (2017). *Is science racist?* Polity Press.

Massumi, B. (2002). *Parables for the virtual: Movement, affect, sensation.* Duke University Press.

Massumi, B. (2012, May 3–5). *Animality and abstraction.* [Plenary session]. The Nonhuman Turn, Milwaukee, WI. https://www.youtube.com/watch?v=YkF9vhsraxE.

Massumi, B. (2014). *What animals teach us about politics.* Duke University Press.

Massumi, B. (2015). *The politics of affect.* Polity Press.

Morrison, T. (1987). *Beloved.* Alfred A. Knopf.

National Research Council. (2012). *A framework for K-12 science education: Practices, crosscutting concepts, and core ideas.* The National Academies Press. https://doi.org/10.17226/13165.

NGSS Lead States. (2013). *Next generation science standards: For states, by states.* The National Academies Press.

Opdal, M. (2001). Curiosity, wonder and education seen as perspective development. *Studies in Philosophy and Education, 20,* 331–344.

Pearce, C., & MacLure, M. (2009). The wonder of method. *International Journal of Research and Method in Education, 32*(3), 249–265.

Schinkel, A. (Ed.). (2020). *Wonder, education, and human flourishing: Theoretical, empirical, and practical perspectives.* VU University Press.

Shalaby, C. (2017). *Troublemakers: Lessons in freedom from young children at school.* The New Press.

Spinoza, B. (2009). *The ethics.* Wilder.

Stern, D. (2004). *The present moment in psychotherapy and everyday life.* W.W. Norton and Company.

Stern, D. (2010). *Forms of vitality: Exploring dynamic experience in psychology and the arts.* Oxford University Press.

Stewart, K. (2007). *Ordinary affects.* Duke University Press.

Thom, K. C. (2016, November 27). *8 steps toward building indispensability (instead of disposability) culture.* Everyday feminism. https://everydayfeminism.com/2016/11/indispensability-vs-disposability-culture/.

Ulmer, J. B. (2017). Critical qualitative inquiry is/as love. *Qualitative Inquiry, 23*(7), 543–544.

Ulmer, J. B. (2018). Composing techniques: Choreographing a postqualitative writing practice. *Qualitative Inquiry, 24*(9), 728–736.

Visvanathan, C. S. (2002). Between pilgrimage and citizenship, the possibilities of self-restraint in science. In C. A. Odora Hoppers (Ed.), *Indigenous knowledge and the integration of knowledge systems: Towards a philosophy of articulation* (pp. 39–52). New Africa (Pty) Ltd.

Vossoughi, S. (2021, April 8–12). *Elsewhere worlds, poetics, and the science of human learning: a reflection in four verses* [Jan Hawkins Lecture]. AERA 2021 Virtual Annual Meeting.

Wallace, M. F. G., & Byers, C. C. (2018). Duo-currere: Nomads in dialogue (re)searching for possibilities of permeability in elementary science teacher education. *Currere Exchange Journal, 2*(1), 59–68.

Washington, H. (2019). *A sense of wonder towards nature: Healing the planet through belonging*. Routledge.

Weber, A. (2016). *The biology of wonder: Aliveness, feeling, and the metamorphosis of science*. New Society Press.

Weber, A. (2017). *Matter and desire: An erotic ecology*. Chelsea Green Publishing.

Weber, A. (2019). *Enlivenment: Toward a poetics for the Anthropocene*. MIT Press.

Whitehead, A. N. (1929). *The aims of education*. Macmillan.

CHAPTER 10

The Salt of the Earth (Inspired by Cherokee Creation Story)

Darrin Collins

DuSable and Son's Personal Legend

"Did you gather the wood?" DuSable spoke to his son as if he were much older. Kana'ti was only 4 months old, though he clearly presented the height and maturity level of a 6-year old. Although Kana'ti's rapid maturation was of great benefit to this nomadic father, he couldn't help but wonder about this divine occurrence. Just four months ago, the two were settled in Georgia with Selu and Wild Boy, DuSable's wife and oldest son. Kana'ti was a newborn, but was already showing signs of prophecy. Within his first month of life, Kana'ti had developed a full set of teeth. No longer was he able to nurse like other newborns. Instead, Kana'ti ate table food and was the smallest person walking in the tribe.

"It was wet from the snow so it wouldn't light, Baba. I tried to ..."

The family was originally located in Florida but had been pushed out of Apalachee territory by French soldiers, then forced out of Georgia by the British as soon as the American Revolution started. Now the two men are traveling north. Selu and Wild Boy didn't make it out of Georgia, a tale we will discuss when appropriate.

D. Collins (✉)
College of Education, University of Illinois Chicago, Chicago, IL, USA

© The Author(s) 2022
M. F.G. Wallace et al. (eds.), *Reimagining Science Education
in the Anthropocene*, Palgrave Studies in Education and the Environment,
https://doi.org/10.1007/978-3-030-79622-8_10

"Go put some oil on it. The oil will burn first and dry out the wood. It'll be ok..."[1] DuSable stated, never looking up. His understanding of the earth and how to manipulate its resources was phenomenal. He was a true student of his environment. Any other time he would have risen up and showed his son how to strike the kindling step by step. However, he was in deep thought about his current situation. They were in the mountains between Tennessee and Kentucky. The territory was hostile at the time. There were thick battles between the English, colonial soldiers, and American Indian tribes. Not to mention the sheer lack of trust for all unknown travelers. There was little safety in the region. DuSable knew how vulnerable he and Kana'ti were. In four months their small family had traveled thousands of miles. He was exhausted. His body, even his spirit wanted to rest. But he couldn't. Kana'ti needed him. And he needed God. Faith was the only paddle he had in the sea of life. In four months his son had experienced so much loss. His rapid physical and mental growth were representations of his forced maturity (Morrison, 2015).

Kana'ti followed the instructions, feeling his father's stress. Children are extraordinarily keen in that way. The boy grabbed the canister of oil from the wagon. It had been stored under the deer and bear pelts that DuSable collected for trade. He was a masterful barterer. He knew the key to any trade was the willingness to walk away. DuSable was one of the original and most successful American merchants. Kana'ti had trouble lifting the pelts. They were cold and very dense. He grabbed the canister and jumped down from the wagon stumbling slightly in the snow. He fell and immediately felt the chill of the cold wet snow on his butt. The canister slipped out of his hand and the pitch black oil poured onto the pure white snow. Kana'ti rushed to grab the canister, but much of the oil had already drenched the snow. The viscosity and warmth of the oil created a depression of blackness in the white snow. The metaphor of American life, especially in the prenatal city of DuSable, was nearly perfect. Except for the fact that Kana'ti wasted most of the oil.

Jean Baptiste Point DuSable saw his son struggling and ran over to the wagon to assist. Snow flung up from DuSable's boots as he trudged over to his son. Kana'ti was embarrassed after his fall. He knew that his father traded two metal flasks full of fresh water for the oil. The boy was simply trying to

---

[1] Indigenous ways of seeing and knowing the world have been marginalized since the onset of European colonialism (Smith, 2012; Fanon, 1963). In this work, the intention is to reimagine the world in which there is common ground for Western and indigenous epistemologies. Throughout this piece, I seek to highlight indigenous science like this example of oil drying damp wood. I apply two primary frameworks to this work of creative writing. The first framework being the concept of "desettling expectations in science" proposed by Bang et al. (2012). In this work, Bang et al. offer validation and confirmation of indigenous epistemologies which have been marginalized in Euro-centric spaces where testability and quantification are supreme. The second framework, which must be simultaneously accepted with the "desettled expectations," is Linda Smith's (2012) "indigenous science." Indigenous science accepts the epistemological distinctions between Euro-centric/Western science through decolonizing the methodological approach. In this work both of these frameworks are applied to offer insight into the experience of both American Indians and Foundational Black Americans within the sciences.

imitate his father's swagger by jumping down from the wagon. He had not yet developed the coordination to do so. DuSable gathered the canister salvaging a corner of the oil for the fire. Kana'ti sat in the snow and cried out of sheer embarrassment. Holding the canister in his left hand, the father scooped his son up by the arm with his right hand. "You always cry on your feet. No matter how hard this life gets, ALWAYS cry on your feet." He tossed Kana'ti on his hip and walked to the future fire pit with his son in tow. Kana'ti felt the love of his father. Even if he didn't understand why he should "cry on his feet."

In his father's arms, Kana'ti stopped crying. He wiped his face on his father's sleeve, and looked DuSable deep in the eyes. "Why did you say that? We have to cry on our feet?" Kana'ti was curious, and his father always encouraged his questions.

"Life is not meant to be easy. Every living creature has obstacles. From the bee to the bear. We all have to struggle through this life, seeking joy, happiness and balance." DuSable paused, thinking of his own imbalances. Kana'ti waited. "The obstacles test our soul. And our desire to reach our personal legends. In the obstacle you do not control winning and losing. All you control is your response to the obstacle. And your response will always be to stand in it. Because your soul is unbreakable. And if you need help standing, I am here. And we will have two unbreakable souls facing the obstacle." DuSable kissed Kana'ti on the forehead and again wiped the boy's nose with his sleeve.

Kana'ti dropped from DuSable's hip with the same bounce as when he jumped off of the wagon. "Baba, can I still light the fire on my own. I hear the oil swishing around in there. It might be enough." The father agreed with his son. It was probably enough. Kana'ti's ears were keen. DuSable nodded allowing his son to start the fire. He knew they would have to find a different camp once they started the flame. *Smoke will draw attention. We need to eat and keep moving. But how much further can the horses go? Que le ciel nous guide. (May the heavens guide us)...* DuSable thought to himself.

The boy had patience beyond his years. He sat rubbing the wood between his hands like he was palm rolling dreadlocks. The wind was picking up, and the snow was blowing in every direction. The two men were cold. The chill touched their marrow. Kana'ti showed great resilience. There was an initial spark. He tried to contain it with his whole body as he had seen his father do so many times. He looked like a rugby player huddling in the scrum. But the wind was strong, and the blinding snow was all around. DuSable sat watching, thinking of his wife, knowing how proud she would be of their son's resilience. Kana'ti continued to palm roll the stick. He blew on the embers out of habit. His breath was instantly absorbed by the strength of the wind. The embers brewed a little stronger. He slowly gathered them and transferred it to the kindling. The oil on the wood caught flame quickly. Kana'ti jumped back. The fire quickly died down and the wood began to take hold of the rumbling burn that could last through the night if they needed it. But DuSable knew that they needed to move and establish a different campsite for the night. Between the wildlife and the war-torn country, there would be nothing to protect them after they had eaten and fallen asleep.

Leaving Kana'ti to tend to the fire, Jean Baptiste Point DuSable walked to the wagon once more to gather the deer steak and onions he and his son were preparing to eat. DuSable poked his head through the side of the Wagon. "Whooooooshhhh!" He heard a very familiar and violent sound whist past his ear. His heart skipped a beat. Without becoming conscious of his movements, DuSable grabbed one of the bear pelts and his feet were already on the ground. He was racing, moving faster than the arrows that began to frame his body. The arrows landed on either side of DuSable's long, muscular, African frame; four of the projectiles landed on the pelt. But the bear fur took most of the damage. Though DuSable felt the daggers in his back, he did not lose any steam. He dove over his son, wrapping them both in the bear fur. *Puis-je attraper le cheval? (Can I grab the horse?)* The father thought to himself. *Je dois attraper le cheval! (I have to grab the horse!)* DuSable thought. "Be calm my son. We are safe."

DuSable sprang up shielding his son from the danger as much as he could. If time was moving quickly at first, everything slowed for DuSable once he felt his son's panicked breathing below him. He was paralyzed. Flee or stay. There was no protection under the bear pelt. Thooomb. Thooomb. Thooomb. Arrows continued to hit the pelt. DuSable decided to emerge from the crouched position. He once again had his son in tow. Kana'ti was caught off-guard. But he trusted his father. DuSable attempted to run but saw a band of Shawnee coming from among the trees; some on horseback, others on foot. The arrows ceased. Looking to the left he saw a man steering his wagon. *Mon dieu, je tout perdu! (My God, I have lost everything!)* DuSable thought to himself. There was nothing to say to his son. All was understood. Power was lost. That bitter truth had to be accepted by father and son as they knelt in the snow with the bear fur draping over them. DuSable raised his hands in submission under the long kinky brown fur coat. His son crouched under him.

The Shawnee appeared intimidating. Kana'ti trembled. He was scared. He was cold. But he was with his father. Kana'ti looked up to see the militia approaching them. They looked like his mother; the child thought they were family. Kana'ti was born on the cusp of two marginalized groups in this prenatal country. He didn't fully belong to either group or country. His mother, Selu, who died tragically during a raid of the British army, was an Apalachee woman. His father an African Haitian of mixed French race. The boy favored his American Indian ancestors. His hair was stereotypically long, wavy, and black. Slightly thicker than most American Indians, but it wasn't all that noticeable. His skin was golden brown with bronze and red highlights.

Even in the harshest of winters his complexion was lovely. His eyebrows and eyes were the same deep black.[2]

The Shawnee recognized Kana'ti's reflection too. He was one of them. They slowed their advance.

Kana'ti broke free from his father's grip. He jogged over the snow to greet his Kentucky cousins. They were not as delighted about the reunion as their Florida relative. Three arrows came flying out to halt the boy's progress. They did not try to hit the boy. They were a warning. Kana'ti was too far for his father to reach. "We are only trying to pursue our personal legend. We have lost everything except our lives and our wagon. Please spare us both." The boy said in eloquent Apalachee, that only his father understood. *Ou a-t-il appris que? (Where did he learn that?)* DuSable thought to himself. Kana'ti's knowledge and maturity at four months of age was divine. As the men watched, Kana'ti grew several inches. The boy continued to speak in Apalachee but soon realized that his cousins did not understand his tongue. He searched his brain to find another way to communicate, but he was at a loss.[3]

The men began speaking in their own tongue among themselves. Kana'ti overheard them and surprisingly understood their language. "He is from the southern tribes." One of the men exclaimed. They discussed his accent and the man he was with. The entire militia looked to a short, long-haired man in the middle of the pack for their instructions. He was silent. He looked beyond the boy at his father who was now behind his son with both hands on the boy's shoulders. DuSable was now keenly aware of the injuries to his back. The Shawnee man assessed them. They were of no harm. But what were they doing alone in Shawnee territory?

Speaking in his tongue, the leader turned to his men saying "They are friends. Friends from the south. It would have been a mistake to harm them. We would have greatly upset the ancestors." Gesturing with his hand extended, he told the men to put their bows down. They obliged.

Kana'ti responded in the Shawnee tongue, "We are friends. I don't know if we are from the south but we are Apalachee." DuSable did not understand

---

[2] In this adaptation of the Cherokee creation story, I use the concepts of housing and food to represent equally important ideas in modern society. To provide a sense of equity and to directly address gender norms in work and responsibility Selu was tasked with the gift of construction. She had a particular gift for assessing materials and engineering structurally sound abodes. I use Selu's death to highlight DuSable and Kana'ti's vulnerability in the wilderness. Just like in the traditional story, the parent's gifts are used to explain the context of the lived experience of the community. Selu's gift was housing. In the city that DuSable founded, as a result of Selu's passing, housing will be precarious. Her death precipitates redlining, gentrification, and mass homelessness in the city of Chicago.

[3] Kana'ti aka Chicagou's rapid growth is representative of the expansion experienced by urban cities as a result of uncontrolled industrialism. Chicagou's ability to acquire new languages is analogous to the cosmopolitan nature of American cities. Modern cities, like Chicago, have expanded and sprawled without limits for over a century. This sprawl, which has been precipitated by industrialization, immigration, xenophobia, and white flight (Loewen, 2018), has caused tremendous amounts of consumption and environmental degradation.

the interaction but was impressed by his son's innate linguistic skills. His heart was no longer fearful, instead it swelled with pride. Kana'ti was of unequaled talent. He could acquire new languages by simply hearing them spoken once. There was something very special about this child.

The men in the militia stood in bewilderment. Then in unison they began to laugh and talk amongst themselves. "Chicagou!" they all said. "Chicagooooooooouuuu" they sang the last syllable together. Holding it like it had some sort of meaning. The ice was broken between these two cultures. The Shawnee welcomed their visitors with open arms. The Shawnee men had DuSable and Kana'ti follow them back to the village, which was approximately three kilometers beyond the Mississippi River. The travel was arduous for all of them and hard on the horses. The Shawnee understood the river and the mountain. They had lived with the river for as long as they could imagine. The men had a particular way of knowing where they could cross the Mississippi. It was a dangerous trek across the frozen body of water. At any point, the ice might cave under the weight of the men, horses, and wagon.[4]

The band of travelers was three-quarters of the way across the river. They moved slowly and efficiently. As the group approached the shoreline, it became more difficult for even the most skilled survivalists among them to identify the barrier between the icy river and the shoreline. KRRRRRRUUUUUKKKK! The resounding and indistinguishable sound of ice cracking shocked everyone. The ice was giving under the weight of the men and horses. The leader raised his hand signaling for the group to stop. He was on foot, he walked slightly ahead of the other men. Kneeling on the ice he became obscured by the whirlwind of snow. Slowly he stood up. Taking an arrow out of his satchel, he situated it on the bow and aimed it between his feet. Releasing the arrow he listened for the sound of the bow on the ice. He gained a sonorous understanding of the earth; as the metal hit the ice he listened freeing up chunks of frozen water. Feeling confident in their chosen path, the leader signaled for the men to continue their journey. His hair draped over his shoulders. He appeared regal. The men trusted him. DuSable and son also grew to trust his judgment. Along the journey, the men spoke sparingly. Mostly they endured the weather. The trek took half a day. They wanted to beat dusk, when the wolves were most active, so they pushed through the fatigue. Once they were clearly on land, they caught a second wind. The danger of falling through icy terrain into a flowing and bone chilling casket has a way of taking the spirit out of a man.

[4] In this work I am actively creating a space for indigenous epistemologies in the Eurocentric/Western science canon. The use of tools to identify the density of ice is science in action. Tachoma, in this example, represents a very keen scientific mind through his understanding of natural phenomena. The goal here is to "decolonize" the many layered structures which prohibit and limit recognition of the many alternate ways of knowing and understanding the world around us (Smith, 2012; Fanon, 1963). This work offers an opportunity for readers to reimagine the world of science without the historical stain of Eurocentric white supremacy. Through this reimagining, other ways of knowing and being can be recognized and humanized.

THE FATE OF SELU AND WILD BOY

Once they arrived at the Shawnee village on the other side of the Mississippi, the men began to open up more. As they related their adventures, Kana'ti worked as translator between his father and the Shawnee. The boy's linguistic skills were flawless. Kana'ti jumped between French, Creole, Shawnee, and Apalachee without accent. As he played with the languages he simultaneously learned and absorbed their nuances. He enjoyed the beautiful flow of the Latin languages. He loved the syllables and accents in the indigenous tongues. Kana'ti thought of where these languages might have come from. He wanted to know the origin of the words. The etymology.

As the travelers shared their tales, the leader disrupted the conversation. Again he raised his hand. He gestured at DuSable, addressing the father and son. Looking toward Kana'ti, the leader spoke very humbly. "I am Tachoma. We are the Shawnee. You all are our honored guests. I really want to apologize for attacking you. But our people come from far beyond the great river. As you do as well. As you all come from the South. We refuse to move again. The land does not belong to the British or the French alone. Nobody owns the land. Rather we belong to it." Tachoma stopped speaking. Again he looked at Kana'ti but gestured with his eyes and head toward his father.

DuSable looked at his son for understanding. Still amazed at the boy's gift, DuSable patiently waited for the translation. Kana'ti now understood that he was the liaison between the two men. Nodding at his son, DuSable began to speak, as Kana'ti relayed the message in Shawnee. "I come from the island of Haiti. My son and I have been through great turmoil like you. His mother, a native Apalachee, was killed in a raid by the British soldiers. They burned our hut to the ground to build a military post. The land does not belong to them. But they took it anyway. They took Selu too." DuSable paused again thinking of Selu and Wild Boy who he found slain in the ruins of their home. He and Kana'ti still hadn't had time to mourn their loss. DuSable was waiting for he and Kana'ti to find safety before asking the boy what happened. He was out hunting game. He came back with two cottontails and a beaver. But there was no need for the dinner. He and Kana'ti had to escape. And they didn't have time to think about the loss. It would be too heavy to carry.

Kana'ti translated. Adding or subtracting what he wanted. He was a clever boy. "We have not been able to keep up with the needs of our people this winter. The patterns of the wildlife have changed since the white foreigners started clearing the lands and burning entire forests." Tachoma spoke genuinely. He too was tired of being displaced in the land of the free. "On your wagon, there were many furs. Deer, rabbit, bear, fox, and animals that I have never seen. You are a great hunter. Can you help us meet our needs?" Tachoma spoke with great sincerity.

Kana'ti translated the message to DuSable. However, he had gotten some of his mischief from his older brother Wild Boy. Kana'ti desperately wanted to know his father's hunting secrets as well. Just before the ambush by the

British soldiers in Georgia, Wild Boy wanted to know his mother's secrets. Selu was a masterful construction worker. No matter how far from Florida they traveled, she could find the resources in her environment to build a shelter for the family. In fact, when they got to Georgia, she showed the locals how to use the clay and tree bark to construct warmer, sturdier shelters that would last through the winter. Selu was especially gifted at manipulating the earth. Just before passing, she had begun to teach DuSable her secrets; he too had begun to teach her his secrets of luring and capturing game. The two brothers, Kana'ti and Wild Boy, were obsessed with their parents' mysterious gifts. The boys wanted to solve the riddle. How did Selu consistently build structurally sound abodes for them, no matter the terrain? How was DuSable readily able to find and capture wildlife for the family?

On the dreadful day when the British came and seized the territory, Wild Boy was determined to find out his parents' secrets. He told Kana'ti, who was only two months old at the time, to follow their father on the hunt and discover the mysteries of his hunting technique, while Wild Boy would stay with Selu and find out how she built the shelters. Once each boy discovered the secret to their parents' gifts, they were to kill the parent and meet at their hut where they would begin a life of independence (Perdue, 1998). As planned, at dawn, when DuSable left his family to chase game, Kana'ti, who hadn't been able to sleep all night, followed him. As he entered the woods, the father was well aware that his son was shadowing him. He didn't understand his motives. He simply thought the boy wanted to be like his old man. DuSable instead went on a very roundabout path in an attempt to lose the two-month-old stalker. Selu stayed back gardening. However, Wild Boy asked "Can we build a hut for me and Kana'ti? I want to show my brother all the things that I learned from my time with the Crocs." Selu was cautious. She knew that Wild Boy was not to be trusted, and she definitely didn't want Wild Boy teaching Kana'ti anything that he learned when he lived with the crocodiles. Wild Boy was the couple's adopted son. In fact, DuSable found Wild Boy naked and orphaned in a swamp three days before he met Selu in her Apalachee village. The boy looked to be about three years old at the time. He was floating in the murky waters like one of the crocodiles. DuSable watched as Wild Boy hunted a great blue heron just like a croc. Floating in his boat, he called to the boy who would later become his son. Wild Boy did not have a language. But he understood DuSable to be his salvation, and he left the heron, the crocs, and the swamp to journey with his new father.

Selu obliged the request; however, she withheld much of the knowledge from Wild Boy, not knowing his intentions. She, like her husband, was trying to mask the secret. Wild Boy watched his mother intently, absorbing each step and asking questions about the process. "Why do you use the red clay?" "How is the soil different from Florida?" "What do you call that tree?" Selu answered his questions, not wanting to reveal too much. As she dug in the soil with her hands creating space for the fourth pillar of the structure, she saw the redcoats on the horizon. "Go, go, go!!!" She shouted at Wild Boy pointing to the family hut. The soldiers instigated a shock of fear that ran through every part of Selu's being. She had heard about the rumblings of a revolutionary war in this prenatal country.

DuSable was unable to shake his tiny tot. He was able to set several small traps for a couple of rabbits and a beaver. But he couldn't perform his routine because of his stalker. He planned on pursuing the hunt again at dusk. When he returned to the hut, he found the territory in flames. He located his son Kana'ti weeping among the trees. They waited in the bush for the fires to stop. It seemed like forever. There was no sound, save the crackle of a fire. The soldiers left. The wildlife had fled. They were just as scared as the father and son. DuSable understood the fate of his wife and son. But he still needed to see them and lay them to rest. Kana'ti understood more than his father. Wild Boy was beside Kana'ti in the spiritual realm. He spoke to the boy of shame and embarrassment. "We were wrong Kana'ti. I am not a crocodile like I thought. Our parents deserved better." Kana'ti only shook his head and cried more.

Secrets of the Hunt: An Exchange of Cultures

Kana'ti continued to translate for his father and Tachoma. But he still wanted to know his father's secrets. Not knowing bothered the boy. Especially after his brother gave his life to discover the secrets. He believed what Wild Boy's spirit told him. But deep down the boy wanted to know the secret. He was in awe of his father's ability to hunt and barter. DuSable had an air about him. Just like when Kana'ti jumped from the wagon and spilled the oil, he wanted to mimic everything DuSable did. The boy told DuSable, "The men say that they want to know the secrets to your hunting. They want to know how you hunt so many different species so efficiently." He told his father what he wanted to know. Tachoma and the Shawnee just wanted help surviving through the winter. Kana'ti needed the secrets of the hunt.

DuSable recognized the mischief in his son. He had not forgotten two months prior the boy was trailing him on the hunt. The father wanted to find a home before he picked at his son about his motives for following him out at dawn. The man also recognized that the Shawnee spared his life and he owed them some gratitude for that. So he was determined to rest and show the men many of his practices at dusk the following day. "I am honored that you think so highly of my skills. I would love to share my ancestral secrets with you. Your people must simply make a trust with me. Do not abuse the earth or the bounty that may spring forth.[5] But my son and I need to sleep. Where can we find shelter?" They were standing in the middle of the village around a large

---

[5] The agreement that DuSable is making with the Shawnee here represents the ontological stance of African and American Indian traditions (Smith, 2012). In both traditions there is an understanding that humankind belongs to the Earth, not the other way around.

fire. Everyone's face glowed in the circle. They all sat attentively watching the exchange of cultures and observing the legendary "Chicagou."

As Kana'ti translated Tachoma nodded. They had indeed traveled a long way. And Tachoma had no clue how far DuSable and Kana'ti had journeyed in the past months. The two had journeyed from Georgia to Kentucky. "We will house you. Do not dismay. But I would be remiss if I did not take you to see the chief. She would never forgive me. She had a dream of your arrival."

DuSable reluctantly accepted the offer. He was beyond exhausted at this point. But again, the Shawnee spared his and Kana'ti's lives. He owed them some lip service. When the three arrived at the chief's tent, they were greeted by the chief's husband, who welcomed them with a pipe. Tachoma passed the pipe to their guest. DuSable was reminded of his family gatherings in Haiti shortly after independence in 1804. He placed his lips on the pipe and pulled hard from his diaphragm. The smoke entered his chest. It was warm. He was reminded of easier times.

"Welcome weary travelers. We have awaited you for many, many moons. I had a dream that the young Chicagooooouuuu would be with us soon." The chief said as she beckoned her guests to sit in her tent. She wore a feathered crown and a flowing and colorful poncho. She was younger than DuSable expected. Prettier too. Her deep brown eyes and high cheeks gave way to the loveliest dimples. He expected the chief of the tribe to be an elder.

Ton mari est un homme chanceux. Your husband is a lucky man. DuSable thought to himself. But he sat silent waiting for Kana'ti to translate. The boy never translated. He was confused too. *What do they mean they were waiting on me? What is Chicagou? Why do they keep calling me Chicagou?* Kana'ti thought in rapid fire. The two guests sat bewildered.

The chief continued. "Chicagou," she said, walking toward Kana'ti. She knelt by the boy touching his hair. "You have the blood of the world in you. I see it in your eyes." The boy understood what she meant. He was on the boundary of so many disparate tribes. Belonging to none of them fully. He was African, French, American Indian, Haitian, American, etc. He was truly American. "You will help your father found the new world. You mark the beginning of something greater than yourselves. Chicago. That's the city. Named after the onion. You are the onion. Chicagou. It is you. Your many layers of culture. The many languages you will learn. You will open the new world for millions of foreigners and immigrants to find their dream. The dreamers will come because of you."[6]

---

[6] I chose to replace the traditional role of Kana'ti (in the Cherokee Creation Story) with Jean Baptiste Point DuSable for two main reasons. First, DuSable, an African man of Haitian descent, is historically recognized as the first settler of the Chicago area and the unofficial founder of the city. In making him a pivotal figure I was giving voice to the historically marginalized role of Africans in the Americas. I also chose DuSable because he and Selu's union represents a true convergence of cultures that makes America the beautifully diverse and unique country that it can be if we accept the contributions of

DuSable looked at his son. He knew something powerful was happening. The chief then grabbed both travelers' hands, and they left to a distant place. A place of marshes and prairies. The grass was taller than their heads. The three walked up the river. The Mississippi River, to a lake. It was an expansive body of water. It was blue and green. Shimmering in the sun. With the tall grass behind them. They saw native men and women gathered water in clay pots. "This is your destiny." The chief looked at the father and son. They were no longer in reverie. They understood better. DuSable puffed the pipe again. He felt anxious about this destiny. He was tired. He wanted to mourn his loss but was being pulled in every direction by life. The lake felt right. He knew that he could settle by the lake. Maybe even garden and trade animal furs. He puffed the pipe once more and passed it to Tachoma who was still standing.

When he and Kana'ti arrived at their own tent, Tachoma gave DuSable a musket. "There have been raids by the westerners. The mountains have protected us. We got this in the last one. I thought you would know how to use it." Tachoma left without providing ammunition or gunpowder. To Tachoma, DuSable represented the West. He believed that he would be able to make the Western tool work.

Inside the tent DuSable turned to Kana'ti, dropped the musket, and hugged his son tightly. They had been through so much. He was just glad they were in one piece. Kana'ti understood the sentiment but was really excited to tell his father exactly what the chief told them. The hug lasted some time. DuSable needed Kana'ti to know that he would always be there. Kana'ti believed that.

When DuSable released Kana'ti the boy was overly excited to discuss their legend. "Baba, did you see the lake too?".

"Yes, son." DuSable explained, wiping his eyes.

"Did you understand what she was showing us?" Kana'ti inquired with a Curious George-type of excitement. "They have been talking about us for over 1,000 years Baba. They say that we will open the new world." Kana'ti could not contain his excitement. "What do you think this new world will be? We get to open it. Baba! We get to open it!" Kana'ti fell to his knees and placed both hands over his mouth, grinning with joy, thinking about the vision. "We get to open the new world to all types of people. It is our destiny."

DuSable placed his hand on his son's head. Looking down at the boy, he said, "We will do all of that in the morning." He then blew out the oil burning lamp that lit up their tent. And the two collapsed from exhaustion.

The next day the men of the village were ready for the hunt. They arrived at DuSable and Kana'ti's tent early. They even brought the wagon. Tachoma offered the two a stew. They ate and were off to teach the Shawnee DuSable's Haitian secrets. DuSable lead the men back to the Mississippi River. They did not cross it this time. Instead DuSable followed the river until the men reached a tributary that branched off into the mountains. They followed the

those groups who have been traditionally othered. In this recasting, Kana'ti, aka Chicagou, is the son of DuSable, representing DuSable's birthing of the Second City.

tributary into a valley that was lush and green. It had been protected from the desertification of winter by the mountain top. The Shawnee were astonished by DuSable's ability to find such an island of life. The father told the Shawnee that they would have to wait until dusk to catch the best of the hunt. The men obliged. Everyone waited with expectation. They had all seen the furs in DuSable's wagon. They knew that he was a masterful hunter. While waiting, the men taught Kana'ti how to shoot a bow and arrow. The boy was quite a natural at that too.

At dusk, the group headed deeper into the tropical island. DuSable halted the wagon with some effort. Even the horses were excited to be out of winter for a period. This was one of the secrets that DuSable was keeping from Kana'ti on the dreadful day they lost Selu and Wild Boy. DuSable hopped off of the wagon with a lion's coat in his hands. It looked like John the Baptist's lion skin. The head and paws were still attached. He handled the fur very delicately, as if he were installing a window frame. Kana'ti jumped off after his father. The boy had grown overnight. His landing was much smoother than the previous day. He now had a mustache and was almost as tall as his father. His clothes no longer fit properly but he didn't mind. He enjoyed his Hulk-like appearance. He was his father's unappointed apprentice. DuSable laid the lion's fur down to his left. The head of the fur faced up. He looked at Kana'ti. The boy had never seen such an animal. Neither had the Shawnee men. They all looked in amazement. Then, kneeling over a fresh patch of green grass, he pulled a spade out of his back pocket. The spade was formed out of the lion's claws. DuSable passed a second spade to his son. The second spade was formed out of the lion's teeth. Then he began to dig. With the little spades and some time, the two men dug a pit that was quite sizeable. The pit was three cubic meters exactly. DuSable instructed Kana'ti to spit in each of the corners of the pit. Then the father laid the lion's coat in the area.

The men waited but nothing happened. DuSable took five deep breaths and then addressed the men. "Get those arrows ready. There is going to be a stampede." He pointed east in the direction that the lion's head was facing. In moments, the lion's fur was ejected from the pit. At first a black and white hooved leg popped out of the pit. Then suddenly three full zebra emerged from the pit running hysterically. Although the Shawnee were ready with their arrows, they were so astonished by the birthing of these animals they had never before seen. Only one was able to fire his bow in time, dropping one of the zebras. Several other Shawnee gave chase on horseback, but the other zebra frantically escaped into the distance.

The Shawnee were all amazed at DuSable's mysterious methods. "What type of horse is that?" One of the men called to DuSable. Kana'ti looked to DuSable without translating. "It is a Zebra." DuSable understood. "That is a lion." He said pointing near the pit. Kana'ti needed more of an explanation than that. *What was the secret? Why dusk? What happened at dusk? Why use the spade? How did he find the lush island?* So many questions rushed through the boy's mind. But DuSable couldn't explain. He looked at his son, who was almost eye to eye, and said, "In due time son." Kana'ti remembered the plan that Wild Boy had devised. He thought about what it would have meant had he followed through on it. He would have been alone. A tear fell from the boy's eye.

DuSable and his son spent one more night with the Shawnee. In that time DuSable taught Tachoma the true secrets of his birthright and left the Shawnee with the lion's fur. Tachoma was incredibly grateful to DuSable for his insights and for revealing secrets of the world to him. Tachoma promised to escort the two men, and by all means Kana'ti was a man at this point, back to the Mississippi River in the morning to continue their journey. He knew that their destiny was great. Both of the men possessed divine gifts, and the city they established and the lineage they created would produce endless possibilities. And Chicago has done just that.

After traveling several kilometers north up the river with DuSable and Kana'ti the next day, the men bid farewell.

"Chicagou. We will never forget the lessons that you both have taught us," Tachoma said with great humility and gratitude.

"Our encounter has been one of great fortune. We have learned about ourselves as well." Kana'ti responded before translating to his father.

"You will accomplish many things with your skills great Chicagou. Your father too." Tachoma paused for a second, trying not to reveal too much. Only the chief can reveal and interpret dreams. Tachoma did not want to overstep. "But remember the two rules of life, great Chicagou. First, you belong to the earth and not the other way around. And if you 'support yourself heaven will help.' You will need those lessons on your journey" (Collins, 2019). Tachoma and the four Shawnee men who accompanied him turned on that note, and left DuSable and Kana'ti to their path. Kana'ti thought long and hard about Tachoma's last words. Again he had not translated them for his father. DuSable assumed that he did not need to know them.

THE SECOND CITY

The two continued their long trek to the lake. In the chief's vision, they understood that the river would lead them to the lake. The lake was where the city would be founded. The city that would open the "new world," as the chief coined it. They traveled for months. The weather changed and so did Kana'ti. He now had a beard similar to his father's. He was even slightly taller than Jean Baptiste. The men looked very much like brothers. The men found a tributary that they followed north. This tributary would later become known as the Chicago River. Named after the legendary Chicagou himself. The river led the men to their destiny.

As they approached their destiny, Kana'ti began to get more and more interested in DuSable's hunting secrets. *Does he know about mom's construction secrets too?* Kana'ti thought to himself. He wanted to know what his father was hiding. *Why didn't he tell me earlier?* Kana'ti was envious of his father's knowledge. *What does a living lion look like?* He thought to himself in Apalachee. He wanted to be able to make the animals appear too. He even practiced performing the exact ritual as his father night after night. But nothing happened. He would follow the order of operations exactly. But no animals would spring forth.

DuSable and Kana'ti camped out less than two kilometers from the super-abundant lake that would serve as the primary resource for their newly founded city. They were the first Westerners to settle the area. It was a rugged and brutal environment. The winters, as the two would find out, were much harsher than those of Georgia or Kentucky. They had arrived in the middle of spring. The territory was beautiful. It appeared just as the chief's vision had shown. Birds like larks and robins fluttered about. "We are close, Baba. Close to our destiny," Kana'ti whispered to his father as they searched for a campsite. Along any journey, when you get closer to the treasure, there is a final test. Kana'ti, aka Chicagou, was going to receive his final test. The test to see if he had learned the lessons from his journey (Coelho, 1988).

Chicagou and DuSable walked the land, scanning their new home. The sun was bright. The two approached the lake. They did not have a name for the lake. But they understood its beauty nonetheless. A large painted turtle rested on a log near the shore. It lay there completely comfortable. Legs and neck outstretched, absorbing the sunlight. Chicagou wanted to possess the turtle as a pet. But he knew that was against nature. "It is time for you to learn the secrets of the hunt my son. We shall discuss the heaviness of our hearts once we get settled." Kana'ti nodded, understanding that he had to finally tell his father about the plot to kill both he and Selu.

Again, the two men followed the Chicago River beyond the mouth of the lake. They traveled until they got to a clearing. The grass was low and greener than most. "You must listen to the earth. She will tell you where to hunt. Can you hear her Chicagou?" DuSable liked his son's new name. He felt like it suited the boy. Kana'ti shook his head indicating no. "You will learn. Now son, it is important that you do everything that I say. I know that you have been practicing at night while I am asleep. But son, remember that our connection with the earth is strongest at dusk and dawn. The spirits of the earth and all who belong to it are oriented to the sun."

"What about the lion, Baba?" Kana'ti had so many questions.

DuSable had so many answers. "It is best to choose a top predator for the hunt. Lions come from the Motherland. Africa. They rule the Sub-Saharan terrain and prey on many of its herbivores." DuSable paused, trying to search for the correct explanation. "I gave you a spade to dig with. The spade has to be made of the teeth and claws of the predator. By digging the pit, we conjured the spirit of that animal." DuSable felt the Ogun spirit of the lion possess him.[7]

---

[7] The Ogun Spirit is a deity who appears in many African and West Indian religious traditions. Ogun is a warrior spirit traditionally.

Chicagou understood the spirit. Wild Boy stood next to him co-signing what the father was saying. He told Chicagou grand stories about lions and their hunting patterns. "They are much stronger than crocodiles. I thought us crocs were top predators," Wild Boy said, slapping his knee. Chicagou looked at DuSable with great intrigue. "How big do you make the pit? Why do you spit in it? Where do the black and white horses come from?" Kana'ti asked rapidly, forgetting the name of the black and white horses.

"I will explain as we dig. We first have to make new spades. I gave Tachoma the lion's relics." DuSable then pulled out one of the brown bear fur coats. Pulling a small cloth from underneath the fur he passed his son a block of wood, string, the bear's claws, and a small knife to fashion a bear claw spade. DuSable took the teeth of the bear and used those objects to fashion the bear tooth spade. The men worked until dawn making the spades and digging the pit. This time the pit was five cubic meters. The two were covered in soil. "The size of the pit," DuSable explained, "determines the number of animals that will spring up from the pit. The predator's spirit is particular in that way. The earth will reward our work, nothing more.[8] Last time it was three meters. This time we made it five."

"Baba? You still didn't answer why we spit in the pit." Kana'ti wanted to know everything about the ritual.

"Oh, the most important part. The earth is a gracious but begrudging provider. You must give her love and offering for her bounty." DuSable paused again. He thought about the spiritual currency that had been spilt on the soil of this prenatal country. He wondered how the earth would seek to balance the atrocities of slavery, racial tyranny, and mass genocides of Black and indigenous people in this land, America. DuSable thought of his own journey as a free Black man traveling throughout this colonized slave state by way of colonized Haiti. "The spit is physical currency for the predator. We give them the authority to return to the physical realm. This is why you need to follow everything I say closely. We give them the authority to return to the physical realm. The earth makes sure that it is only the prey that returns. After we put the bear's fur back in the pit you cannot touch it. Like I said the earth is a begrudging provider."

Chicagou was anxious. He wanted to see the mysterious occurrence again. He had tried so many nights to get it to work on his own. He knew that since his father had walked him through the steps and they had done it at the proper time of day, it was guaranteed to work. "Can we do it, Baba?" Kana'ti said eagerly.

"Hold on son. Like I said you have to follow the ritual completely. The predator's head must always face the east. The sun is our source. We must give him gratitude, just like the earth." DuSable pondered once again. "The

---

[8] This experience of reciprocity with the Earth represents a wrapping of Tachoma's comment: "support yourself and heaven will help." In this example DuSable suggests that the Earth will reward one's work one to one.

earth hears all things and the sun sees all things. The sun communicates with the earth what he sees. And they only give to those who are deserving. If you hold onto any evil deed the earth will not provide sustenance." DuSable ended his thought optimistically, looking forward to his first official hunt with Kana'ti.

Kana'ti wanted to tell his father about the plot that he and Wild Boy conjured before the British raid. He thought that talking about it now would destroy the hunt. But he knew that it was an evil deed that he still carried. Kana'ti was too ashamed to admit his sin. "Ok, Baba. I want to try my new bow and arrow out. I've been practicing since we left the Shawnee." Wild Boy's spirit stood close to Kana'ti and whispered grave warnings.

Kana'ti spit in the corners and placed the bear fur over the pit. "It was five meters you should get five animals out of your pit" DuSable said encouragingly. *My pit.* Kana'ti thought to himself. He waited for the five animals to emerge. They never did. DuSable sat patiently believing in the ritual. He was completely unaware of the heaviness on his son's spirit. The earth was withholding her bounty. The sun told her of Chicagou's plot to kill his father. She would not reward such evil. Kana'ti grew impatient. Though he had grown into a man rapidly, he was still learning life's lessons. One lesson that he had yet to truly learn was the importance of listening. The men waited over two hours. The sun was fully up. Kana'ti grew irritated and rushed to the pit. Jean Baptiste attempted to stop him. Kana'ti ran to remove the bear fur. The earth knew his intentions. She listened deep to his soul. She knew just how mischievous the young man was. But she had a reward in store for his impatience. As Kana'ti ripped the bear fur out of the pit, hundreds of thousands of animals escaped from it never to be returned. DuSable knew what the Earth was communicating, but his son was unaware. No longer would she support such faithless men. Humankind would now be responsible for supporting ourselves; she was not going to give willingly to those who took her for granted.[9]

References

Bang, M., Warren, B. G., Rosebery, A. S., & Medin, E. L. (2012). Desettling expectations in science education. *Human Development, 55*(5–6), 302–318.

Coelho, P. (1988). *The alchemist*. HarperCollins Publishing.

Collins, D. (2019). *Native invisibility*. Mynd Matters Publishing.

Fanon, F. (1963). *The wretched of the earth* (C. Farrington, Trans.). Grove Press. (Original work published 1961.)

[9] In this tale Selu had the gift of construction. When she was slain by Wild Boy, DuSable and Kana'ti became homeless and vulnerable until they reached the Second City. Likewise, Jean Baptiste Point DuSable was gifted with the ability to hunt. Much like Kana'ti in the Cherokee tale. Similarly, DuSable's son Chicagou aka Kana'ti in this rendition misbehaves and releases the animals. This symbolism is representative of the extreme starvation found in urban areas including Chicago. It represents the current epidemic of food deserts, physical starvation, intellectual deprivation, and cultural marginalization that is occurring in modern cities across the globe.

Loewen, J. W. (2018). *Sundown towns: A hidden dimension of American racism.* Touchstone.

Morrison, T. (2015). *God help the child.* Vintage Books.

Niethammer, C. (1977). *Daughters of the earth.* Collier Books.

Perdue, T. (1998). *Cherokee women.* University of Nebraska Press.

Smith, L. T. (2012). *Decolonizing methodologies: Research and Indigenous people.* Zed Books Ltd.

St. Pierre, M., & Soldier, T. L. (1995). *Walking in the sacred manner: Healers, dreamers, and pipe carriers—Medicine women of the Plains Indians.* Simon & Schuster.

Part III

Politics and Political Reverberations

The Science of Data, Data Science: Perversions and Possibilities in the Anthropocene Through a Spatial Justice Lens

Travis Weiland

Statistics often dubbed the science of data (Davidian & Louis, 2012), and data science, the supposed new revolutionary frontier tasked with breaking the bonds of statistics, both have epistemological roots in exploring data—data being numbers in context (Cobb & Moore, 1997). I am a mathematics educator that dabbles in science education through my research focus of statistics education. The inductive investigation of data about our world is the common bond between the different communities/disciplines that I work in, which allows me to move fluidly between mathematics and science education. I love exploring data, particularly data related to sociopolitical issues and education. Exploring data never ceases to excite me when I use it to explore issues in my life, to make sense of things I read and see in the media, or in teaching how to use it to others. Looking at the historical roots of investigating data, we can see the practice of statistics having positive impacts on society, such as Florence Nightingale's work with using data visualizations to convey a statistical argument. Nightingale's now famous polar chart entitled, "Diagram of the Causes of Mortality of the Army in the East," illustrates that the vast majority of deaths in the army were from contagious and infectious disease and were often preventable. Because of this data-based argument, Nightingale is often lauded for influencing vast improvements of sanitary conditions at the time. From Nightingale's story and others in society, we can see why statistics has often been portrayed as a powerful practice that can help society make wise decisions and improve our general conditions and state.

T. Weiland (✉)
University of Houston, Houston, TX, USA

© The Author(s) 2022
M. F.G. Wallace et al. (eds.), *Reimagining Science Education
in the Anthropocene*, Palgrave Studies in Education and the Environment,
https://doi.org/10.1007/978-3-030-79622-8_11

We can also see that the practice of statistics has downright devastating impacts and creates perverse realities. For example, Cathy O'Neil (2016) dubbed the term Weapons of Math Destruction (WMDs) in her book by the same name when she discusses how mathematical and statistical models can create unjust structures. What she is referring to as WMDs are actually statistical and mathematical models developed using data to create "objective" algorithms for decision-making. However, the practice of statistics is not neutral or objective. Statistics is a social construct or discourse that self-perpetuates, privileging certain perspectives while silencing others. O'Neil points out there are serious issues with what models create in terms of the realities they create. She describes some of the characteristics of WMDs: "they're opaque, unquestioned, and unaccountable and they operate at a scale to sort, target, or 'optimize' millions of people. By confusing their findings with on-the-ground reality, most of them create pernicious WMD feedback loops" (p. 12). O'Neil goes on to provide examples such as financial models responsible for the crisis in 2008 and the use of value-added models (VAMs) in personnel decisions (e.g., in teacher evaluation systems). What O'Neil and others in the past have brought up is the issue that statistical and mathematical models are not objective. Models are subjective, and they can be a significant influence in shaping the reality we experience. Furthermore, their influence is at scales not considered possible before big data and data science hit the scene. Even the American Statistical Association weighed in on such issues, arguing that VAMs were not designed to inform personnel decisions and should not be used in education—particularly not as mechanisms for teacher evaluation—other than to help give some sense of what is occurring overall in a school or district (American Statistical Association, 2014). In other words, statistics and data science, and modeling data, are not our great saviors.

Creating a model is a political act. The modeler decides what variables to consider and what weight or influence they have over the model. In statistics, the goals of models are often explanatory or predictive. In other words, we want models to help explain our reality or predict our future realities. The person creating the model has a direct say over what factors will be used to explain or predict our reality. That can be highly problematic because, as educational researchers know, context matters, and people's lived experiences make a difference in how people experience reality. In other words, models are a discourse that shape reality and form regimes of truth. They are also not inherently good or evil; they merely create things. The discursive formations of such discourses, however, can create structures of injustice that privilege some while disadvantaging others (Foucault, 1972). It is such discursive formations that we need to interrogate when thinking about statistics and data science in the Anthropocene.

In the Anthropocene statistics, data science and mathematical models have become a perversion of reality that society has largely chosen to ignore. Data science is embraced as a great savior because people often view numbers as objective purveyors of truth. However, numbers do not interpret themselves, and they do not tell their own story; people do that in all their subjective glory. In this chapter, I will first make some connections between the Anthropocene, in particular the Orbis Spike dating of its beginning, and the human events of that time (Lewis & Maslin, 2015), to statistics and data science, specifically through the context of spatial data. From this discussion, there are two main points I want to elaborate on and then connect to education. The first is that there is a dialectic tension involved in spatial data enquiry between creating new realities using spatial data and using spatial data to make sense of our reality. You might even argue that this is a never-ending cycle. The second point is that people can choose how to investigate and use spatial data based on their ethics. I believe students should have opportunities to investigate and use spatial statistics both to learn about the world around them and to shape the world around them and that they should learn to do so through considering counter-mapping and spatial justice.

The Anthropocene, Spatial Reality, Maps, and Death

The Anthropocene is the geological epoch we are currently in, defined by the influence of humans on the geology of the earth (Crutzen & Stoermer, 2000; Lewis & Maslin, 2015). Many different starting points for this era have been proposed, including the advent of the steam engine, the invention of nitrogen fixing for fertilizer, the height of nuclear bomb testing, and the Industrial Revolution, just to name a few. I choose to use the date of 1610, drawing from the work of Lewis and Maslin (2015) in their identification of the Orbis Spike. As Davis and Todd (2017) discuss, the choosing of a starting date for the Anthropocene is a political decision with significant implications. The Orbis Spike was the time when CO_2 reached historic low levels during the period that Europeans were colonizing America. The drop in CO_2 levels is attributed to the significant population drop of Indigenous peoples, as Lewis and Maslin (2015) describe,

> Regional population estimates sum to a total of 54 million people in the Americas in 1492, with recent population modelling estimates of 61 million people. Numbers rapidly declined to a minimum of about 6 million people by 1650 via exposure to diseases carried by Europeans, plus war, enslavement and famine. (p. 175)

The change in population described is astonishing in a remarkably depressing way. The connection between this genocide and the dip in CO_2 levels globally make a powerful, data-based argument for the Orbis Spike, representing the starting point of a human-driven geological epoch. Using this data also has significant implications, as Lewis and Maslin (2015) discuss:

> The Orbis spike implies that colonialism, global trade and coal brought about the Anthropocene. Broadly, this highlights social concerns, particularly the

unequal power relationships between different groups of people, economic growth, the impacts of globalized trade, and our current reliance on fossil fuels. (p. 177)

The Orbis Spike highlights power relations between peoples and their impact on the natural world. It also makes explicit the insidiousness of colonialism and its influence on the world. Davis and Todd also discuss the implications of the Orbis Spike, "the naming of the Anthropocene epoch and its start date have implications not just for how we understand the world, but this understanding will have material consequences, consequences that affect bodies and land" (p. 767).

The connection of bodies and land that Davis and Todd (2017) make is an important one. This connection highlights the importance of considering a spatial dimension of reality along with historical and social dimensions (Soja, 2010). Soja (2010), who is an urban planner, discusses the importance of a triple dialectic ontology, where space is not treated as just an empty container, but as an integral dimension of reality that is shaped by and shapes the social and historical dimensions. Connecting back to the notion of body and land, Soja considers both to be part of the spatial, "two extremes, the corporeal body and the physical planet, usefully define the outer limits of the concept of spatial (in)justice and the struggles over geography" (p. 31). I bring up Soja's work not so much to focus on his perspective on ontology, but to consider the notion of spatial justice he formulates in the context of its importance in geography and urban planning and connect to issues of justice and the Anthropocene. To illustrate this connection, consider the example of water issues of the American West. Cities like Los Angeles could not exist if it were not for people changing the physical landscape, which did not naturally have the amount of water needed to support the number of people choosing to inhabit the region. To solve this problem they began building a system of diversions, dams, reservoirs, and pipes to take water from the Rocky Mountains, where there was more water. Consequently, the Colorado River now no longer reaches the Pacific Ocean in Mexico. The movement of water is a necessary aspect of urban planning, and we can see the Anthropocene embodied in this example, as the movement of rivers is a major source of geological change, and in this example people have changed the course of the geology of the Colorado River Basin. This example also points to issues of injustice, as the country of Mexico has been deprived of a fresh water source due to people's usage upstream. Furthermore, the change of the flow of the river has changed entire ecosystems and deprived non-human life forms of their habitat.

Geography has consequences. People, however, are capable of interrogating inequities in what geography creates and have agency to change them, as Soja (2010) describes, "their [human geographies'] changeability is crucial, for it makes our geographies the targets for social and political action seeking justice and democratic human rights by increasing their positive and/or negative effects on our lives and livelihoods" (p. 104). It is from this agency that spatial justice holds potential for educational endeavors in creating opportunities for students to interrogate issues of spatial justice and perhaps advocate for transformation. I am choosing to draw from these ideas because they have been taken up in mathematics/statistics education by some scholars (Rubel et al., 2017; Rubel, Hall-Wieckert, et al., 2016; Rubel, Lim, et al., 2016), and I seek to continue to work in the direction they have started. Before I delve into the educational side of this chapter though I want to consider one more example. This time I will relate the notion of models and statistics by considering an example related to one of the most common types of models we consider in spatial statistical analyses, maps, which also relate geography to colonialism and the movement of people that helped cause the Orbis Spike, defining the beginning of the Anthropocene.

Explorers/exploiters in the middle of the last millennium were driven by the desire to learn what was over the horizon and by the greed of what that knowledge would be worth in terms of new trade routes and goods. Imagined geographies impact peoples, societies, and science. As Livingston (2003), points out, fantasies of exotic lands and new peoples made imperialist exploration of the "New World" "at once a moral, an economic, and a scientific event. Much of the early history of this transatlantic engagement depended on Europe's geographical fantasies about the Western Hemisphere" (pp. 8–9). By traversing and measuring these uncharted spaces, Western exploiters gained a better understanding of their world but also created their own images of that world, which they also shared on maps.

To bring ideas full circle let me now connect this discussion back to models and statistics and data science. Maps are essentially models of reality, which means they are inherently flawed, as some amount of abstraction must occur to create a model. Early Western cartographers set out across western America to map out the land. We can see elements of their subjectivity in what they choose to focus their attention on and what they left out. For example, they had a clear focus on objectively rendering the physical landscape, such as rivers, mountains, and forests; however, what they neglected to map were the people that inhabited those lands (Remy, 2018). If we look at one of the few remaining examples of a map made by an Indigenous mapmaker (see Fig. 11.1), what we can see is that the concerns of this mapmaker were not at all on the rivers or mountains, but on the groups of people and the paths connecting them. As a mathematics educator, I also cannot help but see the connection to graph theory here where the distance between the nodes (villages) is not important, but what is, are the connections (paths). The emergence of the reasoning behind graph theory is often attributed to the white European mathematician Leonhard Euler; however, it would be an interesting investigation to see if Indigenous cultures actually came up with the reasoning much earlier.

Now consider the map in Fig. 11.2, which shows the tribes of the same area depicted in the deerskin map except overlaid on a common western map of the Carolinas. One of the powers of modern technology is the ability to consider different layers of graphs. For example, consider Google Earth. The

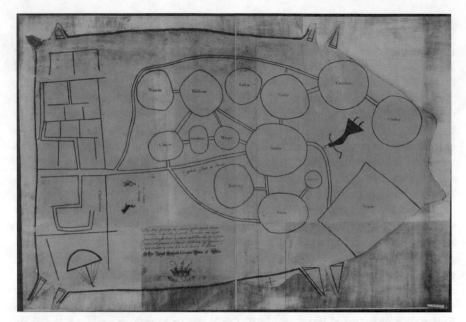

Fig. 11.1 Catawba deerskin map (*Source* Library of Congress, https://www.loc.gov/resource/g3860.ct000734/)

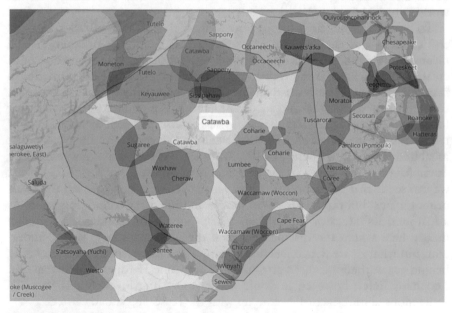

Fig. 11.2 Modern digital mapping of Catawba tribal lands (*Source* native-lands.ca)

user can choose to focus on roads, cities, and political boundaries or they can switch to a satellite view where the user can see a bird's-eye view of the same area. In some cases, the user could also switch to a 3D view to look at the human topography of an area. Hand-drawn maps have inherent constraints in that there is only so much information that can be meaningfully included in the finite space of the map. Early mapmakers chose to use that limited space for conveying the information they found important.

Maps Form Reality in (Un)Just Ways

Now let us think about early mapping endeavors with my first point in mind, which is that there is a dialectic tension involved in spatial data enquiry between creating new realities using spatial data and using spatial data to better understand our reality. If we think of Lewis and Clark's work of mapping the West in North America, they undertook the endeavor to learn more about what was west and how to traverse that space. In other words, they wanted to understand their reality. However, by creating maps that focused only on open physical spaces and ignoring the inhabitants that were already in those spaces, namely Indigenous peoples who had lived there for centuries, they created the "reality" that the west was this vast, uninhabited open space ready to be conquered and owned. The reality created helped fuel the western expansion movement that later occurred and, in turn, the gradual process of dislocating and killing of Indigenous peoples that lived on those lands. We do not know that this was the intention of Lewis and Clark, but it is undeniable that people used their models in such ways and facilitated westward expansion. How we create and how we use models matters and it is a political act.

Now to my second point, which is that people can choose how to investigate and use spatial data based on their ethics. A good example of this point is gerrymandering, which is a hot topic in politics in the United States and intersects with statistics (Honner & Gonchar, 2017; Klarreich, 2017). Gerrymandering is when a political group draws the boundaries for voting districts with their own political interests in mind to disadvantage another group. The state I am writing this in is a prime example, North Carolina. In the most recent election in 2018, Democrats in North Carolina earned 48.3 percent of the total vote cast in House races, but only won three seats. On the other hand, Republicans had 50.4 percent of the vote and won nine seats. The inequities can be seen in spatial graphics in the *New York Times* article by Astor and Lai (2018), entitled "What's Stronger Than a Blue Wave? Gerrymandered Districts." You might ask how such a system of injustice could have been allowed to be created. It all happened when Republican politicians gained control of the governorship, Senate, and House of North Carolina in the 2010 elections, which coincided with the decennial census that is used in the U.S. to draw political boundaries in a "representative" fashion. We also have a new use of technology employed in redrawing the lines, namely GIS. Using demographic data from the census and voter registration databases, the Republican

Party in North Carolina that was in power was able to use this data to create spatial boundaries that favored them. This is an example of how data can be used in unjust ways to create a new reality with spatial models, namely the political power balance based on voting district lines. The issue of gerrymandering is at the center of the current issue around the 45th president of the United States attempting to illegally add a question about citizenship to the 2020 census form, which has been struck down by the courts. This is an example of the use of spatial data based on the ethic of capitalism, greed, and power lust. Later on, I will show an example of how a justice-oriented ethic can create different possibilities.

So how does spatial justice relate to the two points I have made? It relates in that it can be used as an ethic in balancing between better understanding one's realities through spatial data and creating a new reality through spatial data and models. For example, counter-mapping is a strategy that has been taken up by Indigenous groups for several decades now, as described by Remy (2018), to interrogate and trouble common Western maps' narratives. Remy goes on to state: "The development and democratising of the access to digital technologies and online mapping systems opens the possibility for the growth of Indigenous counter-mapping projects and their complete ownership by the community." The idea of counter-mapping draws from the notion of using the master's tools. A similar notion has been put forth by the mathematics education researcher Rochelle Guitiérrez (2016), where she describes using the master's tools as one of the strategies for what she refers to as creative insubordination in mathematics teaching. Drawing from these perspectives, I argue we should be teaching about space and spatial ways of knowing in mathematics classes. With new digital mapping tools and big data, there has been a democratization of access to opportunities to investigate spatial issues that we have never seen before, and it is accessible to not only teachers but students as well.

SPATIAL JUSTICE IN MATHEMATICS/STATISTICS EDUCATION

To start with, I would like to draw from a notion from Marilyn Frankenstein who is arguably one of the first people to consider looking at statistics education through a critical lens. Frankenstein (1994) discusses her goal in teaching to outrage students at the utter ridiculousness of things happening around us through interrogating quantities. For example, let's go back and consider the change in populations of Indigenous people from estimates of 61 million people in 1492 to a minimum of about 6 million people by 1650 (Lewis & Maslin, 2015). That is a 90% decrease in Indigenous peoples in just over 150 years. That is outrageous and helps to highlight the importance of considering such statistics and creating opportunities for students to experience them.

The idea of considering spatial justice in mathematics or statistics education is quite new, with only a handful of scholars investigating the intersection.

The social and historical dimensions of reality have long been important when considering the teaching and learning of mathematics in the social turn (Lerman, 2000), and more recently political dimensions have been highlighted in the sociopolitical turn (Gutiérrez, 2013). Soja (2010) discussed the spatial turn in his work and more recently Larnell and Bullock (2018) created a socio-spatial framework for urban mathematics education, claiming a spatial turn in mathematics education. A recent push in investigating the intersection of mathematics education and spatial justice has been spearheaded by Dr. Laurie Rubel and her team in their work on conceptualizing Teaching Mathematics for Spatial Justice (TMSpJ); (Rubel, 2017; Rubel, Hall-Wieckert, et al., 2016; Rubel, Lim, et al., 2016). In their work, they developed and implemented lessons involving spatial justice themes enacted with critical pedagogy. Their work was situated in a mathematics course in high school and undergraduate classes serving predominantly underserved students in New York City (Rubel, Lim, et al., 2016). In their work, they describe their design heuristic for TMSpJ that included three main considerations: make race explicit, reading the world, and writing the world—the last two considerations coming from Paulo Freire's work on critical literacy and reading and writing the word and the world. With this design heuristic in mind and the points I have made up to this point, I will now show an example of how they can all be incorporated into the classroom through teaching the science of data.

Spatial Justice and Data in the Context of Charlotte, North Carolina

The city of Charlotte, North Carolina, has a history of redlining, particularly after the Civil War. In an attempt to make the city more equitable in the future, the city has undertaken an initiative they refer to as Charlotte 2040, where they are seeking to create a more equitable city by the year 2040. Figure 3 shows how strongly the practice of redlining influenced where people of different races were able to live in the city. The spatial depiction of housing and race data in Fig. 11.3 is part of the Charlotte 2040s interactive slide show to show the issue of housing inequity in Charlotte and is publicly available for people to use. This could serve as a starting point for bringing the notion of spatial injustice and data into a class. This topic has potential for cross-disciplinary connections as well, particularly in terms of connecting social studies to mathematics, talking about the historical context that created such injustices and using statistics and data to highlight such inequities. There are also connections to sciences when you begin considering environmental factors. For example, in the areas where there is a greater density of Black residents, we can see from the data in Fig. 11.4 there is also a higher level of environmental exposure risk, particularly near the airport. There are also possibilities for citizen science in this context, perhaps looking

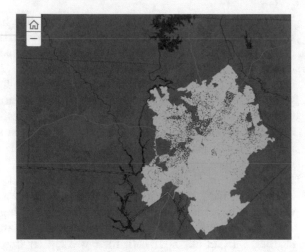

Inequity in Charlotte

Racial Density

The racial segregation catalyzed in the late 1800s and perpetuated, in part, by government policies throughout the next 60+ years is still very prevalent in Charlotte today. The geographic "wedge" south of Uptown, comprised predominantly of white residents (symbolized in teal), aligns with the area categorized as "Best" on the Home Owner's Loan Corporation (HOLC) map. In contrast, several geographies north of Uptown, comprised predominantly of African-American residents (symbolized in pink), align with the area categorized as "Hazardous" on the HOLC map.

Fig. 11.3 Racial density map where each point represents one person and the points are colored based on the race of the person. The light colored dots represent white residents and the darker dots represent non-white, in this case predominantly Black, residents (*Source* https://charlotte.maps.arcgis.com/apps/MapJournal/index. html?appid=0715a3477e484feb9ffc0b9ce88e15d2)

at water quality; measuring soil samples, temperature, air quality, or other environmental factors; and geotagging the data collected to then create spatial representations of the data.

There is also potential to bring issues of spatial injustice across disciplinary lines and into the teacher education curriculum. For example, look at Figs. 11.5, 11.6, and 11.7 that show an interactive data visualization of the multivariate relationship between location, educational attainment, and unemployment rate. You can also notice from these maps that if you compare them to the racial density map in Fig. 11.3 there is a pattern between location, educational attainment, unemployment, rate, and the race of people living in those areas. These types of data visuals can help to show the complex interaction of many different factors that influence and are influenced by education systems and structures of injustice. It is well known that injustices in the educational system are very space based (Kozol, 1991). Furthermore, redlining has caused segregation in schools because school attendance is based on where people live, and the quality of schools is often based on the tax base there. There are also other websites that can be explored that show locations of schools, which could then be considered in regards to the other spatial representations of data. School performance data is also often linked to the factors discussed already. All of this data and the spatial representations of it help to highlight the importance of space in shaping the reality that we experience.

Fig. 11.4 Spatial data map of environmental exposure risk factors in the city of Charlotte (City of Charlotte, 2019). *Note* <1 is a normal, acceptable risk of non-cancer adverse health effects, >1 is likely to increase risk of non-cancer adverse health effects

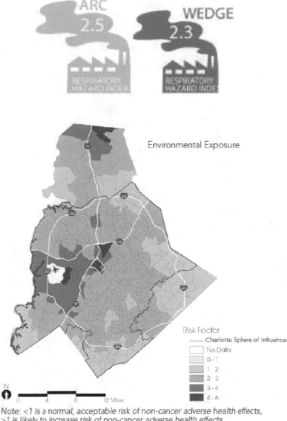

Note: <1 is a normal, acceptable risk of non-cancer adverse health effects, >1 is likely to increase risk of non-cancer adverse health effects

DISCUSSION

My goal in this piece was to try to make some connections between the Anthropocene, science education, statistics and data science education, and issues of spatial justice. I believe there is great potential here to bring new ways of interrogating injustices into not only school classrooms but also teacher education as I tried to briefly highlight in the example of exploring equity in Charlotte. The Charlotte 2040 plan is ambitious and one of the most transparent and significant I have seen to try to transform a community into a more equitable space. However, there is some irony in this plan as well. To bring in one more connection, if you look at Fig. 11.8, you will see a visualization of what Indigenous groups are known to have inhabited the lands where Charlotte is today. The irony of Charlotte's plan is that before redlining was an issue, in fact well before there were inhabitants of European and African ancestry inhabiting the area, there were already many Indigenous peoples that

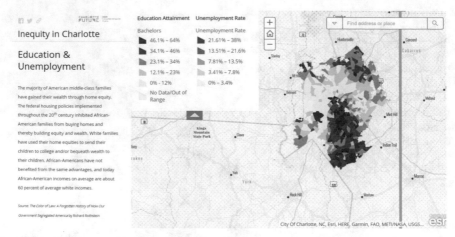

Fig. 11.5 Map of Charlotte region with the graph colored based on the scale on the left for Education Attainment. The vertical bar on the right of the map can be slid across the graph to change the color-coding of the data represented in the graph (*Source* https://charlotte.maps.arcgis.com/apps/MapJournal/index.html? appid=0715a3477e484feb9ffc0b9ce88e15d2)

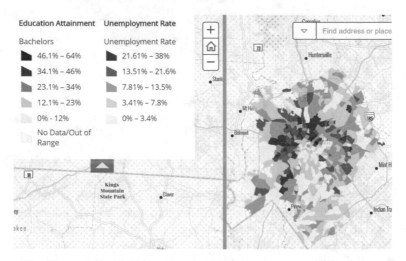

Fig. 11.6 Map of Charlotte region with the graph colored based on the scale on the right for unemployment Rate. The vertical bar on the left of the map can be slid across the graph to change the color-coding of the data represented in the graph (*Source* https://charlotte.maps.arcgis.com/apps/MapJournal/index.html? appid=0715a3477e484feb9ffc0b9ce88e15d2)

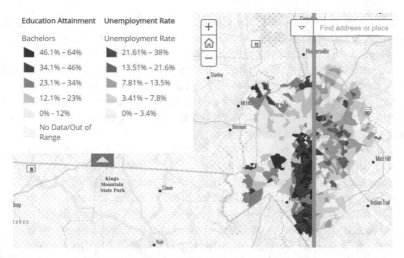

Fig. 11.7 Map of Charlotte region. The vertical bar in the middle of the map can be slid across the graph to change the color-coding of the data represented in the graph with the left side of the bar representing Education Attainment and the right side Unemployment Rate (*Source* https://charlotte.maps.arcgis.com/apps/Map Journal/index.html?appid=0715a3477e484feb9ffc0b9ce88e15d2)

Fig. 11.8 Side-by-side maps of the Charlotte region showing the tribes that inhabited the land on the left and on the right those tribe's lands with labels of the current cities (*Source* https://native-land.ca/)

inhabited the area. Where is the justice for Indigenous peoples in Charlotte's plan? Can spatial justice for all ever truly be realized, or is it a pipe dream?

I would also like to point out the current issues of spatial injustice unfolding in the world in relation to the global pandemic of COVID-19. In the context of the U.S., different states enact different policies toward inhibiting the spread of the coronavirus that at the time I am writing this is responsible for more than 550,000 deaths in my country. Such artificial spatial boundaries then determine where conditions are ripe for the spread of the virus and open travel allows for such policies to impact people across those artificial boundaries. Viruses do not know or respect politics. There are connections between this currently unfolding crisis and the themes discussed in this chapter. A prime example is the case of Brazil, which currently is run by a right-wing president, who denies the dangers of the virus even against his own health advisors' advice, and focuses on capitalistic endeavors by working to focus on the economy rather than saving lives.

Much of the data is visualized spatially. Take, for example, maps of cases per capita in Brazil by the *New York Times* (https://www.nytimes.com/interactive/2020/world/americas/brazil-coronavirus-cases.html) that is updated daily. The map shows that the highest ratios of cases per capita are all in the more rural states in northern Brazil, which are also the locations of many of Brazil's Indigenous peoples. Such examples can also be seen across the Americas, including in Canada and the U.S. (Haig, 2020; Lawrimore, 2020). These are just more examples of capitalist and colonialist ethics leading to the death of Indigenous peoples and how spatial data help investigate such issues.

Reflections

Before ending the discussion, I have a reflection that I believe important to share. I believe it is important for an author to position oneself in their work so the reader can have that information to see how the work is shaped by the author's subjectivity. I see each piece I write as a learning experience as much as it is a form of sharing my own understanding with others. Writing this piece pushed me in several ways and uncovered some tensions in my work that I need to deal with in the future. I typically work at the intersection of mathematics education, statistics education, critical literacy, statistical literacy, and the learning sciences. When Jesse Bazzul asked me if I would be interested in writing a piece in this book, I jumped at the opportunity to push myself and play with new ideas. Playing with new ideas also challenges old ones. Though the Anthropocene is well known in science and science education communities, it has not infiltrated the fields that I generally work in. Furthermore, being a scholar from the United States, I have come across work from Indigenous scholars and seen some educators incorporate it into their work, but still very minimally so, especially when compared to scholars from Canada with whom I have worked. In digging into readings to help me think about what I was going to write for this chapter, I started with one idea. That idea, however, changed significantly over the course of my playing with ideas. Honestly, I ended in great conflict and almost decided to scrap the whole thing and bow out of the endeavor.

Where my greatest conflict came in was in the consideration of who I was citing in my work. In my academic preparation, I was introduced to the notion that who you cite is a political act, and I have tried to always take that to heart in my writing. The conflict I encountered in writing this piece was that though the line of thinking that began to emerge from my reading and the shift in my ideas revolved around reading about Lewis and Maslin's (2015) work, deciding on a date for the beginning of the Anthropocene, and deciding about 1610 because it had both scientific and cultural implications, this was an entirely new concept to me, especially when I was confronted with the quantities they presented for Indigenous populations before and then a century after European conquistadors and settlers arrived. This started me down the path of connecting ideas between the intersection of statistics and spatial justice issues. I had started with the notion of mapping, and how you use the information you gain from data and spatial analyses has implications and can produce systems of oppression and injustice. After having written the chapter in this new line of thinking, what I realized upon reflection and review was that the vast majority of the scholars I was citing were white males. I was discussing the use of the date of the Orbis Spike as an important political act because of how it highlights the devastation brought on by colonization, yet there were likely Indigenous scholars and Indigenous ways of knowing that could be used to speak to these ideas and help to view them through a different lens. I have chosen to still submit this work because the scholars I cite are the sources for where my current ideas as they are come from. Yet, this chapter and reflection serve as a challenge for myself to learn more about the work of Indigenous scholars and ways of knowing. I challenge myself in the future as I continue to write on these ideas to do so from a more diverse array of perspectives, especially those that have been historically marginalized.

References

American Statistical Association. (2014). *ASA statement on using value-added models for educational assessment*. https://www.amstat.org/policy/pdfs/ASA_VAM_Statement.pdf.

Astor, M., & Lai, K. K. R. (2018, November 29). What's stronger than a blue wave? Gerrymandered districts. *The New York Times*. https://www.nytimes.com/interactive/2018/11/29/us/politics/north-carolina-gerrymandering.html, https://www.nytimes.com/interactive/2018/11/29/us/politics/north-carolina-gerrymandering.html.

City of Charlotte. (2019). *Charlotte Future 2040 Comprehensive Plan: Built City Equity Atlas*. http://ww.charmeck.org/Planning/CompPlan/Charlotte_Equity_Atlas.pdf.

Cobb, G. W., & Moore, D. S. (1997). Mathematics, statistics, and teaching. *The American Mathematical Monthly*, *104*(9), 801–823.

Crutzen, P. J., & Stoermer, E. F. (2000). The "Anthropocene." *Global Change Newsletter*, *41*, 17–18.

Davidian, M., & Louis, T. A. (2012). Why statistics? *Science*, *336*(6077), 12. https://doi.org/10.1126/science.1218685.

Davis, H., & Todd, Z. (2017). On the importance of a date, or decolonizing the Anthropocene. *ACME: An International Journal for Critical Geographies*, *16*(4), 761–780.

Foucault, M. (1972). *The archaeology of knowledge and the discourse of language*. Pantheon Books.

Frankenstein, M. (1994). Understanding the politics of mathematical knowledge as an integral part of becoming critically numerate. *Radical Statistics, 56*, 22–40.

Gutiérrez, R. (2013). The sociopolitical turn in mathematics education. *Journal for Research in Mathematics Education, 44*(1), 37–68.

Gutiérrez, R. (2016). Strategies for creative insubordination in mathematics teaching. *Teaching for Excellence and Equity in Mathematics, 7*(1), 52–60.

Haig, T. (2020). First Nations leaders in northern Canada warn of COVID-19 time bomb. *Radio Canada International*. https://www.rcinet.ca/en/2020/03/12/first-nations-leaders-in-northern-canada-warn-of-covid-19-time-bomb/.

Honner, P., & Gonchar, M. (2017, November 30). Investigating gerrymandering and the math behind partisan maps. *The New York Times*. https://www.nytimes.com/2017/11/30/learning/lesson-plans/investigating-gerrymandering-and-the-math-behind-partisan-maps.html.

Klarreich, E. (2017). The mathematics behind gerrymandering. *Quanta Magazine*. https://www.quantamagazine.org/the-mathematics-behind-gerrymandering-20170404/.

Kozol, J. (1991). *Savage inequalities*. Harper Perennial.

Larnell, G. V., & Bullock, E. C. (2018). A socio-spatial framework for urban mathematics education: Considering equity, social justice, and the spatial turn. In T. G. Bartell (Ed.), *Toward equity and social justice in mathematics education* (pp. 43–57). Springer.

Lawrimore, S. (2020, May 8). The impact of COVID-19 on Native American communities. *Harvard Gazette*. https://news.harvard.edu/gazette/story/2020/05/the-impact-of-covid-19-on-native-american-communities/.

Lerman, S. (2000). The social turn in mathematics education research. In J. Boaler (Ed.), *Multiple perspectives on mathematics teaching and learning* (pp. 19–44). Ablex.

Lewis, S. L., & Maslin, M. A. (2015). Defining the Anthropocene. *Nature, 519*(7542), 171–180. https://doi.org/https://doi.org/10.1038/nature14258.

Livingston, D. (2003). *Putting science in its place: Geographies of scientific knowledge*. University of Chicago Press.

O'Neil, C. (2016). *Weapons of math destruction*. Crown.

Remy, L. (2018, December 6). *Making the map speak: Indigenous animated cartographies as contrapuntal spatial representations*. NECSUS. https://necsus-ejms.org/making-the-map-speak-indigenous-animated-cartographies-as-contrapuntal-spatial-representations/.

Rubel, L. H., Hall-Wieckert, M., & Lim, V. Y. (2016). Teaching mathematics for spatial justice: Beyond a victory narrative. *Harvard Educational Review, 86*(4), 556–579.

Rubel, L. H., Hall-Wieckert, M., & Lim, V. Y. (2017). Making space for place: Mapping tools and practices to teach for spatial justice. *Journal of the Learning Sciences, 26*(4), 643–687. https://doi.org/https://doi.org/10.1080/10508406.2017.1336440.

Rubel, L. H., Lim, V. Y., Hall-Wieckert, M., & Sullivan, M. (2016). Teaching mathematics for spatial justice: An investigation of the lottery. *Cognition and Instruction, 34*(1), 1–26. https://doi.org/https://doi.org/10.1080/07370008.2015.1118691.

Soja, E. (2010). *Seeking spatial justice*. University of Minnesota Press.

CHAPTER 12

Science and Environment Education in the Times of the Anthropocene: Some Reflections from India

Aswathy Raveendran and Himanshu Srivastava

ANTHROPOCENE AND THE GLOBAL SOUTH

The word "Anthropocene" is commonly used to denote a geological epoch wherein humans are the central agents driving large-scale and long-lasting environmental changes as manifested in climate change. This chapter uses the term "Anthropocene discourse" to refer to the dominant discourse on the environment and the global environmental crisis and the associated moral and political responsibilities that it places on humankind. Jens Marquardt (2019) points out that this discourse around Anthropocene is West-centric, lending itself to the "ecological modernization" paradigm that emphasizes sustainable development, favoring technological and market solutions to address the ecological crisis. When talking about humankind's collective responsibility in addressing climate justice, it fails to account for the vastly different contributions on the part of the Global North and the Global South in the ecological crisis. Moreover, as Rohan D'Souza (2015) remarks, employing science and technology-based solutions (such as geo-engineering and earth systems science) to the global environmental crisis favors countries of the Global North, which have the advantage of technological know-how and wealth. In another article, situating the notion of the Anthropocene with reference to the history of South Asian environmentalisms, he argues that while

A. Raveendran (✉)
Department of Humanities and Social Sciences,
Birla Institute of Technology and Science Pilani, Hyderabad, India

H. Srivastava
Homi Bhabha Centre For Science Education, Mumbai, India

© The Author(s) 2022
M. F.G. Wallace et al. (eds.), *Reimagining Science Education in the Anthropocene*, Palgrave Studies in Education and the Environment,
https://doi.org/10.1007/978-3-030-79622-8_12

most environmentalisms have defined the environmental crisis in relation to resource scarcity, the Anthropocene discourse has emphasized the reduction of greenhouse gas emissions (D'Souza, 2019). He notes that:

> the Anthropocene framework profoundly disorients the search for the political and economic drivers that historically spurred radical environmental change in South Asia. Instead of debating the ecological gap or processes of unequal ecological exchange between the developed economies and their erstwhile colonies, Anthropocene futurism pushes for a disconnect between an environmentalism for techno-managerial solutions and one that is devoted to uncovering aspects of historical injustice. (p. 247)

The Anthropocene discourse upholds the notion of sustainable development, according to which economic growth is a non-negotiable for our societies but must not occur at the cost of depletion of natural resources. Sustainable development advocates are optimistic about striking a balance between economic growth and resource utilization through technoscientific innovations. In the next section, we historically situate the sustainable development framework and its construal of the interconnection between environment, development, and technoscience.

ENVIRONMENT-DEVELOPMENT-TECHNOSCIENCE: DEBATES IN INDIA

In the Western world, the alarm over dwindling natural resources and the necessity to restrain economic growth first appeared in the 1970s with the publication of the report *Limits to Growth* in 1972. In the very same year, the United Nations Conference on the Human Environment in Stockholm discussed environment and development as interlinked concepts. This conference was also pivotal in promulgating the idea that economic growth could be achieved without necessarily harming the environment (D'Souza, 2012). The term "sustainable development" was eventually introduced in 1987, in the report *Our Common Future*, also known as the Brundtland report (D'Souza, 2012; Kothari, 2014). However, it must be noted that the 1970s also witnessed the emergence of modern environmentalism in the Western world with the publication of Rachel Carson's *Silent Spring*.

The policies of the Indian government responded to these international developments, which is evident in the discursive shifts in the understanding of nature and economic growth in the first to the fifth five-year plan of the planning commission of the Indian government. While the first five-year plan document of independent India discusses both nature and human beings in terms of untapped potential, the notion of resource scarcity and the necessity to acknowledge environmental degradation and pollution are discernible in the fifth plan document (D'Souza, 2012). The shift in the discourse on nature and economic growth also parallels a global discursive shift in the understanding of development, which was conceptualized only in terms of economic growth in the 1800s to an understanding of development as growth that is inclusive of social and environmental indicators (Achuthan, 2011).

The discourses on development in postcolonial India are also important to understand in relation to the visions of three nationalist leaders—Jawaharlal Nehru, M. K. Gandhi, and B. R. Ambedkar. The development model adopted by the Indian state post-independence was inspired by Nehru's scientific socialism, which advocated that controlled economic growth under a socialist state, bolstered by technoscience-based development projects, would save a country struggling to overcome poverty, ill-health, and overpopulation. On the other hand, Gandhi had an entirely different vision of development that sought to sustain rural livelihoods, opposed large deskilling technology and industrialization, and emphasized decentralized governance through the village panchayats (decentralized governance at the village level). Ambedkar, a scholar, anti-caste leader, and the architect of the Indian constitution, emphasized the necessity of state-protected industrialization (Robinson, 2014). His concerns were primarily to protect the people of the marginalized castes. He staunchly believed that industrialization, through the provision of employment opportunities, would put an end to caste-based economic feudal relations prevalent in the primarily agrarian Indian society (Shivaprasad, 2016).

Despite these diverse visions, the Nehruvian imagination of scientific socialism prevailed when Jawaharlal Nehru took over as the first prime minister of independent India. Large technoscientific development projects, such as the atomic energy ventures of the 1950s, the Green Revolution of the 1960s and 1970s, and the White Revolution of dairy technologies of the 1970s and 1980s, were instituted in this period. In these decades, however, people's movements also began to raise critical questions about these technoscientific development projects—whom they benefit and whom they leave behind. For instance, the *Narmada Bachao Aandolan* (Save Narmada Movement) of the 1980s was a resilient movement against the *Sardar Sarovar* hydroelectric power-plant, a big technoscientific development project, raising valid concerns about widespread ecological destruction as well as displacement of the marginalized people who lived near the project site. Thus, the people's movements directly raised critical questions about development—development for whom and at what cost? In other words, while international and national policy documents celebrated sustainable development, people's movements in India through the 1970s and 1980s stressed the political nature of environmental issues, seeking to politicize the notion of development and raising questions on conventional models of economic growth.

Since the 1990s, with the advent of liberalization, the Indian economy opened up to foreign investment, culminating in the integration of the Indian economy with the global economy. Concomitantly, the Indian state receded from investing in essential welfare services such as health and education, resulting in the increased privatization of these sectors. At present, the Indian government has also relaxed environmental laws to favor the private sector. For instance, very recently, laws that protect Indigenous communities that depend on the forests for survival and livelihood were relaxed (Aggarwal, 2019).

In academia, the 1990s also witnessed the emergence of the post-development (PD) discourses (Escobar, 2011). Drawing on post-structuralist theorists such as Foucault, proponents of PD discourses argue that development and its associated terminology need to be understood as a dominant discourse and that this discourse that emerged in the Global North seeks to define, limit, and exploit the Global South. However, critics of the post-development discourse argue that development is not a monolithic, all-powerful entity. The Global South has resisted the development discourse through "local re-imaginations, alternative voices, and different worldviews" (D'Souza, 2012, p. 8).

The Nehruvian vision for postcolonial India forged a strong relationship between science, technology, and development, and therefore, the sustainable development framework has a stronghold in the Indian policy and educational discourse. For these reasons, we argue that it is necessary to dismantle the Anthropocene discourse and work toward transformative educational agendas. More specifically, this translates to: (1) examining the curriculum for how the Anthropocene discourse manifests itself, (2) understanding how the Anthropocene discourse shapes student subjectivities, and (3) working with students to reimagine alternatives that resist or subvert the Anthropocene discourse. Having set the context, in the next section, we switch to a conversational format to discuss what our respective research reveals about the nature of discourses around the environmental crisis, development, and technology in science and environment education.

Educational Discourse on Development and Environment

Aswathy: I will begin by locating myself and my research. I was born and raised in an upper-caste, middle-class family in the south Indian state of Kerala, well-known for its long history of communist politics and its relative prosperity. Religion-based communal tensions are also less pronounced in Kerala than the rest of the country, despite a sizeable minority population constituting Muslims and Christians. Perhaps because of Kerala's uniqueness about the aspects mentioned above and my caste and class privilege, I grew up somewhat oblivious to structural inequalities related to caste and class. However, I faced the full brunt of the misogyny and sexism endemic to Kerala's society in my teen years in the 1990s, when India opened up to the global economy.

After completing my schooling in Kerala, I went on to do undergraduate and master's degrees in biotechnology, after which I switched to science education research. My journey into critical science education research was the culmination of a series of unpleasant experiences with biotechnology research (Raveendran, 2019). Through science education research, I sought to make sense of these experiences and make positive changes to the field that had failed me but had not defeated me. I soon came to realize that science education research offered exciting possibilities to humanize and politicize the natural sciences.

Himanshu: I belong to Kanpur, a crowded city in Northern India, also known to be the industrial hub of the state of Uttar Pradesh, and according to the World Health Organization, the most polluted city in the world in 2018. Kanpur is home to millions of working-class people who migrate from various parts of the country searching for a livelihood. Caste- and religion-based politics have a long history in that region and conflicts between upper-caste Hindus, Muslims, and Dalits are common. Born in a working-class family, I spent my childhood in a neighborhood watching unemployed youth whiling away their time playing street cricket, betting, and fighting. After completing my school education, I got admitted to a prestigious and highly competitive engineering college. However, as a student, I soon began to feel a sense of disengagement with the engineering curriculum, which was depoliticized and disconnected from real life. As part of a voluntary student organization, I spent a significant amount of time with underprivileged children of the locality, helping them with their school education. My engagement with education deepened further when I got an opportunity to work with a non-profit organization in Central India and pursued educational research. In retrospect, I feel that my entry into critical science education was inevitable given my science and technology background and a broader interest in sociopolitical issues.

Aswathy: Now that we have both described what we are about, let's return to the theme of this chapter. At the outset, I think it is important to state that formal science education practices and research in India are steeped in positivism, in terms of restricting its content to disciplinary concerns and marginalising sociopolitical concerns (Raveendran & Chunawala, 2013). Moreover, the textbooks also reveal a commitment to Nehruvian scientific socialism, which forges a link between science, technology, and socio-economic development. Since both of us have done our fair share of analyzing textbook discourses, perhaps we could discuss this a bit?

Himanshu: I will try to respond to this through a discussion of my doctoral work. My primary interest has been to critically examine the discourse on environment and technoscientific development in the school curriculum and explore the ideological commitments of science and environmental studies (EVS) textbooks. As you know, in our context, the mandate is to institute compulsory environment education until grade 12, and this needs to be systematically infused into the curricula of all disciplines (NCERT, 2006b). In practice, however, the significant burden of teaching environment-related topics falls on the science teachers, as they are perceived to be more equipped to deal with the subject.

Aswathy: Where best to insert techno-managerial solutions to the environmental crisis other than the science textbooks!

Himanshu: Yes, and the EVS and science textbooks that I have analyzed do reveal an overarching techno-managerial orientation. In all the textbooks (class 4 to class 10) of the Maharashtra state board, what defines development is the advancement in transport, communication and space technology, big dams, and factories. Occasional references to "welfare" issues such as food security, sanitation, access to water, child labor, and prevalence of superstitions in society exist in the junior classes, but at more advanced levels, the textbooks are quite brazen about pushing the technoscientific model of development. Even the widely criticised Green Revolution is discussed with much optimism!

Aswathy: This has to do with the fact that the curricular emphasis at this stage is disciplinary. In my analysis of science, technology, and society (STS) issues in class 11 and 12 biology textbooks,[1] I noticed that the STS themes are treated in a lackadaisical manner, with very little consistency in the discussion of ethico-political positions across topics. The chapter on environmental issues is of particular interest with its strong techno-managerial focus. For every "environmental issue" discussed in the chapter, there is a proposed technical solution. Even more interesting is the chapter on ecosystems, where the value of natural ecosystems is discussed in the language of economics (Raveendran, 2017).

Himanshu: In my analysis of textbooks too, I find nature conceptualized merely as a "resource"—whether it is land, water, ocean, forests, plants, or animals. This objectification of nature reveals the commodifying tendencies of the economy. Also, the perspective of looking at nature is anthropocentric. Plants, animals, and even microbes are discussed in terms of the useful/harmful binary, as if their lives do not have any other value.

Aswathy: The class 12 biology textbook also discusses the ecosystem in terms of "services," revealing similar anthropocentric tendencies. The class 12 textbook also upholds neo-Malthusian ideologies. For instance, in a chapter on reproductive health, overpopulation is constructed as a burning problem, even when we now have evidence that the fertility rates have declined. Other chapters also advance the population bomb thesis to discuss the environmental crisis and the necessity to enhance food production.

Himanshu: Apart from the Malthusian argument, which posits overpopulation as the root cause of the environmental crisis, some other explanations that are offered by textbooks are people's attitudes toward the environment, lack of awareness, and overexploitation of resources. What is striking to me is the total lack of acknowledgement of differential access and consumption of resources in a highly stratified society like ours. The structural roots of the

---

[1] Education is part of the concurrent list of the constitution of India. This effectively translates to both the central government and all the state governments being responsible for designing, developing, and implementing curriculum at the school level. The biology textbooks that were analyzed were developed by the National Council of Education, Research and Training (NCERT), an apex body responsible for textbook development. The NCERT textbooks are followed in all the schools governed by the central government. However, each state in India has a State Council of Education, Research and Training (SCERT). The textbooks developed by various SCERTs are referred to as the state board textbooks. These textbooks are followed in all the schools affiliated with the state board with no exception.

environmental crisis, such as the profit-oriented nature of economic growth, rapid industrialization, and urbanization, are mentioned only in passing. The measures that are suggested to overcome the situation also reflect the same psyche. Population control, changing people's attitudes, proper management of natural resources, and increasing awareness among people are suggested as solutions in these textbooks—solutions that are managerial, individualistic, and technocratic, reflecting the presence of an eco-capitalist ideology. These findings are consistent with another study that I had carried out in the schools of Madhya Pradesh to examine the environmental philosophies underpinning the teaching of environmental issues (Haydock & Srivastava, 2019).

We pause this conversation briefly here. Textbooks are an essential part of the teaching-learning process in India. Textbook discourses are powerful since they seek to define and limit what can be legitimately articulated and conceived of any topic. Do they speak to everyone the same way? What impacts do the textbook discourses have on their subjects? How do subjectivities get constituted in relation to the textbook discourses? These are difficult questions that educationists in various settings have attempted to answer (see, for example, Bazzul, 2013). Apart from the textbooks, the learners' physical and cultural environment and the broader politico-economic context within which their lives are embedded would also shape their subjectivities, as manifested in their values, attitudes, and aspirations. In the next section, we seek to discuss how student subjectivities are shaped by the Anthropocene discourse.

Students' Values and Aspirations

Himanshu: Let me begin with another part of my research in which I interacted with the children of marginalised communities of the M(East) ward[2] area in Mumbai in out-of-school settings. Part of this interaction focused on their aspirations, attitudes towards environment and development, and political literacy. I also looked at their textbooks and observed classroom teaching in the region. What I gathered from a critical discourse analysis of their textbooks, classroom teaching practices, and my direct interactions with the students is that the school education does not adequately address the lived realities, questions, and concerns of the marginalised communities.

Aswathy: I am not entirely surprised by this finding. I would say that this has more to do with the centralized system of textbook production in India. Could these textbooks ever address the diverse micro-realities of a multicultural country?

---

[2] The M(East) ward in Mumbai is the most neglected ward of the city, with a Human Development Index being merely 0.05. The ward hosts one of Asia's largest landfill sites— Deonar dumping ground, an incineration plant, a slaughterhouse, and oil refineries, as well as a couple of fertilizer plants. Life expectancy at the time of birth in this ward is a mere 39.4 years, which is 30 years less than the national average.

Himanshu: Probably not. The textbooks have to cater to a highly diverse set of learners, but I believe that the teachers can make the necessary connections to the learners' contexts, through a critical and resistant reading of textbooks (Apple, 1992), and if they have to achieve this, they will need to understand what these students are about.

Aswathy: Let me cut to a very different context that I am familiar with. An elite engineering institute with a 100% job placement record. Students who attend this engineering college are from well-off backgrounds. I refer to this context because I too am trying to understand what these students are about—how they have come to embody their neoliberal middle-class subjectivities (Türken et al., 2016) and how they fashion themselves into autonomous, individualised, rational, and entrepreneurial selves, seemingly disconnected from the wider world.

Himanshu: Our contexts are not completely disconnected from one another. The respective student groups that we work with represent two ends of the socioeconomic spectrum. While the students living close to the landfill site face the brunt of the careless production-consumption cycle in terms of severe health and environmental impacts, the student group that you work with will drive the economy, since they will occupy techno-managerial positions in the future. They will drive the innovations that will both destroy and purportedly save the world.

Aswathy: … and what is even more interesting is that both these groups of students share similar aspirations.

Himanshu: Yes, in my conversations with a few adolescents of the M(East) ward, every single one of them unanimously expressed their wish to become engineers—these aspirations also matched with their parents' dreams for them. I recall a parent mentioning once that he would like his child to achieve what he could not achieve for himself. Under no circumstance did he want his child to come back to the same "shit-hole."

Aswathy: This throws open a conundrum—can we fault these students' aspirations, given the conditions in which they live? If becoming engineers and having access to a particular lifestyle is not something these students have experienced, how can we, as educators, deny them that aspiration? What would a transformative science and environmental education mean in this context?

Himanshu: There is no simple answer. But I think, for the M(East) ward students, critical discussions on the living and working conditions of the people in their community are necessary. When I attempted to do this through focus group discussions, one thing that struck me was the shame and humiliation they harbor about who they are and where they come from. I remember a session where I presented some data on the life expectancy in the M(East) ward, which is a mere 39.4 years. They were in denial of it. I also noticed a similar embarrassment when I tried to elicit some discussion on their personal backgrounds. Some of their family members were rag pickers, and they were reluctant to discuss that.

Aswathy: Maybe our task is to facilitate the transformation of these emotions of shame and humiliation to indignation. However, indignation alone cannot be an endpoint because once the students realize that very little can be done to transform their living realities in the short run, frustration would inevitably set in.

Himanshu: I did notice that there were seeds of indignation in some of the students. The groups that I interacted with were surprisingly quite politically aware. Thus, I am not sure if we can be pessimistic about transformative possibilities. Change is bound to take time. In the context of the M(East) ward, there are a few non-governmental organizations, workers' unions, and academics working with people of all age groups—these groups are working on an array of issues, ranging from health and education to the organization of workers. One way forward may be for educators to work with these groups over a more extended period. Transformation cannot happen overnight, and it has to be a sustained effort with a focus on acquiring critical understanding, developing solidarities, and organised, collective action.

Aswathy: Yes, maybe the way forward, as you suggest, is sustained effort in conjunction with other groups. My concerns have revolved around how to work with the privileged student groups. Engaging with this group is also very important—given the fact that they will, in the future, occupy techno-managerial positions and drive the economy. I work with the Department of Humanities and Social Sciences (HSS) in a private engineering institute. As a critical science educator working in this space, it is hard to ignore the fact that engineering practices and education are strongly embedded within the global neoliberal economy. This is evident in how the educational system is structured, the kind of employment opportunities that present themselves to students, and most so in the consumerist lifestyles that they inevitably embrace. Our primary task, as HSS faculty, is to teach "soft" skills such as language skills, personality development, and values to engineering undergraduates. While there are courses that have incorporated elements of critical, Marxist, feminist, and post-structural theory, these are elective subjects, and given the fact that the students that we cater to are gearing to be future technologists, they attribute the least value to the HSS courses. Very few students appreciate courses that introduce critical theory or adopt Marxist/leftist ideological positions.

Himanshu: Probably, this has to do with the numbness towards the harsh social reality that they have come to accept unconditionally.

Aswathy: Yes, a certain kind of numbness or indifference. And, the question is—how does one puncture that? These young people have seen poverty around them, although they have never experienced it firsthand because of their class and caste privilege and, therefore, fail to register it. Our country's social inequality is so stark that two people, even living fifty feet apart from one another, would experience completely different social and physical realities. For example, a person's life in a high-rise would be starkly different from that of a person in a slum a few feet away. However, social consciousness (structured by caste and class) has evolved so that these two human beings would go through life without being severely affected by one another, only, at the most, developing a transactional relationship. I remember a discussion in one of my classrooms where a group of three middle-class girls discussed whether their *bais* (female domestic help) fell ill or not! (Raveendran, 2020).

Himanshu: I find it interesting that you distinguish between social and physical environments. It is also striking that students experience alienation within their social and physical environments, wherein they feel disconnected from each other and their physical environments. I want to focus a little bit on the latter—on the physical environment. While working with students from marginalized communities at the M(East) ward, I have often wondered what developing a connection to the physical environment means in such highly degraded urban environments. For instance, when I would attempt to discuss waste and sanitation (what was salient for me about their context), they would not want to discuss that at length.

Aswathy: What if we begin with *what they like* about their lived environments? To people like us, from relatively privileged backgrounds, what stands out the most about these children's settings is the filth and the waste. However, as educators, it might be worthwhile to work on consolidating and affirming their positive experiences of their environment as well.

Himanshu: That might be a possible way forward, and I have also been thinking along these lines. Environmental education initiatives that seek to establish pro-environmental attitudes through hands-on farming, composting, gardening, etc. do exist in India. Though some exploratory studies (Dutta & Chandrasekharan, 2018) suggest that these activities lead to the development of pro-environmental values in some of the participants, there needs to be a rigorous evaluation of whether the compassion and connection with "nature" extend beyond these built environments (gardens, farms etc.). After all, interventions that use built environments package and present "nature" to participants in a certain way, as thriving, flourishing, enchanting, and clean. The question that I find myself asking is—are there ways to tap into and validate experiences or feelings of connection that children of marginalized communities have with what—in our sanitized perception—is "degraded," "filthy," "unclean" landscapes, which are home to them? This does not mean that we ignore the physical dangers in these spaces—the illnesses, the pollution, the harmful and unpleasant odors. But we often tend to focus only on unpleasant experiences. Probably, a starting point could be to build on the sense of relatedness that these children have to their environments and other people in their communities.

Aswathy: Working with the privileged will be more challenging. How do we get them to question their privileges and develop a sense of empathy with the marginalized and their physical environments? More importantly, how do we convince them that developing these sensibilities is essential?

Concluding Thoughts

India's environmental movement has combined ecological concerns with class and caste politics (Guha & Martinez-Alier, 2013; Kumar, 2016). It is pertinent that transformative science and environmental education in India that is critical of the Anthropocene discourse incorporate these concerns. Our examination of textbooks reveals that the textbooks support sustainable development and the ecological modernization paradigm. Interactions with student groups of different socioeconomic backgrounds also suggest that their subjectivities, as manifested in their values and aspirations, are shaped by the Anthropocene discourse. In our conversation, it emerged that working with both the marginalized and the privileged is necessary. More importantly, we realize that working with their emotions is an essential step toward their politicization, which resonates with other science education researchers' views. For instance, referring to Claudia Ruitenberg, Ralph Levinson (2010) argues that political emotions are distinct from personal insults or feelings of moral outrage. The former entails feeling a sense of anger on behalf of groups less privileged than one's group. Similarly, drawing on the scholarship of bell hooks, Antonio Negri, and Michael Hardt, Bazzul, and Tolbert (2019) argue for the necessity for political love to underpin our critical and transformative educational agendas. They write:

> political love is pursued in the name of collective joy—a refusal of the politics of domination and a recognition that other worlds in common are possible. Love is pedagogical; and so how can we learn to love each other better? How can we learn to extend our love beyond love of self/extensions of self (e.g., our own genetic families) toward a love of/for/with the multitude? (p. 306)

The challenge for educators working with marginalized communities is to find ways to inculcate political emotions that have the power to alter their living conditions. While working with marginalized groups of students living in ecological degradation sites, it might be important to validate their sense of community and their positive relationships with their physical environments. Transformative pedagogical practices involving privileged groups of students would entail getting them to understand how their career choices, lifestyles, and the broader neoliberal power structures are implicated in the sociopolitical and ecological crisis that we are faced with. Undoing privilege is a difficult task, but is a worthwhile goal to aspire for.

References

Achuthan, A. (2011). Re:wiring bodies. *Bangalore: The Centre for Internet & Society*.

Aggarwal, M. (2019, May 25). *What Modi and BJP's return means for environmental laws in India*. HuffPost. https://www.huffingtonpost.in/entry/modi-green-laws-environment_in_5ce7dba1e4b0a2f9f28d7cc4?guccounter=1.

Apple, M. W. (1992). The text and cultural politics. *Educational Researcher, 21*(7), 4–19.

Bazzul, J. (2013). *How discourses of biology textbooks work to constitute subjectivity: From the ethical to the colonial* (Doctoral dissertation). https://tspace.library.uto ronto.ca/bitstream/1807/43477/6/Bazzul_Jesse_T_201311_PhD_Thesis.pdf.

Bazzul, J., & Tolbert, S. (2019). Love, politics and science education on a damaged planet. *Cultural Studies of Science Education, 14*(2), 303–308.

D'Souza, R. (2012). Introduction. In D'Souza (Ed.), *Environment, technology and development: Critical and subversive essays* (pp. 1–16). Orient Blackswan.

D'Souza, R. (2015). Nations without borders: Climate security and the south in the epoch of the Anthropocene. *Strategic Analysis, 39*(6), 720–728.

D'Souza, R. (2019). Environmentalism and the politics of preemption: Reconsidering South Asia's environmental history in the epoch of the Anthropocene. *Geoforum, 101,* 242–249.

Dutta, D., & Chandrasekharan, S. (2018). Doing to being: Farming actions in a community coalesce into pro-environment motivations and values. *Environmental Education Research, 24*(8), 1192–1210.

Escobar, A. (2011). *Encountering development: The making and unmaking of the Third World*. Princeton University Press.

Guha, R., & Alier, J. M. (2013). *Varieties of environmentalism: Essays north and south*. Routledge.

Haydock, K., & Srivastava, H. (2019). Environmental philosophies underlying the teaching of environmental education: A case study in India. *Environmental Education Research, 25*(7), 1038–1065.

Kothari, A. (2014). Radical ecological democracy: A path forward for India and beyond. *Development, 57*(1), 36–45.

Kumar, V. R. (2016). History of Indian environmental movement: A study of Dr. B. R. Ambedkar from the perspective of access to water. *Contemporary Voice of Dalit, 8*(2), 239–245.

Levinson, R. (2010). Science education and democratic participation: An uneasy congruence? *Studies in Science Education, 46*(1), 69–119.

Marquardt, J. (2019). Worlds apart? The Global South and the Anthropocene. In T. Hickman, L. Partzsch, P. Pattberg, & S. Weiland (Eds.), *The Anthropocene debate and political science* (pp. 200–218). Routledge.

NCERT. (2006a). Biology: Textbook for Class XII. National Council of Educational Research and Training (NCERT).

NCERT. (2006b). Position Paper, National Focus Group on Habitat and Learning. National Council of Educational Research and Training (NCERT).

Raveendran, A. (2017). *Conceptualizing critical science education using socioscientific issues* (Unpublished doctoral dissertation). Tata Institute of Fundamental Research Mumbai.

Raveendran, A. (2019). Finding a critical voice. In J. Bazzul & C. Siry (Eds.), *Critical voices in science education research* (pp. 27–36). Springer.

Raveendran, A. (2020). Invoking the political in socioscientific issues: A study of Indian students' discussions on commercial surrogacy. *Science Education*. https://doi.org/10.1002/sce.21601.

Raveendran, A., & Chunawala, S. (2013). Towards an understanding of socioscientific issues as means to achieve critical scientific literacy. In G. Nagarjuna, A. Jamakhandi, & E. Sam (Eds.), *Proceedings of epiSTEME 5 International*

Conference to Review Research on Science, Technology and Mathematics Education (pp. 67–73). Cinnamonteal. http://episteme.hbcse.tifr.res.in/index.php/episte me5/5/paper/view/130/13.

Robinson, R. (2014). Planning and economic development: Ambedkar versus Gandhi. In B. Pati (Ed.), *Invoking Ambedkar: Contributions, receptions, legacies* (pp. 59–71). Primus books.

Shivaprasad, E. (2016). *Ambedkar's perceptions of development an empirical study in Mysore District* (Unpublished doctoral thesis, University of Mysore, Mysore, India). https://shodhganga.inflibnet.ac.in/handle/10603/145202.

Türken, S., Nafstad, H. E., Blakar, R. M., & Roen, K. (2016). Making sense of neoliberal subjectivity: A discourse analysis of media language on self-development. *Globalizations, 13*(1), 32–46.

CHAPTER 13

Rethinking Historical Approaches for Science Education in the Anthropocene

Cristiano B. Moura and Andreia Guerra

> *"Este es un mundo al revés.*
> *Este es un mundo de mierda.*
> *Pero no es el único mundo posible."*[1]
> —Eduardo Galeano, in an interview to #acampadaBCN, a
> social movement occurred in
> Barcelona in 2011

Climate changes experienced in different parts of the planet indicate that the Earth may be rendered inhospitable to human life soon. What once seemed more like a science fiction movie plot is now an urgent reality that is increasingly part of political and academic debate. Political actors, such as the young Greta Thunberg and the Yanomami leader Davi Kopenawa, have made us reflect on fundamental issues on a global scale: What are the conditions of life on planet Earth in a near future? How do we maintain human life in the (paradoxically) so-called Anthropocene era?

At the same time that such questions are regarded as fundamental to contemporaneity, the answers to them seem distant, intangible, under the

---

[1] Spanish for "This is a backwards world. This is a shitty world. But it is not the only world possible".

C. B. Moura (✉) · A. Guerra
Graduate Program in Science, Technology and Education, Centro Federal de Educação Tecnológica Celso Suckow da Fonseca, Rio de Janeiro, State of Rio de Janeiro, Brazil
e-mail: cristiano.moura@cefet-rj.br

© The Author(s) 2022
M. F.G. Wallace et al. (eds.), *Reimagining Science Education in the Anthropocene*, Palgrave Studies in Education and the Environment, https://doi.org/10.1007/978-3-030-79622-8_13

current paradigm. As Eduardo Galeano elaborates in the opening of this text, this seems like an upside-down world: the growing of global anxieties does not seem to produce any major shifts in the ways which we think and live (with) the Earth. Santos (2016) formulated this apparent paradox through what he calls the crisis of the Western society's paradigm, which is directly linked to Western modern science (WMS).[2] The author states that in times of crises like the one we are experiencing, the crisis deepens as strong questions such as those highlighted above have only been responded to with weak answers, that is, answers that do not defy the limits of WMS. According to Santos (2016), strong questions are those that challenge ways of life grounded in WMS, ones that push on, or interrogate, its boundaries. Santos' arguments are not only axiological, but also epistemological: he questions the limits of WMS knowledge systems, and the kind of epistemic responses WMS can produce as it strives to claim a unique position of universality.

The idea that we live in a new geological epoch called Anthropocene has gained momentum since the early 2000s, as the Nobel Prize winner in chemistry, Paul Crutzen, started using the term in his publications (Lewis & Maslin, 2015). The prospect that anthropogenic action has been of such a scale that it has altered the Earth's balance, profoundly changing environmental conditions, is already a consensus in the scientific community (Oreskes, 2007). Despite this, there is still a controversy regarding when the Anthropocene would have started: those who advocate the start date of 1610 claim that the "Great Navigations" (or the Age of Exploration, a period of extensive imperialism and settler colonialism that occurred between the fifteenth and seventeenth centuries) were an unprecedented process of exchange of animal and plant species, at a time when there were also profound changes occurring in the size of the world's population, which had an impact on atmospheric composition. Others state that 1964 should be the chosen date because, added to the repercussions of the great acceleration that occurred in the middle of the twentieth century, a peak of Carbon-14 was identified in the global atmosphere that year, related to the nuclear tests that occurred years before (Lewis & Maslin, 2015).

Turning now to Science Education (SE), those of us who work in the field of History, Philosophy and Sociology of Science (HPSS) and SE note that, despite the seriousness of the crisis that we briefly outlined above and the

---

[2] We are aware of the debates concerning the WMS / Traditional Ecological Knowledge in Science Education and the criticism about using this terminology regarding the bounds they impose (Kim et al., 2017); especially when one take into account the processes of appropriation of local knowledge (Harding, 2015). But, for clarity, what we refer here as Western modern science, based on Quijano (2000), is the way of producing knowledge developed by the end of the Middle Ages from European canons; this way of producing knowledge accounted for the cognitive needs of capitalism, and it "was imposed and admitted in the entire capitalist world as the only valid rationality and as an emblem of modernity" (Quijano, 2000, p. 343). In this way, our choice carries a political stance we sustain as crucial throughout this paper.

urgency of the theme of Anthropocene for science educators, research developed in these fields have given too little attention to such discussions. Our perplexed reaction is the same as that of Bazzul (2012), who points out that this community, gathered at a congress in Greece during a deep political and economic crisis in 2011, held discussions that seemed hermetic to the world that surrounded them at that time. On the other hand, the controversy over the initial date of the Anthropocene seems to point to an important aspect of the Anthropocene as a concept, which is its historical dimension. Choosing one or another marker as the start for this new geological era seems to us to be a decision that goes far beyond deciding on the best stratigraphic indicators—it is a decision with political implications. In summarizing the history of 1610, and the years around, known as the Age of Sail, when the "New World" was "discovered,"[3] as a great movement of species, it is not clear who started this movement, with what intentions, and what the consequences were. The same line of thinking applies to the year of 1964 and the nuclear bombs that increased the concentration of ^{14}C in the atmosphere. Who did it, with what intentions, and what were the consequences? Responses to these questions vary, depending on who is telling the story. Also, as Moore (2017) states, the way we choose to periodize the story completely changes its interpretations. Our intention is not that stratigraphers do the job of considering these factors as criteria for defining the new framework for the Anthropocene, but to point out how important it is for multiple counter-stories to be told. We argue that science education can be fruitful space for grappling with these complex considerations.

The Anthropocene can be used as an analytical lens in history and philosophy of science, though it has not yet materialized as such, given that it is a fairly new concept. Nature of science (NOS) studies have also yet to give much consideration to the Anthropocene. NOS frameworks are used to guide research and pedagogy, but have not explicitly attended to the history of science (HoS). In times of climate crisis and a future at risk, we think that such an omission may have even wider consequences than those related to HPSS in Science Teaching (e.g., Eastwood et al., 2012).

Our objective in this chapter is to underscore the importance of attending to the historical dimensions of the Anthropocene, particularly in HPSS fields of science education. We maintain that it is not possible to act in any field of research—especially in SE—without taking into account the time of strong questions and weak answers in which we live, as well as the role of Western scientific knowledge and diverse forms of knowledge. First, we seek, through a historical case study and with the support of Santos (2002, 2016, 2019), to reframe WMS, bringing other stories and other perspectives into a dialogue about its emergence and establishment. We discuss how Western modern

---

[3] The quotation marks used here intend to express our concerns with these expressions which for a long time have been problematized by postcolonial scholarship.

science as a knowledge system was shaped by the triad of colonialism, capitalism, and patriarchy. Such an analysis not only challenges WMS claims to universality but also honors and uplifts other forms of knowledge that can help to inform solutions for the present moment (Harding, 2015). We argue that enhancing the political-historical dimension of WMS in science education is fundamental to building futures that produce different and potentially less (self)destructive multispecies relationships.

REFRAMING WESTERN MODERN SCIENCE: THINKING ABOUT OTHER STORIES THAT CAN BE TOLD ABOUT ITS EMERGENCE AND CONSOLIDATION

> *"So that is how to create a single story:*
> *Show a people as one thing, as only one*
> *thing, over and over again, and that is what*
> *they become."*
> —Chimamanda Adichie.

In a 2009 lecture, Chimamanda Adichie, a Nigerian novelist, drew attention to the dangers of a single story. For instance, the danger of making a people, a country, or a fact to be identified by only one story. In the lecture, she emphasized how this act of essentializing a story contributed to shaping the global imagination about Africa and Africans, with devastating consequences, such as the dehumanization of Africans and "storying" Africa as if it were a place where good things do not happen. That is, as if it were not a place of creation, beauty, and poetry (Adichie, 2018). Adichie also states that the problem of the single story is closely linked to power and how it is asymmetrically distributed in the world. Therefore, questions such as how stories are told, who tells them, when they are told, and how many stories are told are all dependent on power structures.

Adichie's claims are also useful when it comes to the historiography of WMS. In other words, we still live from a historiography of (if not single) dominant stories, and this is quite evident in science education. Even when it comes to a reinterpretation of episodes that seek to bring down the glorious narratives of the "great geniuses of Science"—the famous fight against a presentist and decontextualized history—theoretical and political commitments that animate these tasks are rarely made explicit (Moura, 2019). In other words, writing a historical narrative is in and of itself an exercise of power and involves political commitments, yet these power dynamics and political commitments are insufficiently addressed in science education. Science education needs to attend in more depth to questions of "who has benefitted and who has suffered in its formation" (Nyhart, 2016, p. 7). Such efforts require a more intentional and in-depth search for other stories about the same events, or stories that reveal other historical events that help to tell different stories about science. These ideas are ones that are also being tackled by such research fields as global history (Roberts, 2009). This is still an incipient movement inside the HPSS and science education research fields, though there are some examples of these new historiographies emerging (Moura & Guerra, 2016; Gandolfi, 2019).

Postcolonial and feminist scholarship similarly challenges Western modern science's claims to universality, exclusivity, and totality. Santos (2002) argues that WMS grounds itself mostly in these three premises and that the obsession with these premises blocks possible understandings of the world that can exceed WMS. Moreover, Harding (2015) contends that WMS appropriated the observation of Indigenous people about their environment and their knowledge from the fifteenth century and marginalized this knowledge by framing it as myth, magic, and superstition that should be replaced by a "universal" WMS. Harding points out how WMS is also full of myth. Her work reveals how science and society are co-constitutive, i.e., producing each other.

The European territorial expansion was also guided by the same three premises of universality, exclusivity, and totality (Santos, 2002). The worldview that places outside Europe could be and should be dominated is also informed by a Christian religious understanding, that God would have given the world to humans ("Fill and subdue the earth," according to the book of Genesis). In this way, the Europeans left the territory they had occupied until the Fourteenth century to colonize others who they considered as non-beings (Santos, 2019), whose knowledge was false, and whose rules of coexistence were illegitimate. The colonization process occurred either through violence, trying to eliminate those who were different, or by assimilating the Other to the precepts of the European world, including WMS.

Harding (2015) and other feminist and postcolonial scholars tell a story about how the exploitation of natural resources and contact with other cultures boosted and helped in the consolidation of WMS, that is, in the consolidation of a universal rationality (Harding, 2015). In other words, although WMS makes claims to universality, this has not prevented WMS from appropriating cultural knowledge through colonization. For example, the specimens collected, the fossils found, the practices, and the knowledge learned in the different colonies expanded the horizons of WMS (Pimentel, 2007). However, the whole process of meeting with the Other in the colonies was understood by colonizers as a mere process of "data collection," extraction of raw material to expand scientific knowledge, rather than a two-way cultural exchange (Livingstone, 2003). Some scholarship has illustrated the importance of contact with the colonies for WMS (Pimentel, 2007; Raj, 2013), contributing to a more diverse array of "origin" stories that can be told about WMS.

In the process of colonizers' masking the epistemologies found and appropriating them as WMS, we reveal what Santos (2002) calls a waste of experience. A range of experiences and knowledge traditions around the world—that were often wider and more varied than the WMS ones—was "wasted" because they were "absorbed" as WMS production or simply neglected or discarded

(Santos, 2002). This waste also refers to what is "worth knowing": the questions and answers usually produced by WMS ended up eclipsing the experiences outside its realm. This means that in dominant stories (even in the non-Whig, non-presentist, and non-decontextualized ones), contact with the colonies is often diminished or relegated to a peripheral role in the production of WMS. If we take into account the aspects that current historians who have engaged in the task of rewriting this story are trying to clarify, we can conclude that: (1) there is more cultural/traditional knowledge from the colonies that was "absorbed" as WMS than what is often communicated in dominant histories (Raj, 2013); thus, we can question how much WMS it actually "European" or actually "universal" (even though it has asserted itself as such); (2) the epistemicide that occurred during the colonization process (and that continues to occur due to various forms of colonialism) may have narrowed possibilities for the future, as much of this knowledge engages with thinking about other forms of existence on earth (e.g., relational forms) and the coexistence with the "more-than-human, other-than-human, inhuman" (Haraway, 2015). That is the so-called waste of experience. To overcome it, · one has to re-inquire into the past (and, thus, the present).

Therefore, when we advocate the need to historicize modern science, it is important to note that we are not dealing with any and all historical approaches. There are those approaches, for example, that can even contribute to further essentialize other ways of knowing, which is not what we seek. Rather we propose the (re)telling stories about the past in ways that render identifiable the power structures inscribed in those histories—that histories, and even histories of science, must be understood as fabricated narratives and not natural ones (Mignolo & Walsh, 2018). As Santos (2019) and Harding (2015) teach us, colonialism, capitalism, and patriarchy are among the main forms of structural inequality that drive contemporary society; therefore, identifying the various forms of colonialism and patriarchy, as well as the hybridizations between capitalism, colonialism, and science, is fundamental in providing us with tools to imagine other futures. Next, we share a case study related to the history of botany that helps to contextualize our argument.

Going Deeper: A Short Case in the History of Botany

Botanical activity developed in Europe had undergone a major change during the eighteenth century when, in the early part of that century, botany began to acquire a strong economic character, prior to focusing on the pharmaceutical function of plants (Sigrest & Widmer, 2011). In this context, the colonizers invested in obtaining natural products from colonies to compete with trade monopolies maintained by other nations (Bravo, 2005). This change in botany increased the demand (and, consequently, the flow) for specimens from the colonies, encouraging the practice of European expeditions to colonies. Spain, for instance, supported expeditions to America and the Philippines to collect, describe, and identify new plants, subsequently publishing these findings and commercializing these natural commodities (Bleichmar, 2011).

The expeditions from Europe to the colonies, with both commercial and taxonomic goals (Bravo, 2005), had the function to locate American plants with potential for commercialization, such as American varieties of cinnamon, tea, pepper, nutmeg, and others. In these expeditions, scientific practices such as the collection and pressing of plants, illustrations, notes on observations of the places where the plants were collected, the transport of plants, activities aimed at adapting them far from their place of origin, and others, were widely developed and mobilized different social actors (Sigrist & Widmer, 2011).

These state-sponsored expeditions were considered an opportunity for young European men to join the scientific *milieu*. Those who wished to accompany these expeditions were trained to be able to take the plants they were looking for to Europe. Thus, in different locations there were book publications, such as the one published in Spain and entitled, in a free trans-lation, "Instruction on the safest and most economical way to transport live plants by sea and land to the most distant countries" (Gómez Ortega, 1779), whose author was a member of the Botanical Society of Florence and of the Royal Medical Academy of Madrid. These publications served as profes-sional conduct manuals for young travelers, in order to reduce the loss of collected specimens, and to train these men as "reliable witnesses" since the observations made by them in colonies could not be verified by botanists in Europe (Livingstone, 2003). The expeditions enlisted naturalists, physicians, clergymen, surgeons, imperial and colonial administrators, and artists. Training the eighteenth-century botanist became a global project (Bleichmar, 2011).

Part of this endeavor, the illustration practices, which were pivotal for the development of botany at that time, enlisted plant collectors, botanists, and also artists (Bleichmar, 2011). To get an idea of the importance of this process, Carl Linnaeus (1707–1778), who published a classification system that stands out in this context, received from only one expedition 250 herbarium speci-mens between 1767 and 1778, and two sizable collections of images produced in the colonies (Bleichmar, 2011). About the work of the artists, one can say that they were not free to represent the plants any way they saw fit. They were carefully trained to produce a version of the plants that was fitting for taxo-nomic description (Daston & Galison, 2007), ensuring that a plant collected in the colonies could be identified within a classification system used for the study of plants in Europe (Bleichmar, 2011). Thus, while the collected plants were recognized as something familiar, the identification of something hith-erto unknown within the system strengthened the premise that the knowledge produced in this process was truly "universal."

In this context, atlases of botany played a central role in the dissemination of the knowledge of botany produced by Europeans in colonies and in Europe (Daston & Galison, 2007). The botanists who participated in the production of plant illustrations guided the hired illustrators so that they represented what the botanists thought that need to be represented (Daston & Galison, 2007).

In this process, the differences between the leaves and flowers of the same plant were not portrayed. In the search for order and regularity, Linnaeus' classification system was built based on "plant sexuality," ordering the specimens in class and order (George, 2006). The number or proportion of male and female parts of each flower, called stamens and pistils, distinguished the classes and the orders. Thus, we find parts of the plants registered in the atlas' images that allow the identification of stamens and pistils (George, 2006). The names of the species and genus who were considered to be named received a Latin name derived from the name of the person who first described the species in publication. In this process, the artists and many other social actors who participated in the botanical processes were made invisible.

Sixteenth-century Europeans came to the Americas and learned about medicine and edible plants unknown in Europe, and they shared this knowledge in the European continent. Physicians and naturalists who never came to America used this knowledge to improve medicine and botany (Bleichmar, 2005), even though Europeans simultaneously disregarded knowledge from the Americas as less sophisticated (Harding, 2015). As Schiebinger (2005, p. 144) argues, "European colonial expansion depended on and fueled the search for new knowledge concerning tropical medicines. At the same time, colonialism bred dynamics of conquest and exploitation that impeded the development of this knowledge." Local knowledge was not only invisible in the illustrations and botany publications, in fact, the value of local knowledge was also disregarded. José Celestino Mutis (1732–1808), a Spanish botanist who lived in Colombia for twenty-five years and whose name was used by Linnaeus to name a specimen, drew much knowledge (which he published) from conversations that he had with local people from various social and ethnic groups. In these conversations, he asked local people about the local knowledge of flora and their medicinal use. He recorded and used these responses in his research while also denigrating local knowledge (Bleichmar, 2011). In 1686, Fontenelle (1686/1993) wrote a book to disseminate the Copernican system and astronomy knowledge to European women. In his book, that had several editions translated to many languages, Fontenelle also denigrated American Indians' knowledge and ways of life. To summarize, a powerful knowledge about plants taken from the colonies was developed, and, at the same time, powerful knowledges and ways of thinking about the same plants and the environment were marginalized, or, in Santos' (2002) words, were "wasted." In Europe and in sites where cultural knowledge appropriation occurred, such as the United States, however, only some stories, perhaps even a single story, about this knowledge construction are taught. If we intend to seek strong answers to the questions that the Anthropocene imposes in contemporary times, it is necessary to recognize these colonial erasures and consider that the neglected knowledge, the "wasted" epistemes, could have built paths to support answers to our current urgent questions. Societies' visions about the HoS are shaped by these hegemonic stories that frequently neglect processes of pillage and erasures that were foisted on native people of the colonies. When these stories are not told, the mythic ideals as to the universality of WMS are reinforced, and one extends the cycle of "wasting" other possible ways of living on the earth.[4] This process of unveiling other stories that can be told about science in schools, we argue, can benefit Anthropocene discussions. In the next section we are going to explore how historicizing the Anthropocene is a way of telling other stories about sciences, and, thus, how it has great potential for reshaping some conversations inside Science Education research and practices.

When Anthropocene and History of Science Meet: Some Insights for Science Education

"Este mundo de mierda está embarazado de otro [...]"[5]
—Eduardo Galeano

The idea of historicizing the concept of Anthropocene is not something only we uphold. Moore (2017) also joins this criticism, who argues that the term "Anthropocene" itself disguises the real story behind how we got to the present moment. Moore (2017) claims that a more appropriate term would be Capitalocene, which would decentralize Man (*Anthropos*) from history, and bring attention to the role of capitalism in our current dire circumstances. According to Moore, during the rise of capitalism, the expulsion of many humans of their homes and from humanity (mainly women, people of color, Indigenous people) provided a material condition for seeing nature as external (Moore, 2017). Haraway (2016) also criticizes the story behind the concept of Anthropocene, which does not make clear the process of looting and plunder involved before the Industrial Revolution, which some advocate is a milestone of the beginning of the Anthropocene:

> One must surely tell of the networks of sugar, precious metals, plantations, indigenous genocides, and slavery, with their labor innovations and relocations and recompositions of critters and things sweeping up both human and nonhuman workers of all kinds. The infectious industrial revolution of England mattered hugely, but it is only one player in planet-transforming, historically situated, new-enough, worlding relations. The relocation of peoples, plants, and animals; the leveling of vast forests; and the violent mining of metals preceded the steam engine; but that is not a warrant for wringing one's hands about the perfidy of the Anthropos, or of Species Man, or of Man the Hunter.

[4] Here we use the words of Ailton Krenak, an indigenous leader that confront the idea of nature and the world, at large, as a place we live in. As Krenak puts it, the indigenous people are the forest itself and not just "live" there. In our vision, the same goes for all the humans and the earth.

[5] Spanish for "This shitty world is pregnant with another world.".

It is this critical view of history that we share, and these are the other stories that we seek to tell about what happened in the past. The stories we tell about the past are not dissociated or dissociable from our present or from our perspectives for the future. As Chakrabarty (2018) puts it, the Anthropocene has challenged us in the sense that if once we constructed our histories about the past with a more or less secure prospect of the human existence in the future, now it is not the case anymore. So, this retelling of the past is not (only) a matter of justice for those who were erased from the dominant histories, but an active and imaginative process of reinvention of this past which could allow us to take care of the present (Haraway, 2015) to move forward and make our existence on this planet possible in the future. We argued, in another work, that the sociology of absences, proposed by Santos (2002) would help in the task of seeing the waste of experience in the past-present, identifying the absences in history and in the present time. We believe, along with the sociology of absences, that the sociology of emergencies (Santos, 2002) can help us to imagine possible futures from the recovery of the past and the present. According to Santos:

> Here, too, the point is to investigate an absence, but while in the sociology of absences what is actively produced as nonexistent is available here and now, albeit silenced, marginalized, or disqualified, in the sociology of emergences the absence is an absence of a future possibility as yet not identified and of a capacity not yet fully formed to carry it out. This is a prospective inquiry operating according to two procedures: to render less partial our knowledge of the conditions of the possible and to render less partial the conditions of the possible. (2002, p. 258)

What are the consequences of this reconceptualization for science education with a historical approach that tells another story about the relationship between modern science, capitalism, and colonialism? To think about this issue, it is important to highlight our standpoint, in order to try to make it clearer why we say what we say, as Chimamanda Adichie teaches us. As woman and man from Brazil, who work primarily in the field of science education, and, perhaps by living in Brazil of 2020 we probably feel more heavily the consequences of social inequalities, climate change, and the rise of political authoritarianism, we consider science classes as a space for social struggle. Therefore, we consider it as a place for the construction of ideas and possibilities for responding to contemporary dilemmas. As science teachers, we believe that the HoS emerges as a possibility to recognize that science education has a fundamental role to play in questioning the exclusivity of WMS. Indeed, "there is no knowledge without practices and social actors" (Santos & Meneses, 2009).

There is a long tradition of research in science education that presupposes that it is essential to promote the study of the nature of science (NOS). Although we recognize the potential of such initiatives in promoting debates about sciences, we consider that, because they do not explicitly consider that WMS was built and consolidated in the light of the triad of patriarchy, colonialism, and capitalism, they do not problematize the concealment, violence, and appropriation processes that occurred in the establishment of WMS. Consequently, the universalism intended by modern science appears as something given and natural rather than as a premise that, in order to be valid, it committed erasures throughout the process of development of scientific knowledge. In this way, the potential of science education to be a space for social struggle, a space for the development of ideas and possibilities for responding to contemporary dilemmas, is also wasted.

We find possibilities within the premises behind HPSS in science teaching:

> [I]t can humanise the sciences and make them more connected with personal, ethical, cultural, and political concerns; it can make classrooms more challenging and thoughtful and thus enhance critical thinking skills; it can contribute to the fuller understanding of scientific subject matter—it can contribute a little to overcoming the "sea of meaninglessness" which one commentator said has engulfed science classrooms, where formula and equations are recited but few people know what they mean; it can improve teacher training by assisting the development of a richer and more authentic epistemology of science, that is a greater understanding of the structure of science, and its place in the intellectual scheme of things. (Matthews, 1992)

Given the current state of affairs and our situation as an endangered species, perhaps it is time to revisit some of these purposes. In the first place, we think that revisiting the HoS such as that we have presented takes us on the path not of humanizing science, but of seeing it as an enterprise that has a certain look at Nature and at the human–nature relationship that is not unique and absolute. This may go beyond humanizing the sciences, in the same sense that understanding our "era" as Capitalocene instead of Anthropocene also does. About thoughtful and challenging classrooms, we certainly cling to this objective. As Harding (2015) points out, the vision that WMS is a science from nowhere should be denied; all sciences are politicized. This concept is still incipient in HPSS & SE research field.

Recalling Santos (2008), our need may not be the refusal of WMS but its reconfiguration "in a broader constellation of knowledge where it coexists with practices of non-scientific knowledge that survived the epistemicide [...] whether or not they have a non-capitalist horizon as reference" (p. 156). We understand that our (urgent) task is to look for ways to retell our stories, having a clear political commitment to the present and the future. Whether in WMS, or in science education itself, we need to reexamine how stories are told, who tells them, when they tell them, and what the power structures underlying these stories are. What could it mean for science education research and practice? Firstly, it is important to remember, as Paulo Freire (1987) states, that there is no education in a political vacuum. The same goes for science education. One important but the overlooked dimension of using HoS in science education is seeking to understand the erasures in our understanding of history. This is a complex reconstruction that will take a time long to be accomplished by historians of science in ways that begin to effectively challenge dominant stories. But, as science teachers, we can seek to unveil the erasures in each history, and maintain a critical awareness of the role of capitalism, colonialism, and patriarchy in History and, thus, in HoS, opening up possibilities of thinking-acting in the present to build different futures. This can help to bring back wasted epistemes to the game of knowledge, reinventing our *histórias*.[6]

To reinvent the future, we need to reinvent the past, which is a task that we should not fear. If we live today, in a *mundo de mierda*, we must remember that it is pregnant with another world—as Galeano teaches us. It is urgent that, following Paulo Freire, we roll up our sleeves to *esperançar*[7] this new world. If there is a certainty in studying HoS, it is that the world has changed a lot throughout history, so that another world is possible. We invite the reader to *esperançar* the world—or the world after the end of the world—together.

References

Adichie, C. N. (2018). *O perigo de uma história única*. Companhia das Letras.

Bazzul, J. (2012). Neoliberal ideology, global capitalism, and science education: Engaging the question of subjectivity.*Cultural Studies of Science Education, 7*(4), 1001–1020.

Bleichmar, D. (2005). Books, bodies, and fields: Sixteenth-century transatlantic encounters with new world materia medica. In L. Schiebinger, & C. Swan (Eds.), *Colonial botany science: Commerce and politics in the early modern world*. University of Pennsylvania Press.

Bleichmar, D. (2011). The geography of observation: Distance and visibility in eighteenth-century botanical travel. In L. Daston, & E. Lunbeck (Eds.), *Histories of scientific observation* (pp. 373–395). University of Chicago Press.

Bravo, M. (2005). Mission gardens: Natural history and global expansion 1720–1820. In L. Schiebinger & C. Swan (Eds.), *Colonial botany science: Commerce and politics in the early modern world*. University of Pennsylvania Press.

Chakrabarty, D. (2018). Anthropocene time. *History and Theory, 57*(1), 5–32.

Daston, L., & Galison, P. (2007). *Objectivity*. Zone Books.

Eastwood, J. L., Sadler, T. D., Zeidler, D. L., Lewis, A., Amiri, L., & Applebaum, S. (2012). Contextualizing nature of science instruction in socioscientific issues. *International Journal of Science Education, 34*(15), 2289–2315.

[6] In Portuguese, the word "histórias" may refer both to History (the hegemonic History with capital H) and stories (the stories we can tell about anything). So, we opted to use "histórias" here to highlight this twofold meaning and also to highlight how different rationales (sometimes expressed by language) can help us to think differently.

[7] *Esperançar* (pronounced is.pe.ran.'sar) is a Portuguese verb, stemming from Paulo Freire's thought, derived from *esperança* (hope) that differs from "being hopeful" because instead of meaning to be in a waiting state, *Esperançar* does not mean to conform, but to pursue and fight for goals.

Fontenelle, B. B. (1686/1993). *Diálogos Sobre A Pluralidade Dos Mundos*. Editora da Unicamp.

Freire, P. (1987). *Pedagogia do oprimido* (17ª· ed.). Paz e Terra.

Gandolfi, H. E. (2019). In defence of non-epistemic aspects of nature of science: Insights from an intercultural approach to history of science. *Cultural Studies of Science Education, 14*(3), 557–567.

George, S. (2006). Cultivating the botanical woman: Rousseau, Wakefield and the instruction of ladies in botany. *Zeitschrift fur Padagogische, 12*(1), 3–11.

Gomez Ortega, C. (1779). *Instrucción sobre el modo más seguro y económico de transportar plantas vivas por mar y tierra á los paises más distantes: Añádese el método de desecar las plantas para formar herbarios*. Por D. Joachin Ibarra, Impresor de Cámara de SM.

Haraway, D. (2015). Anthropocene, capitalocene, plantationocene, chthulucene: Making kin. *Environmental humanities, 6*(1), 159–165.

Haraway, D. (2016). Tentacular thinking: Anthropocene, capitalocene, chthulucene. *e-flux Journal, 75*.

Harding, S. (2015). *Objectivity and diversity: Another logic of scientific research*. University of Chicago Press.

Kim, E. J. A., Asghar, A., & Jordan, S. (2017). A critical review of traditional ecological knowledge (TEK) in science education. *Canadian Journal of Science, Mathematics and Technology Education, 17*(4), 258–270.

Lewis, S. L., & Maslin, M. A. (2015) Defining the Anthropocene. *Nature, 519*(7542), 171.

Livingstone, D. (2003). *Putting science in its place: Geographies of scientific knowledge*. University of Chicago Press.

Matthews, M. R. (1992). History, philosophy, and science teaching: The present rapprochement. *Science & Education, 1*(1), 11–47.

Mignolo, W. D., & Walsh, C. E. (2018). *On decoloniality: Concepts, analytics, praxis*. Duke University Press.

Moore, J. W. (2017). The Capitalocene, Part I: On the nature and origins of our ecological crisis. *The Journal of Peasant Studies, 44*(3), 594–630.

Moura, C. B. (2019). *Educação científica, história cultural da ciência e currículo: Articulações possíveis*. (Doctoral thesis in Science, Technology and Education). Centro Federal de Educação Tecnológica Celso Suckow da Fonseca, Rio de Janeiro.

Moura, C. B., & Guerra, A. (2016). Cultural history of science: A possible path for discussing scientific practices in science teaching? *Revista Brasileira de Pesquisa em Educação em Ciências, 16*(3), 749–771.

Nyhart, L. K. (2016). Historiography of the history of science. In B. Lightman (Ed.), *A companion to the history of science* (pp. 7–22). Welley Blackwell.

Oreskes, N. (2007). The scientific consensus on climate change: How do we know we're not wrong? In J. Abatzoglou, S. Nespor, & N. Oreskes (Eds.), *Climate change: What it means for us, our children, and our grandchildren* (pp. 65–99). MIT Press.

Pimentel, J. (2007). La revolución científica. In M. Artola (Orgs.). *Historia de Europa* (pp. 163–238). Espasa Calpe.

Quijano, A. (2000). Colonialidad del poder y clasificacion social. *Journal of World-Systems Research, 6*(2), 342–386.

Raj, K. (2013). Beyond postcolonialism . . . and postpositivism: Circulation and the global history of science. *Isis, 104*(2), 337–347.

Roberts, L. (2009). Situating science in global history: Local exchanges and networks of circulation. *Itinerario, 33*(1), 9–30.

Santos, B. S. (2002). Para uma sociologia das ausências e uma sociologia das emergências. *Revista Crítica de Ciências Sociais, 63,* 237–280.

Santos, B. S. (2008). *A Gramática do tempo: Para uma nova cultura política* (2nd ed.). Cortez.

Santos, B. S. (2016). *Epistemologies of the South: Justice against epistemicide.* Routledge.

Santos, B. S. (2019). O fim do império cognitivo—A afirmação das epistemologias do Sul. Autêntica Editora.

Santos, B. S., & Meneses, M. P. (2009). Introdução. In B. S. Santos, & M. P. Meneses (Eds.), *Epistemologias do Sul* (pp. 9–20). Cortez.

Schienbinger, L. (2005). Prospecting for drugs: European naturalists in the West Indies. In L. Schiebinger, C. & Swan (Eds.), *Colonial botany science: Commerce and politics in the early modern world.* University of Pennsylvania Press.

Sigrist, R., & Widmer, E. D. (2011). Training links and transmission of knowledge in 18th century botany: A social network analysis. *REDES—Revista hispana para el análisis de redes sociales, 21*(7), 347–387.

CHAPTER 14

Reflections on Teaching and Learning Chemistry Through Youth Participatory Science

Daniel Morales-Doyle, Alejandra Frausto Aceves,
Karen Canales Salas, Mindy J. Chappell, Tomasz G. Rajski,
Adilene Aguilera, Giani Clay, and Delani Lopez

This chapter captures a panel discussion from the 2019 conference of Science Educators for Equity, Diversity, and Social Justice (SEEDS) in Norfolk, Virginia. The panel included two high school students, three high school chemistry teachers, a community organizer, an administrator for a large urban

D. Morales-Doyle (✉) · M. J. Chappell
Department of Curriculum and Instruction, University of Illinois at Chicago, Chicago, IL 60607, USA
e-mail: moralesd@uic.edu

M. J. Chappell
e-mail: mjchappell@cps.edu

A. Frausto Aceves · M. J. Chappell · T. G. Rajski · A. Aguilera · G. Clay · D. Lopez
Chicago Public Schools, Chicago, IL, USA
e-mail: alejandrafrausto2026@u.northwestern.edu

T. G. Rajski
e-mail: tgrajski@cps.edu

A. Aguilera
e-mail: aaguilera@cps.edu

K. Canales Salas
Little Village Environmental Justice Organization, Chicago, IL, USA
e-mail: kcanales@lvejo.org

A. Frausto Aceves
School of Education and Social Policy, Northwestern University, Evanston, IL, USA

© The Author(s) 2022
M. F.G. Wallace et al. (eds.), *Reimagining Science Education in the Anthropocene*, Palgrave Studies in Education and the Environment, https://doi.org/10.1007/978-3-030-79622-8_14

school district, and a university-based science educator. These panelists, the authors of this chapter, had been collaborating on an initiative to support youth participatory science (YPS) projects in high school chemistry classes (Morales-Doyle & Frausto, 2021). We share this lightly edited transcript of our conversation as a way to communicate perspectives about the opportunities and challenges of YPS from viewpoints across these constituency groups.

YPS combines critical and pedagogical elements of youth participatory action research (Cammarota & Fine, 2008) with the disciplinary and democratic elements of citizen science (Irwin, 1995). The project described here is a collaborative effort to engage high school chemistry students in YPS projects about urban heavy metal contamination. In Chicago, like other (post)-industrial cities, there are numerous sources of heavy metal contamination. Communities of color in this hyper-segregated city tend to be disproportionately impacted by this pollution. Lead contamination in the environment comes from decades of leaded gasoline and lead-based paint. Lead and mercury were both emitted as biproducts of two coal power plants that were shut down as the result of a long community struggle in 2012 and a waste incinerator that was shut down in the 1990s. Biproducts of fossil fuels like petcoke and fly ash contributed to contamination as huge piles were irresponsibly or illegally stored outdoors. Besides lead and mercury, there are also metals like manganese and molybdenum, which continue to be emitted by chemical and steel plants in the city. Unlike lead and mercury, these are essential elements, but still have negative health and environmental effects in larger amounts that are less well-understood. Of course, there are also countless miles of lead pipes delivering drinking water to homes throughout the city. In the wake of the Flint water crisis, our group, who had been working together informally for a number of years, refocused on developing YPS projects about heavy metal contamination (Morales-Doyle et al., 2019). As you will read in our comments, different groups of teachers, youth, and scientists have taken up different specific local manifestations of heavy metal contamination in their YPS projects.

We want to emphasize that the work we share has been possible because of long-term involvement and collaboration. Building relationships between students, teachers, university-based scientists, and community organizations is challenging. Establishing trust requires reciprocity, time, humility, and commitment. Some members of our group have collaborated since 2004, with some version of the current project ongoing since about 2012. In 2017, our collaboration became more formal when we secured funding and then held three annual institutes that brought together teachers, young people, scientists, and community organizers to learn from each other and plan curriculum. In 2019, we had a YPS conference that brought about 150 students from neighborhood public schools in Chicago to the University of Illinois Chicago to share our work with each other. The 2020 version of that conference was canceled because of the pandemic.

Over the years that these collaborations have been built, some of our roles have evolved. In the present discussion, Karen Canales shares from her role as a community organizer with the Little Village Environmental Justice Organization (LVEJO). But Karen originally participated in this project as a high school student years ago and has been involved in environmental justice work ever since. Likewise, Alejandra Frausto and Daniel Morales-Doyle, who share from their respective roles as a district administrator and university faculty member, were originally involved in this work as high school teachers. If readers are taking up or planning similar projects, we want to emphasize the length of involvement and the importance of relationships over time. Taking on educational projects that challenge the presumed goals of schooling requires that we be generous and patient with ourselves, and with our collaborators and students.

We also want to be clear in how we are engaging the science. These projects happen in chemistry classes. We study the properties and reactions of metallic elements, asking questions like: why are heavy metals in the environment? how are we exposed? why are they toxic? how can we measure their concentration in the environment? We have been fortunate enough to collaborate with three brilliant and humble university scientists, Alanah Fitch, Shelby Hatch, and Kathryn Nagy. They have helped us learn more about these questions as educators. In our classes, we get into issues of sampling and spectroscopy and the design of environmental studies, but it is really important to us to teach that chemistry is only one way of understanding the problem. We have developed the YPS Curriculum framework, which prioritizes learning to appropriate and appreciate the scientific context. We are learning to critique and change science itself and hopefully to use science as a tool to change society (Morales-Doyle & Frausto, 2021).

The remainder of the chapter is organized around three questions about YPS that structured our panel discussion:

1. What are some of the challenges and possibilities when it comes to engaging with YPS in science classes?
2. How has engaging in YPS exposed both insights and oversights of scientific ways of knowing?
3. In YPS, what are the relationships between learning science and engaging in political and community issues?

We share a lightly edited transcript of the answers that we gave to these questions during the 2019 SEEDS conference with a brief discussion that connects some of our statements with some of the scholarship that informs our work.

Question #1: What Are Some of the Challenges and Possibilities When It Comes to Engaging with YPS in Science Classes?

Giani Clay (Student, George Washington High School):

One of the main challenges is apathy. If students are not interested in the topic that you're teaching, they are only memorizing it for a test and not learning it and keeping the knowledge. I've actually seen this before with one of my closest friends. He used to have poor grades. He would always tell me that we should ditch class, but once we started Ms. Aguilera's class, his behavior started to change, and he became more knowledgeable. He would start to say stuff like, "let's ditch all the first periods and let's go to Ms. Aguilera's class because I want to learn something in her class or tell her something I learned."

We were studying the bioaccumulation of mercury in fish. My friend, he actually taught me how to fish, and I taught him science. As we learned how the mercury accumulates in the fat of the fish, he got more interested, and I also got more interested in the topic. So, while apathy is a challenge to get students interested, once you get them interested it should be a pretty good class.

But another challenge is also trying not to go away from the science. I'm going to be an aerospace engineer and so I wouldn't want to just talk about politics in my science class, I want to learn the science. So, for some students, like my friend, the challenge is apathy and for others, like me, the challenge is not straying too far from the science.

Alejandra Frausto (Project-based Learning Manager, Chicago Public Schools):

I want to talk about the challenges and possibilities from three perspectives; as a high school science teacher that transitioned to middle school, as the lead teacher on the project, and in my new position in the district. As a middle school science teacher, while it was challenging to modify the curriculum that we were co-constructing for a chemistry class, for my sixth graders, it was important that my younger students also had an opportunity to act in our community. Sometimes we weren't able to do all the same labs or activities as the high school students. But we were just as capable of taking on some of the same questions, learning some of the same science, and taking action to address heavy metal contamination in our neighborhood.

As the lead teacher on the project, being very intentional in how Daniel and I designed our time together during the summer institutes allowed us to have more co-constructed possibilities with the teachers, students, community organizers, and scientists. While we had some ideas about things we thought needed to be addressed, like pedagogical needs, we also wanted to position all the participants as the experts that they are. It was important for us that in these institutes, everybody was both a teacher and a learner. The challenge was making the time to co-construct those learning moments together, especially with the unexpected work you don't think about. For example, we held a YPS student conference and we learned there are a lot of logistics with getting together 150 students.

Currently, I lead the district's service-learning project requirement and civic-oriented project framework that supports this requirement across about 105 high schools. This creates an opportunity for more YPS projects to happen in science classrooms as school leaders look to provide opportunities for students to meet the requirement. But the challenge now lies in the willingness to make space and time to make the pedagogical shifts needed to co-construct projects like these with students and community. Last year, 243 service-learning projects were completed in the district and only about 20 of them were in science classes. As we work to expand civic-oriented projects across disciplines, science class will also need to focus on who's making the decisions around specific environmental policies that impact our communities. I hope to better understand how to integrate YPS projects like these into science classrooms at a larger scale while making sure the projects are still meaningful, community responsive, and student led as the work spreads.

Tomasz Rajski (Teacher, Hubbard High School):

I predominately teach chemistry. A project like this can seem very intimidating to a new teacher. Being in my third year proved to be a real challenge that I faced when engaging with YPS. I was relatively new to social justice in science, even though I might have had some experience from my student teaching placement.

However, one of the good things about our project is that it does become more like a professional learning community. Not only in the fact that we are working with other teachers, but working with students, community organizers, professors; and so, there is a lot of different input of information that helps a new teacher develop their skills.

So, what we have noticed is this project grows with the people who take part in it. One of the things we did was a tuning protocol of a lesson that I have taught as part of the YPS program and that is similar to the wonderings that we had here at the SEEDS conference. One of my students was there; some of my colleagues were there, including other members of this panel. They listened to my presentation of the lesson, asked questions, and offered me a lot of advice and critical feedback on how to improve it.

After the protocol, my student who participated in this lesson a couple of months before told me that she felt awful because they were tearing my lesson apart by adding this and changing that. It is important to realize that was an experience for my student and myself. We had to understand that they were not just tearing the lesson apart or that they had a negative view of it, but what was actually happening was that they were giving critical feedback. I was learning from it, and I did not feel bad after hearing the changes that needed to be made. I felt confident that the next time I teach it, it will be improved. So, it is a moment of vulnerability that you end up learning quite a bit from.

Mindy Chappell (Teacher, North-Grand High School):

I teach students who are in 10th grade chemistry at North-Grand High School. The pipeline discourse, from Sputnik to Nation at Risk to No Child Left Behind, commodifies students who are well adapted to science, technology, and math. It positions science education as a place to build knowledge capital for a global market. This type of science education continues to produce functioning beings with the knowledge and skills to *exist* in the current system rather than pose questions about it or add methods to *transform* it. If we're being honest, STEM education continues to be a site of continual racial stratification of the U.S. workforce and associated differences in economic wealth.

Science education should allow students to see science as a tool for interrogating their world and lived experiences and evoking social change. For me as a teacher, I am not concerned with whether my students choose to be in STEM once they leave high school. That's their choice. What I want them to walk away from the classroom with is that their knowledge is valuable, and they *can* engage in rigorous science classes, regardless of their desire to study STEM. YPS gives me the opportunity to create this type of learning environment, which promotes the usefulness of scientific knowledge outside of the pipeline discourse.

Daniel Morales-Doyle (Assistant Professor, University Illinois Chicago):

One of the possibilities of YPS is connecting science classes with organizing efforts in the neighborhood. When I started teaching in the Little Village community in Chicago, I went to the Little Village Environmental Justice Organization and humbly asked them to teach me about the issues around which they were organizing. And they were gracious enough to do that. I had the instinct to ask because I came from a politically engaged family where my dad was a community organizer. So I had this sense that I needed to learn from community organizers if I was going to be an effective teacher in the community.

But one of the related challenges of doing YPS is constantly being engaged with complexity and contradictions. In this project, folks in schools, community organizations, and universities are working together. All of these can be contradictory, problematic spaces in and of themselves, but there are also power and resource differentials between them. Universities have tons of power and resources the schools don't. And even public schools often have power and resources that community organizations don't. So, the relationships we have built with each other are good and trusting and rich and long term. But, especially in terms of the relationships between institutions, they're definitely not unproblematic.

Question #2: How Has Engaging in YPS Exposed Both Insights and Oversights of Scientific Ways of Knowing?

Adilene Aguilera (Teacher, George Washington High School):

What I learned from leading YPS projects over a few years is the importance of recognizing each other and our families as sources of valuable knowledge who are capable of leadership. For example, as a Mexican woman with a family who was raised around agriculture, I tell my mom:

> Amá, you're a scientist, you just weren't labeled a scientist. You know all of this information that people who are labeled as scientists also know, the only difference is that you do not have the title, but you definitely have a lot more experience.

I share this interpretation with my students and encourage them to think about their families this way. I tell them, "Your parents know more than you think, or more than they show you. Therefore, you must ask them." The students themselves don't always view their parents as scientists. This is where we begin to learn how to acknowledge our families as a resource. In our school, the parents are resources because in the Southeast side of Chicago, they were either working for the industry that was polluting our areas or knew someone directly affected by the contamination. There was a year where we had about three students whose parents passed away from some type of cancer, so these students became more involved. They said, "I wonder if our air and our soil is causing our parents to pass away." These were their ideas. They were looking for information on their own to tackle these issues. This allowed us to have authentic student engagement. Even if we only searched through different databases for, let's say, asthma rates in the Southeast Side, students were engaged and leading this work in one way or another.

When our students took leadership, they went back home to tell their parents about these issues of heavy metal contamination going on in the Southeast side of Chicago. But, just as students do not always see their parents as a source of knowledge, parents often do not believe their children. I remember this dynamic in my house growing up. And this is especially true when students' analysis places blame for their circumstances on power structures. Unless some parents see their children's analysis confirmed on the news, a high school student is not a credible source. However, this changed with YPS; students became more vocal about issues of power and pollution, and they felt their voice validated.

The YPS conference that we organized as part of this project inspired me to organize an annual community conference at my high school. At this conference students present their work and parents came to show support for their children and their research. The fact that this conference is held yearly for students creates a sense of fulfillment for the students and it validates them as scientists. Something as simple as the conference program with their name and project title with abstract on it has students saying, "wow look at this, I did this." Now, at this very moment, Giani is experiencing the bigger picture. He is here. It's real to him and he's feeling that agency and fulfillment!

Tomasz:

So, I would like to expand on that a little bit. I feel that YPS can expose a lot of insight into student prior knowledge. In framing this, I looked at myself as a student learning with my students. Daniel presented to my class, and we started looking at maps of the neighborhood. Part of the process was identifying what might be areas in the neighborhood that we might consider industrial, what are the parts of the neighborhood where people live, what parts of the neighborhood might be considered beautiful, what parts of the neighborhood might be considered ugly. Interlocking that prior knowledge gives students that opportunity to take back ownership of their community. These are their observations. They live there. They know it better than I can ever try to know it, and I do really try. That is where the conversation begins, but oftentimes, that conversation then grows when you actually go on site. So, we had identified an area that we were interested in sampling, in the residential areas and parks close to a chemical plant down the street from the school. As community experts, we devised a plan to sample soils. We figured out our grid, where we want to go and why. Then we hopped on the bus and set off to collect samples.

As community experts, the first observation that we made on site was that the facility looked appalling. We were standing right in front of it, this is not a picture anymore, this is not a map—it is right in front of our faces. Not only did it look dreadful, but it also smelled horrible. There was this lingering smell that a lot of us attributed to whatever was coming out of the smokestacks. Looking at EPA documentation, that company does release a toxic inventory report for the heavy metal, molybdenum. This gave us a starting point to focus on.

The process of developing that prior knowledge from the community perspective began by observing how the facility looks, how it smells, what the area around it is—and led to students fusing that prior knowledge into a scientific perspective. We collected these soil samples. We looked at the EPA Toxic Release Inventory reports. We looked at what heavy metals are being released and the concentrations of the heavy metals released. Suddenly we are going beyond just experts in the community and becoming science experts within our community.

Alejandra:

I want to mention some insights and oversights as I share some of the work my middle school students did. Two years ago, I introduced our study of heavy metal contamination with an assignment that is similar to what Tomasz described. Students were asked to take four pictures in their working-class community where most of the residents are of Mexican descent: something that's ugly, something that's beautiful, something that's clean, and something that is contaminated. We noticed that many students took pictures of the viaducts that many of us walked under on our way to school. These viaducts, underneath the railroad tracks, have rust and chipped paint that flakes off and can be found in the soil in the next block. So, we decided to collect soil samples near the viaducts. The first year our results weren't really conclusive. So, an oversight that arose was needing to be explicit that one experiment won't usually answer a question conclusively. My middle school students that year realized addressing the paint chips of the viaduct was going to take more time.

In the second year, Daniel and I collected paint chips ourselves and found lead levels 170 times higher what national standards should be. We quickly realized science wasn't the *only* thing we had to learn; we had to also look at what policies existed and who made them. My students decided to present our findings to our local elected official, which in Chicago is our alderwoman or city council member. We contacted her and invited her to come and listen to my students' presentation and recommendations. Since that day, we have gotten excuses for why the city has not addressed the viaducts and an informal invitation to present to the commissioner of public health. But our scientific data was not enough. We have proof that high levels of lead are there and yet we have to keep fighting to convince the city to properly remove the lead that is toxic.

Question #3: In YPS, What Are the Relationships Between Learning Science and Engaging in Political and Community Issues?

Delani Lopez (Student, North-Grand High School):

I began working on this project my sophomore year, with my wonderful teacher, Ms. Chappell. For me, the goal of learning science is to further engage students in political and community issues so they can relate and empathize. So sometimes not all social and political issues are obvious, especially if you don't know the reason behind it. For example, I was not aware of heavy metal contamination in general. So, to look at maps and all of this information on how there were high contamination levels in predominately Black and Latinx communities is something that made me really passionate about this project. Alejandra mentioned the paint chips in the viaducts. That's a good example of how a lot of people don't realize what chemicals they might be exposed to in their everyday lives. So, what made me prioritize learning about this was that it affects me and the people in my communities. And I don't want to just get this knowledge and then sit back and do nothing about it when I know that there's something that needs to be done. Because it doesn't just affect me, it affects the people around me and other people that I don't know. You have to be able to stand up for that.

The learning I did in this project also made me curious about what other scientific, political, or social issues I am unaware of and how can we not only carry on this knowledge but build from it and find ways to face these issues head on. YPS gives students a sense of encouragement and empowerment: "This is wrong, and I have the science and proof to tell you this is wrong, so you can't knock what I say." A lot of times, youth are seen as inferior, like they don't know as much. That's not true. We know as much and we have the proof to fight against injustice. I also feel like knowing the science about these situations, and how it leads to them, it sparks an interest in students, but not only in the scientific field but in political issues as well. Now you see how something that didn't seem as big of a problem to us, because we're unaware of it. Now it's a bigger issue. It affects me. It affects other communities. It affects my family. We need to do something about it.

Mindy:

For me as a teacher, Delani's last point is exactly what it's about for many communities, but especially those that are marginalized and largely impacted by environmental injustice and racism. We need scientifically literate experts who become the engineers, scientists, science teachers, and advocates who determine our future. We need experts that protect us from the harms that get a pass in the name of science. YPS positions science education as a tool for students to understand the world and use science to transform the experiences and conditions in their communities. Maintaining the United States' place in the global market is removed from the conversation about science learning and replaced with pedagogy regarding science knowledge as a means of addressing environmental injustice and other socioscientific issues. This critical and social perspective to STEM education promotes the development of students' STEM knowledge through social issues that directly and indirectly impact their lives. The students are engaged through their knowledge of STEM to challenge and fight against oppression in multiple strands such as environmental racism and economic exploitation. Young people are not mere spectators, but they are actors who use their science agency to promote necessary change. And so for me, as a teacher I'm not concerned with whether or not my students go on to be scientists; if they choose to, by all means. But I want them to use their scientific literacy to advocate and evoke change in their communities and know that they are capable and deserving of their seat at the table.

Karen Canales Salas, Little Village Environmental Justice Organization (LVEJO):

As was mentioned above, I have been involved with this project since 2012. Now as a community organizer, I've noticed that young people generally want to take a step back from talking about politics. We've noticed that at times, it can feel as though politics is something that you talk about when you're able to vote, so it can feel intimidating to young people or people who are not citizens. However, through the popular education work that we do, we try to show them that everything around us is political. From the clothes that we wear, the way we wear them, to the businesses we're supporting. Every decision is a political one.

In YPS, it allows us to do science politically without even realizing it. Students are actively engaged in a social justice issue without even realizing how impactful the work is in the moment. In my case, I was doing the science and wasn't necessarily thinking I'd become a community activist or an organizer, but it happened naturally. YPS gives students a direct way to engage more meaningfully with science because of the social justice issue. I transferred to my neighborhood high school, which ironically enough is named, Social Justice High School, for the last two years of my high school career. I had failed general chemistry as a sophomore at a magnet high school; the only reason why I took AP Chemistry is because I didn't want to disappoint my environmental science teacher. Because I had failed general chemistry, I had a lot of self-doubt before I even started the school year. I thought to myself, "How can I go into AP Chemistry after not even knowing what a mole is?" Regardless of this self-doubt, I ended up doing really well in AP Chemistry and part of that was because I was actually engaged in something that meant more than just chemistry problems.

My neighborhood was *my* classroom. That was really important for my success in the class and as a high school student doing the project. In this context, I was actually more into the science content instead of the social justice issue. However, it wasn't until mid-way through the school year, or even a little bit later, that I started making the connection that it truly is my responsibility now that I have gained these new skills and new knowledge to be able to communicate it with my community. A lot of the language that we use when we talk about heavy metal contamination can go over our heads sometimes. It can be very empowering for me, and students, to be able to communicate the scientific language and jargon to community members that maybe don't have access to that. We had a community science night at the end of the school year where LVEJO had the Illinois Environmental Protection Agency present, and their presentation was very technical, and I think went over a lot of our heads. I remember feeling as though people were more engaged when the group of us students presented our work. Seeing young people there, listening to their own children and actually able to understand a lot of the stuff that the EPA did mention but in a different way... it was really empowering. It was because of this class and through my walks and drives around the neighborhood that I started noticing that there was a lot of industry in my community. What's my responsibility to that?

Now that I work with LVEJO, I'm able to go back to classrooms and provide the sociopolitical context that I didn't want to talk about as a student myself. It's also important for me to talk about and teach community resistance and the agency that community members have, whether or not they have degrees, to create change. Little Village is also predominately a Latinx community, and earlier I mentioned how and in what ways community folks been able to fight against powerful, huge, polluting industries, and win. Our base members may not all be citizens, but that's never stopped them from being civically engaged in campaigns. They know that whether or not they have legal documentation, they're still able to make change. Politicians don't ask a crowd of protestors whether they can vote or who they'll vote for, they just see and hear a big crowd, and they're pressured to listen to community members. We recognize that community still has power whether or not they're able to vote, or whether or not you choose to vote. I'd like to believe that a lot of the ways that we encourage folks to be civically engaged is through resistance. We can share grassroots victories from our cities to show students how being civically engaged strengthens our community resistance.

Daniel:

The relationship between learning to use scientific evidence and becoming politically engaged is complex for me, even though there have always been connections between the two in my life and my work. Communities with political clout don't have to find evidence that an unwanted polluter is doing harm—they just say they don't want the polluter there. So, by even participating in the collection of evidence of pollution, we're legitimizing that it's up to the community to find that evidence. To take an example from neighborhood where I live and Tomasz teaches, we don't want a plant that manufacturers catalysts for the petrochemical industry and spews out molybdenum compounds with uncertain health ramifications along the way. Why is the burden on the largely Mexican, working-class community to find evidence of harm? Why doesn't the multinational corporation that owns this plant have to provide evidence to the surrounding community that what they're doing is safe and sustainable?

CONCLUSION

Among various ways to conceive of equity in science education, we find the most promise in connecting science learning with movements for social justice (Philip & Azevedo, 2017). At the same time, our work has taught us that forming these connections is complex, especially within schools. One complexity is navigating equitable ways to deal with differentials in power, resources, and constraints between schools, community organizations, universities, and the people who learn or work within them (Tolbert et al., 2018). A related complexity is considering the ways in which studying problems of environmental justice in science class reproduces "regimes of evidence" (Liboiron et al., 2018). We engage those responsible for pollution, or those who have failed to properly regulate it, on their terms. This strategy is often more reproductive than transformative.

Despite these challenges, our panel concluded the session with words of encouragement for others who might take up this sort of work. Among the panelists, one common reflection is how YPS has given us all an opportunity to be more responsive to the communities where we live and/or work. YPS is one way to value the ways of knowing and lived experiences our students bring into the classroom. It is one way to push back against the ways that environmental racism has shaped the landscapes of our city. During the Q&A session, we emphasized that projects like ours require patience, optimism, and determination, but also that they can inspire hope. We try to create spaces for learning, acting, and reflecting that build on each other's expertise in meaningful ways. A central part of this work is to flatten differences and imposed hierarchies between participants. We view sharing in this format as a small step in that direction.

Acknowledgements We would like to acknowledge the youth, teachers, and scientists who have contributed to our work but do not appear in this chapter, including Kenneth Booker, Tiffany Childress Price, Darrin Collins, Maribel Cortez, Alanah Fitch, Shelby Hatch, Elizabeth Herrera, Amy Levingston, and Kathryn Nagy, among others. We also want to acknowledge the advisory roles of Julio Cammarota, Ashaki Rouff, and Sara Tolbert. Sara graciously served as the facilitator for the panel discussion captured in this chapter.

Funding The work discussed in this paper has been supported by the National Science Foundation Grant #1720856. Any opinions, conclusions, or recommendations expressed in this material are those of the authors and do not necessarily reflect the views of the National Science Foundation or collaborators who are not listed as authors.

References

Cammarota, J., & Fine, M. (2008). *Revolutionizing education: Youth participatory action research in motion.* Routledge.

Irwin, A. (1995). *Citizen science: A study of people, expertise and sustainable development.* Psychology Press.

Liboiron, M., Tironi, M., & Calvillo, N. (2018). Toxic politics: Acting in a permanently polluted world. *Social Studies of Science, 48,* 331–349. https://doi.org/10.1177/0306312718783087.

Morales-Doyle, D., Childress-Price, T., & Chappell, M. (2019). Chemicals are contaminants too: Teaching appreciation and critique of science in the era of NGSS. *Science Education, 103*(6), 1347–1366. https://doi.org/10.1002/sce.21546.

Morales-Doyle, D., & Frausto, A. (2021). Youth participatory science: A grassroots curriculum framework. *Educational Action Research, 29*(1), 60-78. https://doi.org/10.1080/09650792.2019.1706598.

Philip, T. M., & Azevedo, F. S. (2017). Everyday science learning and equity: Mapping the contested terrain. *Science Education, 101*(4), 526–532. https://doi.org/10.1002/sce.21286.

Tolbert, S., Schindel, A., & Rodriguez, A. J. (2018). Relevance and relational responsibility in justice-oriented science education research. *Science Education, 102*, 796–819. https://doi.org/10.1102/sce.21446.

Part IV

Science Education for a World-Yet-to-Come

Learning from Flint: How Matter Imposes Itself in the Anthropocene and What That Means for Education

Catherine Milne, Colin Hennessy Elliott, Adam Devitt, and Kathryn Scantlebury

James Lovelock in *Gaia: A New Look at Life on Earth* (1995/1979), described how his idea for Gaia emerged when he was working at the Jet Propulsion Laboratory in California in 1965. Collaborating on a team that was tasked with developing a series of experiments to look for life on Mars as part of the Voyager project, the planned experiments consisted of an automated microbiological laboratory designed to sample the Martian soil to see if it was suitable for simple microscopic life and exploratory examinations of Martian soil to see if the soil contained chemicals that might indicate "life at work," including amino acids and optically active substances (p. 2). After a year, Lovelock

C. Milne (✉)
Department of Teaching and Learning, New York
University, New York City, NY, USA
e-mail: catherine.milne@nyu.edu

C. H. Elliott
Department of Instructional Technology and Learning Sciences,
Utah State University, Logan, UT, USA
e-mail: che217@nyu.edu

A. Devitt
California State University, Stanislaus, Turlock, CA, USA
e-mail: adevitt@csustan.edu

K. Scantlebury
Department of Chemistry and Biochemistry, University of Delaware,
Newark, DE, USA
e-mail: kscantle@udel.edu

© The Author(s) 2022

M. F.G. Wallace et al. (eds.), *Reimagining Science Education in the Anthropocene*, Palgrave Studies in Education and the Environment, https://doi.org/10.1007/978-3-030-79622-8_15

245

started to ask questions that had not specifically been a focus of the scientists collaborating on this project including, "How do we know that life on Mars will reveal itself to tests based on life on Earth? What is life and what tests will tell us that there is life?" When other scientists there asked him how one might answer these questions, he suggested they would look for entropy reduction, since the ability to organize is a feature of living things. For Lovelock, this discussion initiated an ongoing exploration of the relationship between the material and the living in the world. In the 1960s, Lovelock was already grappling with the issue of the entanglement of the geo and the bio, matter and life, which has become of the mainstream focus of scientists like Robert Hazen (2012), who argue that life and matter are entangled and that the evolution of matter and life have shaped the Earth—an Earth which humans often seek to claim for themselves. Lovelock's thinking might lead you to think that science has not only been considering the entanglement of life and matter but has begun to move in that direction. However, as we show in this chapter, drawing on the effectivity of water as a solution, humans tend not to notice matter unless it brings an effect upon them.

EFFECTIVENESS, AGENCY, AND THE ANTHROPOCENE

In 2000, Paul Crutzen (1995 Nobel Prize co-winner in chemistry for his research on ozone depletion) and Eugene Stoermer (2000) published a statement, "an affirmation that our species belongs amongst other biological and geological forces" and with that claim, they proposed "the term 'Anthropocene' for the current geological epoch" (p. 17), superseding the current geologically accepted Holocene ("Recent Whole") epoch. Their rationale for a new epoch was based on an analysis of the impact that the expanding *human* population was having on earth's resources. Simon Lewis and Mark Maslin (2015) claim that it was Crutzen and Stoermer's argument that the Anthropocene had already begun, which galvanized an increase in the usage of this idea and term. As Stacy Alaimo (2019) notes, the Anthropocene, as a claim for human exceptionalism, is both illuminating and misleading; illuminating, because it acknowledges the implications for global phenomena of taking seriously the intra-action between living (human) and nonliving and calls into question the separation of nature and culture that structures modernity, and misleading because it grants humans an exceptional status that is not assigned to any other living thing.

Of course, the presentation of "the Anthropocene" as part of the geological timescale is easy to critique, especially if it seems to come from a position of human hubris. Lewis and Maslin (2015) acknowledge that from a geological perspective, human activity is fleeting and recent. However, there can be no denying that human activity is speeding up the rate of change in a number of different contexts. If we accept that the issue is not only that science is coming around to the idea that the ontology of the Earth is an outcome of bio, geo, and chemo intra-actions but also that humans are not central, exceptional, or separate from nature, then what are the implications of such a stance?

Isabelle Stengers (2010, p. 7) notes, the category of human "has never been neutral because it entails human exceptionalism at its crudest.... From this standpoint, the very drastic opposition between humans and nonhumans would then itself be the witness of the unleashed power of this (nonhuman) idea that made us humans, as it allowed us to claim the exception, to affirm the most drastic cut between those beings who 'have ideas' and everything else, from stones to apes." So, it was the unleashed power of the non-human that allowed humans to claim this undeserved exceptionalism? According to Sarah Whatmore (2006, p. 602) the "livingness" of the world shifts matter, "from indifferent stuff out there, articulated through notions of 'land,' 'nature,' or 'environment' to the intimate fabric of corporality that includes and redistributes the 'in there' of human being." In *Inhuman Nature*, Nigel Clark (2011, p. 15) argued that the proposal for the Anthropocene constituted for Crutzen and Stoermer (2000) requires more than an appreciation that humans are part of the biological foment. It requires that scientists invest in a version of relationality in which observable realities are understood in terms of entanglements that are inescapable and inseparable, mutually interdependent, and relationally co-constitutive. However, as Melinda Benson (2019) notes, in science and in the everyday world, "the dominant ontology reinforces a familiar binary—one in which humans are separate *from* and doing things *to* nature" (p. 252). This ontological stance, which emerged during the Enlightenment, underlies environmental laws in the United States. As Benson observes, this ontology, and the associated laws, make assumptions about agency, which is assigned to only humans, and exceptionalism, which we have already noted. If humans persist with this ontological stance then cultural, government, and political strategies designed to address environmental challenges are likely to be limited in their effects.

Benson (2019) argues for the need to engage in practices and processes that "more accurately reflect our lived experience of the material world" (p. 254). Indeed, since the 1980s Earth system modeling has recognized the interrelation between bio (including humans), geo, and chemistry elements. However, when researchers argue for the need for Earth system models "to address explicitly the inter-actions between the ecosphere and the 'anthroposphere'" (Scholze et al., 2012, p. 131), one is left to acknowledge that the underlying stances of agency and exceptionalism remain unexamined and unproblematized in much of this work. In an exploration of the Rio Grande Forest System, Benson explores the need to decenter the human and "rather than seeing wildfires, flooding and other events as problems to be solved" (p. 276) or as "disasters," beginning with an exploration of the roles they play as actors in the system with humans in relational engagements with all elements and processes. In this case, we take Jane Bennett's (2010) notion of agency as effectivity rather than intentionality seeing agency as relational, when entities, material and human, intra-act, engendering phenomena that may be unpredictable. Benson sees the proposal of the Anthropocene not just as a geological epoch but as an opportunity to restructure human intra-action in the space time mattering of the Earth (Barad, 2007).

A MOLTEN, MORE-THAN-HUMAN WORLD

However, for many people matter only comes to their attention when they encounter an ontological disturbance where matter forces itself on human experience and thus is noticed. The more-than-human world becomes "molten" and leaks into everyday experiences as the material world imposes itself on the everyday world of most human agents (Whatmore, 2013). In this chapter, we use the water crisis of Flint, Michigan, to explore when the material leaked into humans' everyday world to cause an ontological disturbance through water and other elements of the material more-than-human world, such as lead ions, legionellosis (Legionnaire's disease), corrosion control, and water pipes.

Flint is a city in the "rust belt" of the United States with a majority Black population. It was once a thriving economic city (although people knew little of the ecological costs) in an area that was the industrial heartland, where steelmaking and associated industries fueled the country's economic growth in the mid- to-late twentieth century. At one time, Flint was a focus of industry. In 1908, General Motors (GM) was founded in Flint, but by the end of the twentieth-century GM had closed most of its factories, which had a flow-on effect to other industries associated with the auto industry and small businesses supporting the community (Clark, 2018a). Nearly 50% of the populace left Flint, and one consequence was a decline in tax revenue for local government to run utilities and maintain infrastructure. Concurrent with a decline in local tax revenues, the state of Michigan reduced fiscal support to its cities, and by 2008, the Great Recession had a devastating impact on government budgets.

THE COST OF WATER

Not too long after the financial crisis hit the state of Michigan hard, Rick Snyder—a venture capitalist with no governmental experience—was elected governor in 2010. Early in his term, he declared the City of Flint in financial crisis and appointed an emergency manager, which effectively got rid of local governance for any fiscal decision-making. In early 2013, the managers of Flint decided to change the source of its water from the Detroit Water and Sewage Department (DWSD) to a pipeline from the newly formed Karegnondi Water Authority (KWA), which planned to extract water from Lake Huron and pipe the water to cities in order to save money. Centering the water in analysis of the cascading policy moves reveals that the consequences of this decision were not just a material-human connection but unfolding agency of the Flint River water infrastructure, which co-constitute each other in the toxic phenomena that followed.

The cities were responsible for water treatment. However, the construction of the new pipeline was not scheduled until 2016, and the state-appointed emergency manager for the City of Flint decided that rather than continuing with the expensive DWSD, the city would use Flint River water treated at the defunct water treatment plant (Olsen & Fedinick, 2016). In 2014, the defunct plant was reactivated and the switch from DWSD to Flint River water was made (Dingle, 2016). There were two main issues with this decision. Firstly, the Flint River is more corrosive than the water from Lake Huron, raising issues of corrosion control which should have been addressed before the water was released to citizens, and secondly, the treatment plant required refitting and updating and the appointment of skilled staff. Within weeks of the switch to Flint River water, Flint residents reported that their water had turned brown, tasted metallic, smelled foul, and people were experiencing skin rashes and hair loss (Clark, 2018a). Further, some pediatricians like Dr. Hanna-Attisha began recognizing high lead levels in Flint children's' blood. Yet, only Flint residents in the direct path of the contaminated water recognized the agency of the water and their entanglement with it. As Dr. Hanna-Attisha argues in her book about the crisis (2018), testing the children's lead levels is actually a test of their lived environment, including the water infrastructure. This intra-action led to a recognition of a debilitating problem that required immediate action. Yet the decision-makers failed to act. It was not just lead in the water that flowed out of Flint residents' tap but failed governmental policy and woeful disregard for the agency of residents. The events illustrated environmental racism. It took local activism and national recognition through President Obama's eventual declaration of an emergency to spur local officials into action.

For Flint residents, the water was leaking into their everyday experiences in unexpected but real ways. As Clark (2018b) noted, when the people of Flint complained that their tap water was odorous and made children sick, it took officials 18 months to accept that there was a problem, even as doctors, led by Hannah-Attisha, began noticing higher levels of lead in children's blood. The phenomena that emerged from the intra-actions between the dissolved chemicals and humans was initially rejected by state and local officials (Olsen & Fedinick, 2016). Although high lead levels can exist without a change in water color, Flint's brown water was possibly the first indication of lead contamination (Torrice, 2016). A timeline of key events helps to communicate just how quickly the material world imposed itself on Flint (Flint Water Advisory Task Force [FATF], 2016). The FATF laid the blame for the Flint water event with replacing of local representative decision-making with state-appointed emergency managers. In April 2014, the City of Flint began distributing treated Flint River water to its customers[1] and soon after residents begin complaining

---

[1] According to the Flint Advisory Task Force (FATF, 2016), the Michigan Department of Environmental Quality (MDEQ) determined that corrosion control, as legally required by the Federal Environmental Protection Agency's (EPA) lead and copper rule, was not

about its odor, taste, and appearance. On July 1, 2014, Flint started its first six-month monitoring period for lead and copper in the drinking water. By August 15, 2014, discovery of *E. coli* bacteria violation led to a local boil water advisory and a month later, the Michigan Department of Environmental Quality requested an evaluation for trihalomethanes (disinfection byproducts) in the water. By October 1, 2014, the Genesee County (the county in which Flint is the largest urban center) Health Department communicated its concern to Flint Public Works regarding the detected increase in Legionnaires' disease cases since April 2014. In October 2014, General Motors announced that it would cease to use water from the Flint WTP, citing corrosion concerns related to chloride levels in water from Flint WTP. Of course, for Flint one of the issues was that the material which was implicated in the phenomena that Flint residents reported is submicroscopic and therefore difficult to engage with. Also, the phenomenon that residents and the water constructed could easily be ignored by city and state managers because, effectively, they were not elected officials and not directly answerable to the residents.

The agents that were responsible for all the observed phenomena are still not completely understood. In some cases, even if residents lived in relatively new houses that had plastic piping into the house, the piping that distributed water from the treatment plant used lead or iron pipes. These pipes require the formation of a phosphate preservation layer to stop the lead or iron from the pipes dissolving into the drinking water. This preservation layer is created through the continued use of a phosphate corrosion inhibitor (PO_4^{-3}), which has the goal of stopping lead or iron from the pipe dissolving into the water as lead ions (e.g., Pb^{2+}, Pb^{4+}) and iron ions (e.g., Fe^{2+}, Fe^{3+}). The phosphate does this by reacting with the iron or lead in the pipe to form lead (II) phosphate or iron (III) phosphate that forms a crust on the inside of the pipe. This inhibitor needs to be added as an ongoing process because without the continual supply of phosphate, the crust corrodes. When Flint's city officials decided to use water from the river, they failed to add the phosphate corrosion inhibitor, even though they were legally bound to do so because of the federal Copper and Lead Rule, which limits the levels of copper and lead in residential tap water.

Compared to the water from Lake Huron, Flint River water contained more particles, ions, organic matter, and microbes. Flint River water was naturally high in chloride (Cl^-) ions (Dingle, 2016). As soon as Flint water was used, any iron pipes in the water distribution system began corroding as the iron phosphate in any remaining preservation layer reacted with the chloride ions, releasing iron (Fe) ions into the water. These iron ions then reacted with chlorine (Cl_2), (different from chloride ions) that was added as a disinfectant to kill any microorganisms in the water. The chemical reaction between

required immediately but instead the Flint Water Treatment Plant would carry out 2 six-month monitoring periods and then the MDEQ would make a determination about the need for corrosion control.

iron ions and the chlorine in solution meant that the amount of chlorine was reduced, which forced the humans "overseeing" water quality to add more chlorine to control the coliform bacteria detected in the water (the presence of which indicated contamination from human and other animal fecal waste). While additional chlorine did control the bacteria, it also initiated the formation of trihalomethanes. The chemistry of trihalomethane formation is a little complex. When chlorine gas dissolves in water it forms a hypohalous acid. In water, this acid exists in equilibrium with hypohalite ion. In the case of chlorine, this would be hypochlorous acid ($HOCl$) and hypochlorous ion (OCl^-). At low concentrations of the halide ion (Cl^-), formed when chlorine reacts with water, the reaction stops with the formation of $HOCl$ and OCl^-. The formation of the trihalo ion, Cl_3^-, which can form when more chlorine is available, does not take place to a significant degree (Boyce & Hornig, 1983). Both $HOCl$ and OCl^- are powerful disinfectants for removing coliform bacteria. However, the addition of more chlorine in a more basic environment initiates a reaction between chlorine and chloride with the resulting formation of trihalomethanes, in this case trichloromethane. The same types of reactions occur with all halogen elements, most commonly chlorine and bromine. The presence of trihalomethanes was concerning because of their association with negative health effects and reproductive health (Hood, 2005). Specifically, with everyday tap use, absorption through the skin and inhaling these chloromethanes can strongly affect the level of trichloromethanes in the blood.

Flint city officials did not take matters seriously, resulting in negative consequences for the material and the people of Flint. Officials did not attend to the possible issues this water was creating for the people of Flint, the same way that General Motors did for its engine parts. General Motors, because it was of interest to their business, listened to the phenomena that their materials were producing and they switched their water supply. The state and city officials in charge of making decisions for the residents of Flint did not listen to the ways the materials in the water and their residents were speaking to them, even as the material world had become molten for residents, imposing itself on their everyday lives.

At each stage of the process from the water that was chosen, the treatments applied to the water, the pipes that transferred the water to homes, and the people and other animals that used the water as a source of nutrition and cleanliness, the matter was taken for granted. Human responses to the materials coursing through the water pipes of Flint constituted phenomena depending on how the material-discursive practices were enacted and how cuts were made to establish the boundaries. If the management had acted on water treatment as a phenomenon in which agents come together, this might have resulted in a more holistic approach to water delivery and treatment. As such, the goal of water delivery and treatment would have acknowledged the role played by ions, water, pipes, residents, and their bodies as belonging to a system of water functioning instead of prioritizing management efficiency and fiscal prudence. As this example illustrates, the submicroscopic agents in water have their own ideas about how they will react under certain circumstances. Perhaps this case is also suggestive of how science educators might rethink their construction of systems thinking and the Anthropocene more specifically. An earth system approach offers the possibility of removing humans from the center of the narrative, while engaging in the intricacies of the systems that produce phenomena like water running out of a tap.

IMPLICATIONS FOR SCIENCE EDUCATION

The so-called Flint Water Crisis is an illustrative microcosm of the increasing number of global challenges delineating human-Earth-chemistry intra-activity. The Flint Water Crisis, in combination with other examples, such as the Australian wildfires, Amazon deforestation, and most currently, COVID-19, epitomize the onto-epistemological stance of human exceptionalism under-girding human activity and worldviews. At the same time, we need to accept that all participants, living and nonliving, have agency and therefore participate in a complex and dynamic system that is distributed across multiple and overlapping bodies (Bennett, 2010). As we have argued throughout this chapter, we are living in a new epoch, the Anthropocene, where nature and culture are inherently entangled—yet, science and current visions of science education fail to engage with this worldview. Rather than constructing Flint and other examples as emergencies or crises that need to be solved, in a worldview where nature and culture are entangled, education would explore the dynamic nature of these events and the intra-actions of all elements. We need to transform what science education looks like for both developing scientists and everyday citizens.

In the USA, the public K–12 educational system uses national standards to set the vision leading both curricular and instructional reforms. The current enacted vision of science education in K–12 education is dominated by the Next Generation Science Standards (NGSS) (Lead States, 2013). In the NGSS, disciplinary science ideas and the practices of scientists (i.e., human activity) are presented as dis-entangled dimensions of science learning. NGSS aims to connect humans with nature through an exploration of disciplinary core ideas, the practices of scientists and engineers, and the relationships of crosscutting concepts among traditionally separated disciplines (Lead States, 2013). While NGSS attempts to reconcile the role of humans in the natural world, it still promulgates a unidirectional impact of humans on the earth and maintains human exceptionalism in that the standards only focus on science that centralizes the human and human agency in the world (Maslin, 2015). For example, consider the report, *Using Phenomena in NGSS-Designed Lessons and Units*, which begins with the statement, "Natural phenomena are observable events that occur in the universe and that we can use our science knowledge to explain or predict" (NGSS Lead States, n.d., p. 1). Note that, in this description, phenomenon exists naturally to be observed and explained by humans using science. We advocate a need to understand the perspective of the phenomena; in other words, rather than looking for phenomena from pre-defined constructs, understanding how phenomena emerge out of situations like the Flint Water Crisis and how phenomena is emergent as humans and the material world are entangled.

A new vision of K–12 education needs to incorporate the role of intra-activity, nature, and human activity entanglements, making the emergent nature of phenomena explicit rather than telling learners what the phenomena is and then expecting them to recreate that example of phenomenon. For example, the human-nonhuman inputs in the phenomena of the Flint Water Crisis unfolded as the community intra-acted with the water and its unseen infrastructure, which only became "molten" and seeped into the aware-ness of the humans when the issues became visible and tangibly changed humans' lives. Flint water and the submicroscopic agents that were intra-acting with human systems and initiating a range of emergent phenomena had negative implications for human health as a consequence of how the water came together with the pipes and how the political structures in the state of Michigan created positions for people to impose a new water source without community accountability. This series of intra-actions culminated in the contaminated (poisonous) water (sometimes brown in color, but also sometimes undetectable by the naked eye) that came out of Flint residents' taps.

Student learners under the present conception of NGSS, where life and physical science are dis-entangled from human activity, will most likely be inad-equately prepared to understand the complexity of such challenges. While an important goal of NGSS is to prepare students to be problem-solvers (Lead States, 2013), essentialized knowledge and practices will deprive them of the criticality, knowledge, experience, care, and ethic they need to analyze the nuances involved in this situation and prioritize action. Learning continues to be theorized and practiced as a cognitive activity—where the "object" has an essence to become known by the "subject." By separating object and subject, this epistemology establishes the need for "representations" or concepts that are as like the object as possible but which need to be learned. This approach to education continues to influence how K–12 science education and teacher education principally theorize how science is taught. The goal for students is to provide symbolic representations of requested phenomena. Particularly from anthropocentric epistemology, we no longer can pretend that phenomena "exist" in an objective form nor that students can learn about pre-determined phenomena in an objective totality.

Hetherington et al. (2019) take a material-dialogic stance for reframing science learning and, subsequently, suggest nuanced explanatory rationales for explaining why common K–12 teaching tools and strategies are fruitful for teaching science. Explanatory frameworks for student learning, such as "visual learning," "making learning hands-on," or "adds motivation" saturate K–12 teachers' and teacher education discourses, whereas explanatory frameworks rooted in Barad's (2007) work would emphasize the role of material agencies intra-acting in the formation of student knowledge and pedagogical decisions. To illustrate, for example, the necessary intra-actions between phenomena and humans, Hetherington et al. (2019) rearticulate the role of students' gesture-making during activity is evidence of material agency, "talking-back" (Schoen, 1996) shaping students' thought processes prior to typically desired symbolic representations of the phenomena (i.e., definitions, descriptions, formulas). From a dialogic-materialist perspective, gestures, which reverberate the voice of material, are recognized as significant communicative devices, whereas according to current assessment practices, only written, and in rare circumstances verbal, symbolic forms are worthy of evidencing meeting learning objectives.

Ultimately, the teaching and learning, and subsequently the materials used in assessment practices, matter. Likewise, experiments and activities are considered supportive of student learning because of the "hands-on" component but would offer more if the agency of the participating materials were acknowledged. A common experiment for students to explore conservation of mass is through observing ice cubes melting in a closed container and measuring the mass before and after the melting takes place. Yet rarely are learners asked to listen to what the ice cubes and the water and the container are saying to them. We argue that the framework of "hands-on" undermines any consideration of the role of scientists in how they engage in scientific work in real-world conditions, such as Flint Water, because this framework fails to provide a context in which consideration is given to the dynamic interplay between living and non-living in the emergence of phenomena that typically only become acknowledged when they become molten and seep into our awareness as they impose themselves.

Understanding the phenomena created in Flint necessitated a broad understanding of history, racism, classism, politics, and geography to appreciate what was pouring through the tap of residents of the Flint community. It was, in fact, everyday citizens, particularly mothers, who noticed how the phenomena of Flint water "responds back" to the human (Barad, 2012; Hetherington et al., 2019). They were the first to notice the discolored water, the smell, and the resulting rashes and hair loss of using said water. Likewise, auto workers working for General Motors (GM) in 2014 noticed that engine parts were corroding and swiftly switched water sources—an option that was not possible without the economic and political means. Rather than teaching facts, formulas, and demonstrations, we can learn a lot of scientific expertise and practices from people who are most impacted from crises, like the residents in Flint. Material-dialogism values the learning of new onto-epistemological frameworks that include social configurations, spatial relations, purposes, and the agential role of material resources. Such additional frameworks drastically enhance how teachers could reconfigure and be reconfigured when teaching science.

LISTENING TO THE WATER AND THE PEOPLE

The officials' failure to respond to the challenge of Flint water was because they devalued the lives of the city's African American residents. Moreover, those residents whose economic lives were already impacted by the city's declining fortunes were excluded from the decision-making process, which included brutal financial accounting and fiscal austerity. Ironically, in late August 2020, the state of Michigan proposed at $600 million settlement for Flint's residents (Bosman, 2020).

Bullard (2001) shows us that this kind of inequitable policy decision-making is consistent throughout environmental policy in the United States and around the world. Consistently, lower income neighborhoods and neighborhoods that are majority Black are exposed to more pollution and more detrimental environmental factors due to governmental policies and corporate decision-making. The Flint Water Crisis is another example of environmental racism (Bullard, 2001) and classism that emerges from the intra-action between histories of racial injustice, outdated infrastructure, and corporate/government decision-making for fiscal interests over humane conditions. Therefore, we believe a systems approach that deepens perspectives on how the human and material agents produce unfolding phenomena would actually lead to a more humanizing perspective for these kinds of policy decisions and further scientific practice. This would inform data collection systems for lead testing of humans that would offer easier integration across providers (Hanna-Attisha, 2018), as well the ways we teach and learn scientific knowledge making processes. So instead of just building "hands-on" science pedagogies, science education and education in general should be building pedagogies that critically engage in material agencies intra-actions and the political nature of knowledge together. What the Flint Water Crisis also shows us is that scientific perspectives are important for building critical narratives about ongoing policy and infrastructure issues, which the residents and activists took up because their health depended on it. Science standards, like the NGSS, therefore must take into account the unfolding, complex, and intricate ways that science, politics, and ethics unfold currently and a vision of change for a more onto-ethico-epistemologically focused future (Barad, 2007).

REFERENCES

Alaimo, S. (2019). Your shell on acid: Material immersion, Anthropocene dissolves. In R. Grusin (Ed.), *Anthropocene feminism* (pp. 89–120). University of Minnesota. https://doi.org/10.5749/j.ctt1m3p3bx8.

Barad, K. (2007). *Meeting the universe halfway: Quantum physics and the entanglement of matter and meaning*. Duke University Press.

Barad, K. (2012). Nature's queer performativity. *Women, Gender and Research, 1*(2), 25–53.

Bennett, J. (2010). *Vibrant matter: A political ecology of things*. Duke University Press.

Benson, M. H. (2019). New materialism: An ontology for the Anthropocene. *Natural Resources Journal*, *59*, 251–280.

Bosman, J. (2020, August 19). Michigan to pay $600 million to victims of Flint water crisis. *The New York Times*. https://www.nytimes.com/2020/08/19/us/flint-water-crisis-settlement.html.

Boyce, S. D., & Hoernig, J . F . (1983). Reaction pathways of trihalomethane formation from the halogenation of dihydroxyaromatic model compounds for humic acid. *Environmental Science & Technology*, *17*, 202–211.

Bullard, R. (2001). Environmental justice in the 21st century: Race still matters. *Phylon*, *49*(3–4), 151–171.

Clark, A. (2018a, July 3). 'Nothing to worry about. The water is fine': How Flint poisoned its people. *The Guardian*. https://www.theguardian.com/news/2018/jul/03/nothing-to-worry-about-the-water-is-fine-how-flint-michigan-poisoned-its-people.

Clark, A. (2018b). *The poisoned city: Flint's water and the American urban tragedy.* Picador.

Clark, N. (2011). *Inhuman nature: Sociable life on a dynamic planet*. Sage.

Crutzen, P. J., & Stoermer, E. F. (2000). The "Anthropocene." *International Geosphere Biosphere Program Global Change Newsletter*, *41*, 17–18. http://www.igbp.net.

Dingle, A. (2016, December). The Flint water crisis: What's really going on? *ChemMatters*. https://www.acs.org/content/acs/en/education/resources/highschool/chemmatters/past-issues/2016-2017/december-2016/flint-water-crisis.html.

Flint Water Advisory Task Force. (2016). Final report. Accessed on July 14, 2021 at http://flintwaterstudy.org/wp-content/uploads/2016/03/Flint-task-force-report_2438442_ver1.0.pdf.

Hanna-Attisha, M. (2018). *What the eyes don't see: A story of crisis, resistance and hope in an American city*. One Worlds.

Hazen, R. (2012). *The story of Earth: The first 4.5 billion years, from stardust to living planet*. Viking.

Hetherington, L., Hardman, M., Noakes, J., & Wegerif, R. (2019). Making the case for a material-dialogic approach to science education. *Studies in Science Education*, *58*(2), 141–176. https://doi.org/10.1080/03057267.2019.1598036.

Hood, E. (2005). Tapwater and trihalomethanes: Flow of concerns continues. *Environmental Health Perspectives, 113*(7), A474.

Lewis, S. L., & Maslin, M. A. (2015). Defining the Anthropocene. *Nature*, *519*, 171–180. https://doi.org/10.1038/nature14258.

Lovelock, J. (1995/1979). *Gaia: A new look at life on Earth*. Oxford University Press.

NGSS Lead States. (2013). *Next generation science standards: For states, by states*. National Academies Press. www.nextgenscience.org/next-generation-science-standards.

NGSS Lead States. (n.d.). *Using phenomena in in NGSS-designed lessons and units.* https://www.nextgenscience.org/resources/phenomena.

Olsen, E., & Fedinick, K. P. (2016). *What's in your water? Flint and beyond*. Natural Resources Defense Council.

Schoen, D. (1996). Reflective conversation with materials. In T. Winograd, J. Bennett, L. Deoung, & B. Hartfield (Eds.), *Bringing design to software* (pp. 171–184). Addison-Wesley.

Scholze, M., Allen, J. I., Collins, W. J., Cornell, S. E., Huntingford, C., Joshi, M. M., Lowe, J. A., Smith, R. S., & Wild, O. (2012). Earth system models: A tool to understand changes in the Earth system. In S. E. Cornell, I. C. Prentice, & J. I. House (Eds.), *Understanding the earth system: Global change science for application* (pp. 129–156). Cambridge University Press.

Stengers, I. (2010). Including nonhumans in political theory: Opening Pandora's box? In B. Braun & S. Whatmore (Eds.), *Political matter: Technoscience, democracy, and public life* (pp. 3–33). University of Minnesota Press.

Torrice, M. (2016). How lead ended up in Flint's tap water. *Chemical and Engineering News, 94*(7), 26–29.

Whatmore, S. (2006). Materialist returns: Practicing cultural geography in and for a more-than-human world. *Cultural Geographies, 13*(4), 600–609.

Whatmore, S. (2013). Earthly powers and effective environments: On ontological politics of flood risk. *Theory, Culture & Society, 30*(7–8), 33–50.https://doi.org/10.1177/0263276413480949.

Resurrecting Science Education by Re-Inserting Women, Nature, and Complexity

Jane Gilbert

INTRODUCTION

I am a female Pākehā[1] New Zealander who has worked in science education for four decades, first as a high school science teacher and later as a university teacher and researcher. However, I have never really "belonged" in science or science education. In earlier years, this was just a feeling; however, it was a feeling I set out to explore in postgraduate study, first in linguistics, then feminist theory, political theory, and science education. Here I again found myself on the margins, an outsider to the intricacies of academia, but by then I had decided to see this marginality as a strength, a space from which to see things differently. I have a long-standing interest in what we now call "diversity issues" in science and science education, but I am critical of conventional strategies for attracting women and/or other marginalised groups into science. I am old enough to have seen the same strategies rolled out repeatedly with little discernible effect on the problem. In this field the issues tend to be conceptualised at the surface-level, and the "other" question in science

---

[1] The Māori word Pākehā is used in New Zealand to describe the descendants of the European settlers (mainly British) who have come to New Zealand over the last 150 or so years. Māori are Aotearoa-New Zealand's *tangata whenua* (indigenous 'people of the land').

[2] By the "other" question, I mean work exploring how science *conceptually* excludes or "others" certain major classes of human. See, for example, the work of Evelyn Fox Keller (1985, 1992).

J. Gilbert (✉)
Auckland University of Technology, Auckland, New Zealand

M. F.G. Wallace et al. (eds.), *Reimagining Science Education in the Anthropocene*, Palgrave Studies in Education and the Environment, https://doi.org/10.1007/978-3-030-79622-8_16

has received little attention.[2] In this chapter I want to argue that the coming of the Anthropocene could—and should—change this. Picking up the editors' invitation to think differently about science education as it is now, this chapter's starting point is that science-as-we-know-it can't provide solutions to the issues we now face because it is part of the problem. Re-imagining science education for the Anthropocene needs to involve much more than improving public understanding of science, especially climate science, or political activism, conceived of within the current conceptual system. This chapter argues against these strategies. Proposing deconstruction as a frame for envisaging—and resurrecting—science education for the Anthropocene, it advocates a pedagogical approach based on deconstructing science-as-we-now-know-it. The chapter argues that if we are to think our way out of the situation we're now in, we need to "unpack" the conceptual system that led us into it.

THE ANTHROPOCENE

The term Anthropocene came to prominence in the first years of the twenty-first century, when the atmospheric chemist Paul Crutzen proposed, at a geologists' conference, that planet Earth has left the Holocene and entered a new geological epoch that is defined by the effect of human activities, not just on other living things, but on the Earth's deeper physical processes. This new epoch, dubbed the Anthropocene (from the Greek "anthro" meaning "human"), is the result of the widespread burning of fossil fuels since the time of the Industrial Revolution in Europe. Burning carbon sequestered over hundreds of millions of years from the atmosphere, via living processes, has vastly increased atmospheric carbon dioxide levels, which has in turn triggered a steady rise in mean global temperatures. This is expected to have a major impact on world sea levels, weather systems, and ecosystem stability, which will affect the habitability of the planet for humans and have major implications for human social, political, and economic life (Hansen, 2009; Klein, 2014; Kress & Stine, 2017; McNeill & Engelke, 2014; Scranton, 2015). These changes are already happening, but, as widely discussed elsewhere, we have not yet managed to put in place measures that could reverse or delay these trends, nor have we developed strategies for adapting to or mitigating their likely effects (Flannery, 2005; Hamilton, 2010; Jamieson, 2014; Oreskes & Conway, 2014). Many now argue that we are in a "climate emergency"—that urgent action is required if we are to avert abrupt catastrophic change.

The Anthropocene discourse originated in science. Groups of scientists, using scientific language and evidence, used the term to persuade non-scientists to put in place policies and protocols to address its causes. However, the concept was quickly taken up by scholars in the arts, humanities, and social sciences. Analysis of the intellectual implications of the Anthropocene is now well underway. Arguments are being made for new "post-carbon" philosophies (e.g., Irwin, 2010) and for new social, political, and economic theories (e.g., Elliott & Turner, 2012; Klein, 2014; Newell & Patterson, 2010; Urry,

2011). However, some scholars argue that, because the crisis we're now in is a direct consequence of capitalism, Capitalocene is a more appropriate term (e.g., Moore, 2016).

The Anthropocene's arrival has significant implications for education (and science education in particular), but scholarly exploration of these implications is only just beginning (the present volume notwithstanding). This chapter looks at how the Anthropocene challenges science education and explores how it could catalyse change. However, first it examines how the Anthropocene is portrayed by scientists in their interactions with policymakers, arguing that the story they tell is not a helpful basis for re-imagining science education.

The scientists' Anthropocene story emphasises evidence-based predictions, targets, and demands for urgent action. This rhetoric is accompanied by apocalyptic stories of collapse if something isn't done, or alternatively, by "it's already too late" stories. While there are good reasons for using this language in this context, this construction of the Anthropocene, if it is picked up and used "as is" in education, has several problems.

The first problem is that this story constructs the Anthropocene as an engineering and/or a policy problem that can be solved using existing ways of thinking. This construction misses the point. The Anthropocene names a new epoch in the history of planet Earth. The term was invented to denote a significant rupture with the past. It signals the advent of new systems and processes that quite possibly will not be comprehensible using current ways of thinking. However, more importantly for the present purposes, the circumstances the Anthropocene names have been *caused* by actions that arise from and are informed by current ways of thinking. The second problem with the scientists' Anthropocene story is that it reinforces the widely held idea that science and technology *are* the future: they are what will "save" us from the problems we face. But science and technology don't, in themselves, shape our future: they are guided by human values, choices, and actions (Slaughter, 2012). And while technological mitigations for climate change will undoubtedly be developed (Kolbert, 2018), thinking this way sends us down *one* possible pathway to the future, closing off other options (Facer, 2013; Inayatullah, 2008).

The third problem with the prevailing climate change story is that it is profoundly anthropocentric and Western-centred. It exists in a filter bubble which puts humans front and centre and reifies their agency. Humans are constructed as separate from nature, active, autonomous subjects who can make meaning about, act on, and master an essentially passive nature. This draws on—and reproduces—the thinking system that is the source of the problem. The story also expresses the interests and worldview of particular humans in particular countries, obscuring the interests and worldviews of other groups of humans, as well as those of non-human living things and the non-living things with which/whom we share planet Earth (Haraway, 2016). It obscures the fact that most of the world's humans play a very limited role in contributing to climate change,[3] and it reinforces the idea that while we live "on" planet Earth, we are not part of it, that we are entitled to take what we want and to conquer and control it.[4]

A fourth aspect of the prevailing Anthropocene story that is unhelpful is that it sets up either/or choices. *Either* we succeed in saving the planet and human life on Earth can continue, *or* we don't and humans (and a great many other species) are totally eradicated. There are of course other possibilities. Catastrophic events causing mass extinctions of other species have happened many times in the Earth's history.[5] Humans on planet Earth have experienced catastrophic events before and survived, often inventing completely new ways to be human. This could happen again, and it is possible that this could actually be positive.[6] Completely new ways of thinking about what it means to be human could emerge, ways that we can't imagine from within existing thinking systems. The prevailing Anthropocene story allows only two possibilities—success or failure. If we don't do X, Y will inevitably follow. Why only two possibilities? What *other* possibilities are we avoiding thinking about in the present circumstances? What other possibilities can we *not see* in the present circumstances? Can we think *outside* the prevailing story?

In this chapter I argue that these "urgent action or collapse," "science will save us" stories are underpinned by the set of assumptions that have created the problem[7] and, because of this, they aren't a helpful basis for re-imagining science education. Using these stories uncritically will produce science education with a focus on teaching students *about* climate change, engaging them in "climate action," and/or encouraging them to "contribute to the cause" by considering science-related careers.[8] These approaches will reproduce the thinking systems that have produced the problem. In the current circumstances, this would be deeply *mis*-educative, in the sense meant by Dewey

[3] According to a recent Oxfam study (Gore, 2015), 50% of the world's carbon emissions are produced by 10% of the world's population.

[4] This is in contrast to the reciprocal relationship with nature assumed by many non-dominant groups of humans (e.g. Kimmerer, 2013).

[5] A well-known example of such a catastrophe is the asteroid strike sixty-six million years ago that sparked global firestorms, followed by a nuclear winter-like cold (caused by the smoke), that caused the mass eradication of 75% of the planet's species, including the dinosaurs. A new, radically different, world order eventually emerged: the age of mammals and birds replaced the age of the dinosaurs (Lee, 2020).

[6] At the time of writing, we are in the grip of the global COVID19 pandemic. While at this point in time it is hard to see the positives, there are, in my country anyway, perceptible shifts in thinking. There is talk, not of going "back to normal," but of a "new normal" in which deep expertise and "essential workers" are newly appreciated while mis-information and social inequalities are not. The ubiquity of international travel is being challenged, as are many long-held assumptions about educational "delivery.".

[7] These assumptions are outlined later in the chapter.

[8] See, for example, https://educatorsdeclare.org/resources/.

(1938).[9] Other stories are needed, stories that allow us to see science, the Anthropocene, and ultimately science education differently. Drawing on work in the philosophy and social studies of science (not science itself), in the next section I explore two alternative Anthropocene stories.

SEEING SCIENCE AND THE ANTHROPOCENE DIFFERENTLY

Bruno Latour, in his 2013 Gifford Lectures, argues that the Anthropocene's arrival is a significant challenge to science-as-we-know-it. It requires a major shift in how we think about science: what it *is*, what it is *for*, with what and whom it should *engage*, and how it should do this (Latour, 2013).[10] Building on his long-term work on how scientists think about—and "do"—science (e.g., Latour, 1993), Latour argues that scientists need to rethink their relationship with nature, to see it, not as something to be tamed, objectified or "deified," something we are "apart from," but rather as something in which we are inextricably entangled, embedded, and connected. This shift, he argues, will require completely new ways of thinking, new tools that allow us to investigate nature, not as a set of "entities" to be understood and controlled, but as constructed and reconstructed in reciprocal relationships with science (and scientists). He argues for a focus on this relationship, on the spaces or "crossings" between science and nature.

Donna Haraway, on the other hand, argues against using the Anthropocene concept (or the Capitalocene) to think our way out of the situation we're in. She argues that both discourses assume—and reproduce—the binaries of Cartesianism[11] and using them can only, as she puts it, "end badly." For her, both terms too easily lead to cynicism and defeatism, to "game over, too late" thinking (Haraway, 2016). Instead, she proposes a new concept, the Chthulucene,[12] as a positive way forward. In her new Chthulucene age, human entanglement with all other living and non-living things on earth is acknowledged, not denied.

> Unlike the dominant dramas of Anthropocene and Capitalocene discourse, human beings are *not the only important actors* in the Chthulucene, with all the other beings able simply to react. The order is reknitted: human beings are

---

[9] In *Experience and Education* (1938), Dewey argues that "educative" experiences are those that open up possibilities for active, ongoing intellectual growth, that is, the capacity to think in increasingly complex, abstract ways. M*is*-educative experiences, on the other hand, constrain, distort or arrest intellectual growth (p. 25).

[10] See also: http://www.modesofexistence.org.

[11] The next section has an explanation of Cartesian binaries.

[12] The term Chthulucene, invented by Haraway, is derived from the name of a Californian spider (*Pimoa cthulhu*), which in turn comes from the language of the Goshute people of Utah (Haraway, p. 31). For Haraway the Chthulucene signifies entanglement, everything's connection to everything else.

with and of the earth, and the biotic and abiotic powers of this Earth are the main story. (Haraway, 2016: p. 55)

Haraway's Chthulucene concept denotes a new way of thinking, a new way of doing/making things. It rejects anthropomorphism and the *anthropos*, the autonomous, rational, outcome-focussed possessor of agency and knowledge. Instead, subjectivity, knowledge, and agency are seen as emerging in multi-species collaborations, in what she calls "sympoiesis," or "making-with," a process of breaking down and re-making the old, using it in new ways, to do new things. For her:

> The unfinished Chthulucene must collect up the trash of the Anthropocene, the exterminism of the Capitalocene, and chipping and shredding and layering like a mad gardener, make a much hotter compost pile for *still possible* pasts, presents, and futures. (Haraway, 2016: p. 57)

Latour, Haraway, and many other contemporary theorists[13] make a strong case for the urgent need to find ways to think *outside* the old paradigms, to develop new modes of thinking that can allow new ways of doing things to emerge. However, this is incredibly difficult. Our thinking has been formed, structured, and colonised by the existing conceptual frameworks, to the extent that it appears as though this is just "how things are," that this is "all there is." Anything that can't be shoehorned into the existing frameworks can't be thought. Because it is unrepresentable, "uncomputable" (Bridle, 2018), the "left-over" material is treated as though it doesn't exist. It is invisible to the system, unwanted "excess" or "waste" (Irigaray, 1985, p. 30). We can't simply reject and/or replace the prevailing conceptual frameworks: we are part of them, and we can't think outside them. All we can do, to use Derrida's (1991) term, is to put them "under erasure": signal that they are problematic, that they may eventually need to be erased, while at the same continuing to work with—or around—them.

These difficulties are further compounded in educational contexts. A key goal of education is to foster intellectual growth, traditionally achieved by exposing learners to increasingly complex forms of knowledge. Knowledge is generally regarded the "raw material" for thinking: we "think with" knowledge (Willingham, 2019). But, if the knowledge we are exposed to "formats" our thinking in certain very specific ways, our intellectual growth is channelled and constrained in ways that make it very difficult to think "other-wise." Moreover, whatever—or whoever—was "excess" to this knowledge is excluded right from the start. These are of course not new problems: however, attending to them is now urgent as we try to imagine what being educated might look like in the Anthropocene.

---

[13] For example, Braidotti (2013), Barad (2007).

The purpose of this chapter is to oppose the use of the prevailing Anthropocene story in our attempts to re-imagine science education. Instead, drawing on the alternative readings proposed by Latour, Haraway, and others, I want to make the case for an approach that is based on deconstructing science-as-we-now-know-it. In the next section I outline how I plan to use the term deconstruction and provide an example of how science-as-we-now-know-it could be deconstructed. I then propose a pedagogical approach based on deconstruction, arguing that this approach, unlike business-as-usual science education, could be genuinely "educative."

Deconstructing Science-As-We-Know-It: How Women, Nature, and Complexity Were Left Out

The deconstruction concept, while common in the humanities and social sciences and occasionally found in education, is rarely used in science-related contexts. Deconstruction's purpose is change, particularly in relation to idea-systems, and in situations where these idea-systems are seen to be oppressive. It is a process for trying to break out of, and see beyond, the conceptual categories that, at a very deep level, structure the way we think. Deconstruction involves looking below the surface of a conceptual system to examine its key concepts and how they work together to form a coherent narrative. It also involves looking at what these concepts were built on or from, and what has been excluded or disallowed to make the system work. Doing this, its protagonists argue, is enough to produce change (Grosz, 1989; Lather, 1991; Davies, 1994). Deconstruction is different from analysis or critique. Its aim is *not* to take apart, refute, or destroy existing conceptual systems: rather, it is to *work with* these systems, but in new ways. The purpose is to open up spaces between the existing categories from which it is possible to see the system—and think—differently. In what follows I draw on scholarly work in feminist theory, political philosophy, economic history, and the history and philosophy of science to attempt a deconstruction of science-as-we-know-it. This material forms the background to the pedagogical approach that follows.

The development of capitalism and then science over the last five hundred years or so has produced a very specific way of thinking about—and organising—the relations between humans and the rest of nature (Patel & Moore, 2018). The success of both capitalism and science rests on the idea of humans as separate from—and superior to—nature. This idea first appeared in Western European thought in the 1600s and is an organising principle of much of modernist thought, including the sciences, politics, economics, and the other social sciences. It is so embedded in modern thought that it seems self-evident, obvious, and "natural." However, while it is an abstraction, invisible to most, this idea has deep material effects. It has affected how humans have thought about, organised, and dominated each other, how they have lived on planet Earth, and how they have affected other living and non-living things, including, now, the Earth's fundamental geological processes. One of these effects is to exclude, at the conceptual level, women, nature, and complexity from this thought system. This removes their agency and power, and allows them to be, as Patel and Moore (2018) put it, "cheapened" to serve the interests of a particular class of humans. The human-nature split idea has been hugely successful for this class of humans, but it has also produced the planetary emergency we now face.

The idea of humans as not-nature originates in the work of the seventeenth-century philosopher René Descartes (1596–1660).[14] For Descartes, reality is made up of "thinking things" (*res cogitans*) and "extended things" (*res extensa*). Humans are "thinking things," and nature is made up of "extended things." However, for Descartes, not every human was a "thinking thing." Specifically excluded were women, indigenous/colonised peoples, slaves, and servants. For Descartes, these classes of person were not fully human, but part of nature. "Thinking things" are rightfully the masters and possessors of nature (which includes those considered not fully human), and nature is something to be controlled, dominated, and known. Thinking, in Descartes' schema, is reason, the exercise of "pure intellect." It is the functioning of a mind defined by its ability to radically separate itself from the bodily substrate that nourishes and supports it *and* from the matter it contemplates. At around the same time, Francis Bacon (1561–1626), a philosopher widely characterised as the "father" of modern science, was arguing that understanding nature is achieved by "attending to" or "courting her" so that "she reveals her secrets." This understanding, he thought, would allow man to exercise his rightful dominion over nature (thought of as feminine).[15]

Bacon's empiricism—looking for patterns in nature—and Descartes' rationalism—pure reasoning in a mind radically divorced from nature—are foundational to modern science. However, the thinking on which these foundations sit (and the implications of this thinking) is only really visible to historians and/or philosophers of science. The invisibility of science's conceptual foundations to most working scientists, science educators, science policymakers, and the general public has allowed science to be widely thought of not only as "representing" nature, but as if it *is* nature, while scientists are thought of as "not-nature," able to master, control, and use it. This view of science, originating as it does in the Baconian/Cartesian worldview, is based on some important exclusions. One of these, important for the present purposes, is that only "thinking things" can be the "subjects" or "knowers" of scientific knowledge. All non-thinking things—that is, the classes of human listed earlier, non-human living things, and non-living things (soils, rocks, rivers, oceans, weather, and so on)—are thought of as part of nature and therefore

[14] The following account of Descartes' thinking draws material from Lloyd (1993) and Tuana (1993).

[15] This outline of Bacon's ideas is informed by Merchant (1980, 2008), Keller (1985) and Lloyd (1993).

the "objects" of scientific knowledge. These "extended things" have no independent agency: they are non-rational matter, to be studied, understood, acted on, controlled, and dominated (Keller, 1985; Irigaray, 1987; Sartori, 1994).

Modern political and economic thought is founded on the same principles. Because, at the conceptual level, the abstract individual actor of the political/economic sphere is a Cartesian "thinking thing," women, nature, and complexity are excluded. As Anna Yeatman puts it "some individuals are more individual than others" (Yeatman, 1988). The rational, autonomous, choice-making individual of the modern public sphere is, at the conceptual level, a white, male, property-owning individual. All other categories of person—women, non-property-owning men, servants and indigenous/colonised peoples—are conceptually part of the domestic sphere and/or nature. They are not "fully" individual: they are part of, subsumed into, and controlled by the male head-of-household (or coloniser) who is the political/economic individual. While two hundred or so years of activism has produced formal, surface-level equality and women (and indigenous/colonised peoples) can now participate in science and public life, at the deepest conceptual level, they cannot "really" be the "knowers" of science or actors in the public sphere.[16] Neither can non-human living things and non-living things: like women, indigenous, and working-class peoples, they are Descartes' "extended things." conceptually part of nature.[17]

The Cartesian "revolution" was also crucial to the success of capitalism. Patel and Moore (2018), in their history of capitalism, identify four transformations that formed the world we know today and continue to shape thinking.

> First, either-or binary thinking displaced both-and alternatives. Second, [the Cartesian revolution] privileged thinking about substances, things, before thinking about the relationships between those substances. Third, it installed the domination of nature through science as a social good. Finally, the Cartesian revolution made thinkable, and doable, the colonial project of mapping and domination. (p. 54)

[16] This section draws on material in Pateman (1988, 1989); Yeatman (1988); Flax (1990); Gutman (1980).

[17] It might seem odd to argue for formal "human" rights for non-human living things and non-living things: however, in Western-influenced political systems, this is the only available strategy for redressing claims of injustice. This strategy is being actively pursued by some indigenous and environmental groups: for example, in 2017 the New Zealand Parliament officially recognised the Whanganui River as a living being with legal personhood status and is soon to do the same with Te Urewera forest and Taranaki mountain. See: https://www.nationalgeographic.com/culture/2019/04/maori-river-in-new-zealand-is-a-legal-person/. Similarly, Lake Vattern in Sweden was recently recognised as a living being with legal personhood rights. See: http://www.naturensrattigheter.se/2019/05/12/verdict-for-the-tribunal-of-the-rights-of-lake-vattern/. However, it could be argued that this strategy reproduces the Cartesianism that created the injustice.

For Patel and Moore these four transformations are far from innocent. They are "undetonated" forms of symbolic violence that reflect the interests of the already powerful and licence them to organise the world in ways that suit those interests (2018, p. 47). Patel and Moore show how the mapping of nature, organising the world into grids which then became reality, allowed it to be measured, enclosed, known, conquered, and, importantly, owned. Knowledge of nature was authored and authorised by European men, and all other forms of knowledge of nature were classified as witchcraft or folklore.[18] The privatisation of land, along with the proletarianization of human labour (turning human activity into labour that can be bought and sold) were central to capitalism's success. Nature, land, and human (and non-human) labour were turned into "cheap things" to be exploited, turned into money, then capital. This cycling of nature into money and then capital has brought us to the point in history that Patel and Moore call the Capitalocene.

Cartesianism also underpins the "computationalism" that has produced the digital technologies and the internet that organise today's world (Bridle, 2018). Bridle defines computational thinking as the

> extension of what others have called solutionism: the belief that any given problem can be solved by the application of computation. . . . Computational thinking supposes—often at an unconscious level—that the world really is like the solutionists proposes. It internalises solutionism to the degree that it is impossible to think or articulate the world in terms that are not computable. (Bridle, 2018, p. 4)

Describing its origins in the mapping of nature, meteorology, and the antecedents of today's digital platforms, Bridle shows how computational thinking now structures nearly everything, so much so that anything that cannot be computed is excluded and effectively invisible. For him, this invisibility is the most striking—and dangerous—feature of today's computational "regime." Reality has been replaced by digital models of it, which because they simulate reality by simplifying it, selecting certain elements to include and leaving others out, are flawed. Leaving out the complexities of the situation being modelled (many of which are unknown) inevitably means that models are not especially successful in predicting the future. However, Bridle argues, their ubiquity in today's world has led to an inability to distinguish between simulations and reality. Models are now so pervasive that they are effectively reality, just "how things are." They no longer stand for, frame, or shape today's culture: operating beneath our awareness, they *are* culture.[19] Bridle's concern is that as we think more and more in the channels provided by machines,

---

[18] Linnaeus's system for naming and classifying living things, now taken as reality, as naturally ordained, is a paradigm example of Cartesianism.

[19] Bridle here cites the extent to which knowledge is now defined by Google and relationships are now defined by Facebook.

we are losing our capacity to think deeply, or even to think at all.[20] He also worries that computationalism is constructing futures that fit *its* parameters, modelled on (a selection of) past events. Excluding the uncomputable narrows our field of possible futures: it *colonises* our futures with current thinking. This is not a good way to think about, in, or for the Anthropocene.

If, as I am arguing here, the conceptual system we are embedded in, in particular, its invisibilising of the "uncomputable," has produced the present crisis, and if it is not actually possible to think outside this system, *is* there a way forward? In the final section of this chapter I propose a pedagogical approach that I think could help to navigate these difficulties.[21] This approach draws on the work of the Belgian-French feminist philosopher and psycho-analytic theorist Luce Irigaray. Irigaray's work offers an approach that is very different from most Anglo-American feminist theory, and, while its focus is the conceptual exclusion of women, it can be used to think about the exclusion of other groups.

READING "BETWEEN THE LINES"

Irigaray argues that the conventional representation of sex and sexual differ-ence is *not* actually a system of difference: rather it is what she calls "a Logic of the Same." Within this system there is *one* sex, *one* sexuality, *one* form of subjectivity, and so on. The category "woman" is defined in relation to the category "man" as whatever "man" is not. The result of this is that it is not possible to think of "woman" as a separate, self-defining, independent category, and it is not possible to simultaneously be a woman *and* the author-itative subject of knowledge (Whitford, 1991; Grosz, 1989). While women can contribute to knowledge, their contributions must be authorised by the "real" subjects of knowledge. Women cannot be "the one": they are always "the other," occupying a position "next to" or in support of male authority, a "substitute" for the "real thing" (Sartori, 1994).

Much of Irigaray's work focuses on how "woman" can be thought inde-pendently of "man," as a completely separate category, developed and defined by women. For her, the problem with developing a separate category is that, at the very deepest level, and from the very earliest stages, our psyche and our thinking are entirely structured by the masculine Symbolic order.[22] Women *and* men only have access to a Symbolic order structured by the

---

[20] See also Wolf (2018).

[21] When I started thinking about this pedagogy I had high school science education in mind. However, this model is applicable wherever the primary purpose is *educational* (in the Deweyan sense of fostering ongoing intellectual growth), as opposed to where the focus is pre-professional training and/or developing specific skills. Suitably modified, the model could be used in elementary, university or informal education contexts.

[22] The Symbolic order concept comes from Lacanian psychoanalytic theory, which posits three (largely unconscious) orders through which human existence is structured. The Symbolic is the realm of language, signs, culture, law and so on; the Imaginary is the

male Imaginary. Because femaleness is not, and cannot be, represented here, there is effectively no foundation on which a specifically female subjectivity could develop. Irigaray argues that if female subjectivity and female authority are to become possible, strategies designed to develop a female Symbolic and a female Imaginary are needed. For her, there are two aspects to this, each of which depends on, and is necessary for, the other. One involves *relationships*: developing new ways for women to relate to, and work with, each other *as women*.[23] The other involves *knowledge*: developing ways to analyse, deconstruct, and refuse the fantasies of the male Imaginary. In educational contexts, this implies developing ways to teach, while simultaneously *also* deconstructing, the traditional subject matter. To address the second of these two aspects, Irigaray proposes a strategy she calls "reading as a woman" (Irigaray, 1985). In what follows I describe this strategy and explore how it could be used to develop teaching approaches designed to make visible what is currently "uncomputable."

Irigaray's "reading as a woman" involves reading the "texts" of a given knowledge system at two levels. Drawing on the psychoanalytic concepts of interpretation and transference, she distinguishes between what she refers to as the "male" and the "female" readings of a text. In interpretation, the analyst/reader "masters" the text and is able to explain the analysand/writer's intended meaning. Interpretation, for Irigaray, is the "male" (or positive) reading of a text. Its aim is to produce coherent, transparent, and verifiable statements that can be applied to other situations. Transference, on the other hand, is a (negative) position of *non*-mastery. It involves paying attention, not to the writer's intended meaning, but to the reading's *effect* on the self of the analyst/reader. It requires the reader to recognise and identify these effects (as an analyst does). This kind of reading involves interaction between the analysand/text and the analyst/reader and it generates *new* meaning. These new meanings will be specific to the situation they were generated in, necessarily contingent, temporary, and ungeneralisable. This, for Irigaray, is the "female" reading. It involves reading "between the lines," looking for the blanks, the negatives, for what has been left out in the masculinist search for "positivity." "Reading as a woman" acknowledges that coherence is an illusion, an illusion produced by eliminating all that cannot easily be defined, quantified, and computed. Irigaray does not claim that the "female" reading is superior. She says that we should engage both, "one with the other," to develop new forms of *genuine* partnership (Irigaray, 1993a; Whitford, 1991).

realm of the ego and unconscious fantasy; and the Real is the pre-linguistic biological substrate we leave as we enter language.

[23] I don't discuss the relationship aspect of Irigaray's model at all here, but for work on this, see Piussi (1990), Cicioni (1989).

Drawing on these ideas I want to suggest an approach to science education that is based on reading the texts of science at three levels.[24] The first level is the "male" or "positive reading" proposed by Irigaray. The aim of this reading is to decode the key concepts of a particular area of science, to comprehend them as they are represented in the current paradigm, and to explore how these concepts are connected up to form this paradigm. For example, if the area of focus was genetics, this reading might focus on the meaning, significance, and connections between cells, chromosomes, mitosis, meiosis, DNA, RNA, transcription, protein synthesis, and so on, as they are understood by biologists. This first-level reading resembles current practice, but its purpose is different. Rather than being an end-in-itself, it is the groundwork on which the second and third readings become possible.

The second-level reading would look "underneath" the concepts examined in the first-level reading. It would explore the wider historical, philosophical, and cultural contexts in which these concepts were developed, and to which they contribute. It would aim to find—and deconstruct—the assumptions and metaphors on which these concepts rest, and through which they are connected to their origins. Using genetics as the example again, this reading might focus on the ways the cell is commonly represented using the metaphor of a hierarchical, command-and-control system or a corporate organisational chart. The cell nucleus, or more specifically, the DNA, is represented as "in control of" and/or "directing" cell processes. These processes are represented as the linear, one-way transmission of information that has come directly from instructions inherited on the parental chromosomes. DNA is thus the "master molecule" of life, exercising its "authority" over cell processes to provide genetic stability, much as the leader of an authoritarian organisation, government, or family might. Similarly, the relationship between the cell's nucleus and its cytoplasm is routinely represented in gendered terms. The nucleus is the masculinised "mind" or "head of household" of the cell, while the cell constituents are its feminised "body," charged with executing decisions made in the nucleus. These representations are metaphors that, while they are easily traced back to science's roots in the philosophies of Descartes and Bacon, persist in today's thinking.[25] They are a way of thinking about biology that comes from the cultural contexts in which biology developed. But embedding these metaphors into biology's conceptual system "naturalises" them. It allows the metaphors to be thought of as if they are not metaphors, but "facts of nature," which in turn disallows other possible metaphors.

The third-level reading corresponds to Irigaray's "female" or "negative" reading. Aiming to read "between the lines" of the apparent "positivity" of

---

[24] This three-level approach draws on the "critical literacies"/"multiliteracies" field, in which literacy is much more than simply the capacity to decode existing texts (see: New London Group, 1996).

[25] This is still the case, despite today's science's acknowledgement of the complexities of the interactions and feedback loops between internal and external cell processes, between genes and their environment, and so on.

the first-level reading, it would search for the negatives, for what is left out: in particular, the hidden relationships and interdependencies that make the first reading possible. The third-level reading's purpose is to disrupt the apparent coherence of the level-one narrative. It also aims to explore the effect of this narrative on the self of the reader (Irigaray's "transference"). A third aim is to explore how the concepts examined in the first-level reading could be read "other"-wise, and how, possibly using a kind of "science fiction" approach, these concepts might be re-presented differently, if they had different foundations. Providing an example of what this third-level reading might look like is less straightforward than it is for the first two readings. However, continuing with the genetics example, reading genetics "other"-wise might explore representations of genes, cells, and so on, not as entities acting on other entities, but as complex, fluid, continuously re-negotiated relationships of exchange, mutual construction, and reconstruction, as Irigaray herself puts it, in partnership, "one *with* the other".[26] Irigaray's "reading as a woman" is different every time it occurs. What emerges from the reading depends on the situation, the interaction between the participants, and the effects of this interaction on the participants. In the introduction to this chapter, I mentioned my younger self's "feeling" of being excluded from science. Noticing, acknowledging, and using this kind of feeling, in interaction with others, is a starting point for generating new meanings, new narratives, and new spaces to be. We can't know in advance what these will look like, but this (almost) doesn't matter. The point of reading in this way is not to replace old narratives: it is to expand our intellectual capacities for life in the Anthropocene.

For Irigaray, this deconstructive work is important because it refuses the fantasies of the masculine Imaginary. It opens up new symbolic spaces in which women, nature, and complexity can represent themselves *as themselves*, not in relation to, subsumed or defined by another, but in partnership, "one *with* the other." It seems to me that this kind of work is needed to make it possible to, as Donna Haraway puts it, "reknit" things so that humans can *conceptualise* themselves as embedded in and connected to nature, as *able to* engage in the kind of multi-species collaboration, the "sympoiesis" Haraway envisages. This work needs to begin in educational contexts, especially, but not only, in science education.

REFERENCES

Barad, K. (2007). *Meeting the universe halfway: Quantum physics and the entanglement of matter and meaning.* Duke University Press.

Braidotti, R. (2013). *The posthuman.* Polity Press.

Bridle, J. (2018). *New dark age: Technology and the end of the future.* Verso.

[26] Irigaray uses biological examples in her discussion of the placenta as a space belonging to "neither one, nor the other", but "one with the other" (Irigaray, 1993b; p. 38–9).

Cicioni, M. (1989). "Love and respect, together": The theory and practice of *Affidamento* in Italian feminism. *Australian Feminist Studies, 10,* 71–83.

Davies, B. (1994). *Poststructuralist theory and classroom practice.* Deakin University Press.

Derrida, J. (1991). Of grammatology. In P. Kamuf (Ed.), *A Derrida reader: Between the blinds.* Harvester Wheatsheaf.

Dewey, J. (1938). *Experience and education.* Collier MacMillan

Elliott, A., & Turner, B. (2012). *Society.* Polity Press.

Facer, K. (2013). The problem of the future and the possibilities of the present in education research. *International Journal of Educational Research, 61,* 135–143.

Flannery, T. (2005). *The weather-makers: How man is changing the climate and what it means for life on Earth.* Text.

Flax, J. (1990). *Thinking fragments: Psychoanalysis, feminism and postmodernism in the contemporary West.* University of California Press.

Gore, T. (2015). *Carbon emissions and income inequality.* Oxfam International. https://oxfamilibrary.openrepository.com/bitstream/handle/10546/582545/tb-carbon-emissions-inequality-_1;jsessionid=1FAA69761CF420EA43BDE6599F7 2DB3C?sequence=2.

Grosz, E. (1989). *Sexual subversions: Three French feminists.* Allen & Unwin.

Gutman, A. (1980). *Liberal equality.* Cambridge University Press.

Hamilton, C. (2010). *Requiem for a species: Why we resist the truth about climate change.* Earthscan.

Hansen, J. (2009). *Storms of my grandchildren: The truth about the coming climate catastrophe and our last chance to save humanity.* Bloomsbury.

Haraway, D. (2016). *Staying with the trouble: Making kin in the Chthulucene.* Duke University Press.

Inayatullah, S. (2008). Six pillars: Futures thinking for transforming. *Foresight, 10*(1), 4–21.

Irigaray, L. (1987). Is the subject of science sexed? *Hypatia, 2,* 65–87.

Irigaray, L. (1985). *This sex which is not one* (C. Porter, Trans.). Cornell University Press (original work published 1977).

Irigaray, L. (1993a). *An ethics of sexual difference* (C. Burke & G. Gill, Trans.). Cornell University Press (original work published 1984).

Irigaray, L. (1993b). On the maternal order. In *Je, tu, nous: Towards a culture of difference* (A. Martin, Trans.). Routledge (original work published 1990).

Irwin, R. (Ed.). (2010). *Climate change and philosophy: Transformational possibilities.* Continuum.

Jamieson, D. (2014). *Reason in a dark time: Why the struggle against climate change failed and what it means for our future.* Oxford University Press.

Keller, E. (1985). *Reflections on gender and science.* Yale University Press.

Keller, E. (1992). *Secrets of life, secrets of death: Essays on language, gender and science.* Routledge.

Kimmerer, R. (2013). *Braiding sweetgrass: Indigenous wisdom, scientific knowledge, and the teachings of plants.* Milkweed.

Klein, N. (2014). *This changes everything: Capitalism vs the climate.* Simon & Schuster.

Kolbert, E. (2018, November 15). Climate solutions: Is it feasible to remove enough CO_2 from the air? *Yale Environment, 360.* https://e360.yale.edu/features/negative-emissions-is-it-feasible-to-remove-co2-from-the-air.

Kress, J., & Stine, J. (Eds.). (2017). *Living in the Anthropocene: Earth in the age of humans*. Smithsonian Books.

Lather, P. (1991). *Getting smart: Feminist research and pedagogy with/in the postmodern*. Routledge.

Latour, B. (1993). *We have never been modern* (C. Porter, Trans.). Harvard University Press (original work published 1991).

Latour, B. (2013, February 25). *The Anthropocene and the destruction of the image of the globe*. Gifford Lecture No. 4. The University of Edinburgh. http://knowledge-ecology.com/2013/03/05/bruno-latours-gifford-lectures-1-6/.

Lee, M. (2020, January 6). Bushfires have reshaped life on Earth before: They could do it again. *The Conversation*. https://theconversation.com/bushfires-have-reshaped-life-on-earth-before-they-could-do-it-again-129344.

Lloyd, G. (1993). *The man of reason: 'Male' and 'female' in Western philosophy* (2nd ed.). Routledge.

McNeill, J., & Engelke, P. (2014). *The great acceleration: An environmental history of the Anthropocene since 1945*. Harvard Belknap Press.

Merchant, C. (1980). *The death of nature: Women, ecology, and the scientific revolution*. Harper & Row.

Merchant, C. (2008). Secrets of nature: The Bacon debates revisited. *Journal of the History of Ideas, 69*(1), 147–162.

Moore, J. (Ed.). (2016). *Anthropocene or Capitalocene? Nature, history and the crisis of capitalism*. PM Press.

New London Group. (1996). A pedagogy of multiliteracies: Designing social futures. *Harvard Educational Review, 66*(1), 60–92.

Newell, P., & Patterson, M. (2010). *Climate capitalism: Global warming and the transformation of the global economy*. Cambridge University Press.

Oreskes, N., & Conway, E. (2014). *The collapse of Western civilisation: A view from the future*. Columbia University Press.

Patel, R., & Moore, J. (2018). *A history of the world in seven cheap things* (2nd ed.). Black.

Pateman, C. (1988). *The sexual contract*. Stanford University Press.

Pateman, C. (1989). *The disorder of women: Democracy, feminism and political theory*. Polity Press.

Piussi, A.-M. (1990). Towards a pedagogy of sexual difference: Education and female genealogy. *Gender and Education, 2*, 81–90.

Sartori, D. (1994). Women's authority in science. In K. Lennon & M. Whitford (Eds.), *Knowing the difference: Feminist perspectives in epistemology*. Routledge.

Scranton, R. (2015). *Learning to die in the Anthropocene: Reflections on the end of a civilization*. City Lights Books.

Slaughter, R. (2012). Welcome to the Anthropocene. *Futures, 44*, 119–126.

Tuana, N. (1993). *The less noble sex*. Indiana University Press.

Urry, J. (2011). *Climate change and society*. Polity.

Whitford, M. (1991). *Luce Irigaray: Philosophy in the feminine*. Routledge.

Willingham, D. (2019). How to teach critical thinking. *Education: Future Frontiers Occasional Paper Series*. https://apo.org.au/node/244676.

Wolf, M. (2018). *Reader come home: The reading brain in a digital world*. Harper Collins.

Yeatman, A. (1988). Beyond natural right: The conditions for universal citizenship. *Social Concept, 4*, 3–32.

CHAPTER 17

Watchmen, Scientific Imaginaries, and the Capitalocene: The Media and Their Messages for Science Educators

Noel Gough and Simon Gough

SCIENTIFIC IMAGINARIES AND SCIENCE EDUCATION IN THE CAPITALOCENE

We write this essay collaboratively because, as father and son, we have first-hand experience of different generations interpreting popular media texts in different ways. However, like Deleuze and Guattari (1987, p. 3), we understand the limits and opportunities of writing together: "Since each of us was several, there was already quite a crowd." Each of us is indeed several, but we agree that some things might be better said with one voice.

Notwithstanding this book's title, we are reluctant to call the present epoch "the Anthropocene." We prefer *Capitalocene*—the age of capital—because, as T. J. Demos (2016, n.p.) writes, "it names the culprit, locating climate change not merely in fossil fuels, but within the complex and interrelated processes of global-scale economic-political organization." With respect to alternative epochal names, Elizabeth de Freitas and Sarah Truman write:

> Concepts like the Capitalocene and Plantationocene remind us that the Anthropocene has been manufactured by a portion of humanity invested in accelerated capitalist accumulation and white supremacy. Science has played a crucial role

N. Gough (✉)
La Trobe University, Bundoora, VIC, Australia
e-mail: N.Gough@latrobe.edu.au

S. Gough
University of Melbourne, Parkville, VIC, Australia
e-mail: simon@simongough.net

© The Author(s) 2022
M. F.G. Wallace et al. (eds.), *Reimagining Science Education in the Anthropocene*, Palgrave Studies in Education and the Environment, https://doi.org/10.1007/978-3-030-79622-8_17

in shaping this "global" condition as a legacy of European imperialism. And yet it would be foolish to deny scientific knowledge as simply serving the white establishment, particularly today under neoliberal post-truth conditions. Science denialism is on the rise, allied with nationalist anti-establishment movements and libertarian free market interests. (2020, pp. 1–2)

Regardless of which term best encapsulates our current crises, we share de Freitas and Truman's (2020, pp. 1–2) interests in "foregrounding speculative fiction as a way to open up scientific imaginaries... to think through the many pasts, presents, and futures of science."

"Scientific imaginaries" are abundant in popular media, especially (but not exclusively) in the storytelling genres signified by "SF."[1] Some audiences see popular artistry as ephemeral and/or inconsequential, but as J. G. Ballard (quoted in Vale & Juno, 1984, p. 155) observes, "pop artists deal with the lowly trivia of possessions and equipment that the present generation is lugging along with it on its safari into the future." We focus on *Watchmen* not only because the novel and film speak to us of issues in science education, but also because they are among the "lowly trivia of possessions" we are "lugging... into the future"—"equipment" that connects us with the world and helps us to make sense of it. We argue that familiarity with (and informed appreciation of) the "lowly" artworks that *Watchmen* exemplifies should be understood as key indicators of a science educator's "cultural literacy."[2] We also suggest that critically appreciative readings of scientifically inflected popular media texts (such as we demonstrate here) are particularly relevant to the multigenerational practice of science teacher education, which typically involves professors, trainee teachers, and the learners they anticipate teaching.

From Clockwork to Complexity: (Re)Connecting Science and Fiction

Modern Eurocentric science (beginning with Copernicus, Brahe, Kepler, Galileo, and Newton) was constructed on the assumptions of empiricism and experimentalism. By the mid-nineteenth century it was typified by Newtonian

---

[1] Donna Haraway (1989, p. 5) writes: "In the late 1960s science fiction anthologist and critic Judith Merril idiosyncratically began using the signifier SF to designate a complex emerging narrative field in which the boundaries between science fiction (conventionally, sf) and fantasy became highly permeable in confusing ways, commercially and linguistically. Her designation, SF, came to be widely adopted as critics, readers, writers, fans, and publishers struggled to comprehend an increasingly heterodox array of writing, reading, and marketing practices indicated by a proliferation of 'sf' phrases: speculative fiction, science fiction, science fantasy, speculative futures, speculative fabulation.".

[2] Coined by E. D. Hirsch (1988), "cultural literacy" refers to the ability to understand and participate fluently in a given culture—in this case, the culture of science education as a multicultural, multigenerational, cross-class, and cross-disciplinary activity. If readers are not already familiar with the two graphic novels to which we refer here (*Watchmen* and *Animal Man*) we strongly recommend doing so before reading further.

physics and mathematics, with the universe likened to a gigantic mechanical clock, continually ticking along with gears governed by Newton's laws of motion and universal gravitation. Scientists and educators alike assumed that science was chiefly a matter of patiently seeking the "facts" of nature and accurately reporting them.

In the late 1880s, the discovery of radioactivity led to a revolution in the goals and structures of physics. As Joseph Schwab (1964, p. 198, his italics) observes, "The new physics... did *not* come about because direct observations of space, place, time, and magnitude disclosed that our past views about them were mistaken." Rather, assertions about these matters changed because physicists found it productive to treat them in a new way—not as matters for empirical verification but as principles of inquiry—conceptual structures that could be revised whenever necessary. Schwab concludes:

> Today, almost all parts of the subject-matter sciences proceed in this way. A fresh line of scientific research has its origin not in objective facts alone, but in a conception, *a deliberate construction of the mind*. On this conception, all else depends. It tells us what facts to look for in the research. It tells us what meaning to assign these facts. (1964, p. 198, our italics)

In other words, many of the interpretations and explanations that constitute "reality" and our experience of it are not "facts" (as empiricist science conceived them) but meanings fashioned by human agents: that is, they are fictions. "Science" and "fiction" do not exist in separate domains but are culturally connected. This is not simply a matter of science and literature finding common meeting places in SF, or of scientific imaginaries being translated into literary themes, a practice that long preceded the early twentieth-century emergence of SF as a distinctive literary mode in popular culture.[3] In her study of scientific field models and literary strategies, Katherine Hayles (1984, p. 10) concludes that "the literature is as much an influence on the scientific models as the models are on literature," there being a two-way traffic in metaphors, analogies, and images between them.

The emergence of chaos and complexity as foci of scientific speculation provides a relatively recent example of SF incorporating the leading edges of scientific inquiry. Ilya Prigogine's investigations during the 1960s and 1970s explain how complex, far-from-equilibrium systems spontaneously transform themselves into new levels of complex organization. Prigogine's model of self-organizing systems as "dissipative structures" reconciles several contradictions in twentieth-century science, including the divergent models of physical function provided by entropy versus evolution and the different roles and attributes of time in physics and biology. The cosmological significance of Prigogine's work was recognized by his receipt of the Nobel Prize for chemistry in 1977, but his work was not published in English in any popular form until 1984,

---

[3] For example, Copernican cosmology permeates the poetry of John Donne (1572–1631). For other examples see Brian Stableford (2003).

when *Order Out of Chaos* (Prigogine & Stengers) was translated from the French.

Stories about chaos and complexity began to appear in popular media during the mid-1980s (e.g., Atkinson, 1985), at about the same time as their applications in educational philosophy and theory were beginning to be explored (e.g., William Doll, 1986). James Gleick (1987) popularized chaos theory with *Chaos: The Making of a New Science*. Given this chronology, we agree with David Porush that.

> it is a tribute to the general intuition of SF, and in particular the long-distance imaginative radar shown by A. A. Attanasio, that in his extravagant and lavishly imagined tour de force, *Radix*, Prigogine's theories make a crucial, if cameo, appearance. Attanasio must have seized very quickly upon Prigogine's work . . . in order to have abstracted some of its essential implications . . . in a novel that was published as early as 1981. (1991, p. 372)

Other SF authors to give imaginative form to Prigogine's work and its successor projects include Bruce Sterling (1985, 1989), William Gibson and Bruce Sterling (1991), Lewis Shiner (1988), and graphic novelists Alan Moore and Dave Gibbons (1987).

Prigogine's thinking catalyzed highly original interdisciplinary work in astrophysics, biology, biophysics, chemistry, ecology, economics, education, management, neurology, particle physics, thermodynamics, and traffic studies. However, it had little or no effect on the "textbook science" of late-twentieth -early twenty-first century school science curricula, despite many of Prigogine's ideas and their implications being accessible through SF.

WHY COMICS/GRAPHIC NOVELS?

Our engagement with comics/graphic novels began with Noel's longstanding interests in the potential contributions of SF to curriculum studies (Gough, 1991) and in particular to science and environmental education (Gough, 1993a). His monograph, *Laboratories in Fiction: Science Education and Popular Media* (Gough, 1993b) offered a vision for science education that rejected the "textbook science" of late-twentieth-century schooling, which retained a nineteenth century conceptualization of science as a study of the material structures of simple systems. *Laboratories in Fiction* was received enthusiastically by US science educators and curriculum scholars (see, e.g., Appelbaum, 2019; Weaver, 1999, 2019). Noel did not argue that SF is what Catherine Hasse (2015) calls "a motivating fantasy," that is, as "bait" on a "hook" that lures and lands learners in the flawed representations of late-twentieth-century textbook science. Rather, he demonstrated that SF gives imaginative form to the limits of our socially constructed knowl-edges (including whatever might lie *beyond* those limits) and thereby opens up conceptual territories in which to explore scientific imaginaries in more accessible ways than conventional science textbooks.

Seeking examples of SF beyond adult-oriented books and films, Noel looked for works oriented to younger audiences. Simon's fascination with the 1987–1996 children's animated series, *Teenage Mutant Ninja Turtles*, which premiered in Australia during 1989 (the year after Simon's birth), led Noel to the graphic novel on which the series was based (Eastman & Laird, 1986), which in turn led him to other comics and graphic novels that interrogated scientific imaginaries, including *Animal Man* (Morrison et al., 1991) and *Watchmen* (Moore & Gibbons, 1987).

For example, *Animal Man* reimagines superhero myths as it chronicles the adventures of a sometimes over-zealous, sometimes self-doubting animal-rights activist with the power to take on the capabilities of animals with whom he comes into contact (e.g., flight). Early issues rework the conventions of superhero comics in a sinister tale of scientific research corrupted by agents of the capitalist military-industrial complex, and includes an implicit comment on Ruth Bleier's (1988) question: "Lab coat: Robe of innocence or klansman's sheet?" Bleier identifies the contradictory meanings of the white coat as a scientific imaginary:

> It is the lab coat, literally and symbolically, that wraps the scientist in the robe of innocence—of a pristine and aseptic neutrality—and gives him, like the klansman, a faceless authority that his audience can't challenge. From that sheeted figure comes a powerful, mysterious, impenetrable, coercive anonymous male voice. (1988, p. 62)

In one episode (Morrison et al., 1991, Ch. 1, p. 27), a scientist's spectacles appear to be opaque, contributing to his "mysterious, impenetrable" presence, but later in the story (Ch. 3, p. 17), when challenged to tell the truth, he *removes* this symbol of detachment and objectivity, thereby undermining the mystique of what scientists actually *do* in pursuing "truthful" accounts of the world.

WHY *WATCHMEN*?

Watchmen initially appeared as 12 issues of a comic-book series between September 1986 and October 1987. It features a self-contained narrative requiring no prior knowledge, with new characters and a setting separate from other publications in the DC Comics universe. When packaged as a "graphic novel" it was widely acclaimed as a groundbreaking work of SF (Van Ness, 2010, pp. 8–15).

Watchmen is a dark satire on superhero[4] mythologies and US politics. One of its central characters is Dr. Manhattan, an aptly named superhuman physicist. Manhattan is the reincarnation of Jon Osterman, a nuclear scientist who is materially "disassembled" when accidentally irradiated by subatomic particles. In a sequence of three pages (1987, Ch. IV, pp. 118–120), *Watchmen* depicts Osterman's reconstruction through two intertwining—and to some extent contradictory—metaphors, one of which borrows from Albert Einstein's (alleged) rueful reflection on his role in the release of atomic power: "if only I had known, I should have become a watchmaker" (Ch. IV, p. 138), despite his role in overthrowing Newtonian mechanics. Osterman (whose father is a watchmaker) repairs a friend's wristwatch shortly before his demise, and his resurrection as Dr. Manhattan (foreshadowing and remembering his transformation by reference to watchmaking) is depicted as "just a question of reassembling the components in the correct sequence" (Ch. IV, p. 119). But other visual and verbal cues suggest that his transformation can be understood as a metaphor of emergence in complex systems, insofar as Osterman's disassembled components can be interpreted as a dissipative structure that progressively reorganizes itself (shown as an emerging sequence of a neural network, a circulatory system and a partially muscled skeleton) into a higher level of complexity represented by Dr. Manhattan's superpowers. In this sequence of words and images, Dr. Manhattan's ambiguous genesis can be interpreted as symbolizing the contesting paradigms of modern and contemporary science: of the deterministic mechanics of Newton's "clockwork universe" versus the unpredictable dynamics of complex self-organizing systems.

Watchmen's brief (but conceptually rich) interpretation of competing scientific paradigms is a stark reminder that although the explanatory power of complexity has transformed many disciplines, it has had relatively little impact on science education. For example, the current specifications for Australia's national science curriculum (ACARA [Australian Curriculum Assessment and Reporting Authority], 2020a) make no mention of complexity as a key *scientific* concept, as distinct from suggesting it as a criterion for judging the quality of a student's learning relative to achievement standards: "Inferences can be drawn about the quality of student learning on the basis of observable differences in the extent, complexity, sophistication and generality of the understanding and skills typically demonstrated by students in response to well-designed assessment activities and tasks" (ACARA, 2020b).

---

[4] Few comic-book "superheroes" actually possess superpowers. Those that do are likely to be aliens (Superman), mutants (X-men), or physiologically altered by a laboratory accident (Spiderman, The Flash). Most have deliberately enhanced their physical powers by harnessing fictional and/or advanced technologies (Batman, Iron Man, Wonder Woman). As crime fighters, they often function as masked vigilantes. In the alternative USA of 1985 depicted in *Watchmen*, masked vigilantes have been outlawed and only one character, Rorschach, continues to wear a mask in defiance of the law. *Watchmen's* "heroes" refer to themselves as "adventurers".

The Sciences of *Watchmen*

Brent Fishbaugh (1998, p. 191) argues that the significance of the sciences is signaled in *Watchmen's* first chapter, wherein Moore and Gibbons subtly depict the advanced technologies existing in their alternative 1985, where people drive electric cars (we later learn that Dr. Manhattan made this possible) and airships travel the skies between buildings:

> It is in the heroes themselves, however, that Moore proposes his primary question: Is humanity responsible and humane enough to properly use science? As such, he personifies the sciences within the major characters and through the text, asks the reader if placing the power of various sciences in the hands of the subject morality and wisdom of human beings is a wise idea.

Fishbaugh (1998, p. 194) interprets the major characters as "exact personifications" of various sciences. For example, Rorschach, the first character that readers encounter as he investigates the Comedian's murder, is linked to the sciences through psychology, as his name (and mask) implies:

> Rorschach is the epitome of soft science not only in his obvious connection to psychology but in his subtle connections to it as well. Two easily recognized examples of this link are revealed in his relationship with his psychiatrist and in the way he is shaped by his environment. (p. 194)

Fishbaugh (1998, p. 194) considers Dr. Manhattan to be "at the opposite end of the science spectrum; where Rorschach represents the soft, personal, somewhat subjective sciences, [Manhattan] represents the cold, hard, true mathematical and chemical sciences." This is illustrated in a sequence (Ch. I, p. 29, emphasis in original) in which Rorschach asks Manhattan if he is "concerned about Blake's [the murdered Comedian's] death," who replies, "a live body and a dead body contain the same number of *particles*... structurally there's no discernible *difference*... life and death are unquantifiable *abstracts*... why *should* I be concerned?".

Fishbaugh (1998, p. 196) also argues that Adrian Veidt (previously known as masked adventurer Ozymandias) personifies a melding of the hard and soft sciences by embodying soft sciences (especially history), while manipulating hard ones (especially genetic engineering) to achieve his plans. However, we suggest that from the standpoint of science educators, Fishbaugh's exaggerated "personifications" risk reinforcing stereotypes of "hard" and "soft" science, and see little merit in debating whether (or not) any of the major characters' personifications of science are "exact." We see more educative possibilities in examining the ways in which *Watchmen* illustrates dilemmas of science education. For example, the novel includes expository materials (appended to chapters 1–11), fictitious documents from within *Watchmen's* world which help readers to understand the chronology of events or reveal details of the adventurers' private lives. One such document is an article

"reprinted from the [non-existent] *Journal of the American Ornitholog-
ical Society*" (Ch. VII, p. 241) ostensibly written by Daniel Dreiberg (an
ornithologist who was the masked adventurer Nite Owl II). The article begins:

> Is it possible, I wonder, to study a bird so closely, to observe and catalog its
> peculiarities in such minute detail, that it becomes invisible? Is it possible that
> while fastidiously calibrating the span of its wings or the length of its tarsus, we
> somehow lose sight of its poetry? . . . I believe that in approaching our subject
> with the sensibilities of statisticians and dissectionists, we distance ourselves
> increasingly from the marvelous and spell-binding planet of imagination whose
> gravity drew us to our studies in the first place.

In its entirety, the mock scholarship of Dreiberg's article (it mocks many
subjects including itself, science, and the discursive forms it both emulates
and criticizes) is a lucid critique of identifying-naming-collecting-measuring-
classifying-dissecting[5] approaches to bird-watching (and other "scientific"
studies of nature). Moreover, unlike a "real" scientific journal article, it is
written in language that is accessible to people of all ages.

SIMULTANEITY: THE MESSAGE
IN *WATCHMEN*'S (1987) MEDIUM

The comic/graphic novel *Watchmen* exemplifies sequential art's ability to
convey the experience of *simultaneity*—a key imaginary in contemporary
physics—by representing objects and events assumed to be isolated in space
and time as coexisting in the "fourth dimension" of a space–time continuum.
As Mark Bernard and James Bucky Carter (2004, p. 1) explain, the fourth
dimension "refers to a special relationship with space and time wherein the two
conflate such that infinite multiple dimensionalities become simultaneously
present":

> When the reader's interaction, his or her own space-time, is accounted for, this
> evocation of space-time becomes quite literal and expands exponentially. The
> fourth dimension is bridged by human experience and interaction. The sponta-
> neous, real-time interplay of all these forces at once create an ethereal dimension
> of its own. . . . Therefore, the fourth dimension is defined as simultaneous,
> multitudinous dimensionality deeply entwined in and part of individual experi-
> ence. There is special artistry in sequential art and narratives in the relationship
> of this metaphorical and literal space-time continuum.

In the novel *Watchmen*, simultaneity is clearly illustrated by the ways in
which Dr. Manhattan experiences every moment simultaneously. In 1959, he
knows what will happen in 1969 because he is already there; he knows about

---

[5] See Gough (2002, pp. 118–121) for a critique of dissection and associated practices
in school science classrooms..

JFK's assassination at his rebirth because he was already experiencing it, unable to change the course of history (see Ch. IV, p. 128).

Watchmen as sequential art not only represents simultaneity as something that Dr. Manhattan experiences, but also provides the reader with their own experience of it:

> From the very opening pages of *Watchmen*, it is clear that the reader is in for a virtuoso bridging of space and time, made all the more complete by his or her own role. For example, while two detectives investigate the apartment where the Comedian [Edward Blake] ... has been murdered, we are able both to hear about the murder through their dialogue and to see it through Gibbons' graphic illustrations of the crime that are spliced in-between the detectives' examination of the murder scene. Through the combination of texts and visuals, we, the readers, are truly in both places at the same time as well as in our own space. (Bernard & Carter, 2004, pp. 4–5)

Bernard and Carter (2004, p. 5) argue that this bridging of the space–time continuum is particular to sequential art, and that comics/graphic novels constitute a twentieth-century culmination of the artistic goals of pivotal modern and postmodern genres such as cubism and futurism, which they illustrate with Michel Duchamp's (1912) cubist *Nude Descending a Staircase No. 2*.[6] Watchmen also gestures toward surrealism in panels showing Dr. Manhattan perusing a reproduction of Salvador Dali's (1931) *The Persistence of Memory* (Ch. IV, p. 128).

ADAPTING SIMULTANEITY AND SCIENCE IN *WATCHMEN* (2009 AND BEYOND)

Watchmen now represents other forms of simultaneity, insofar as its existence includes both a sequential art novel and a filmic text, Zack Snyder's (2009) Watchmen and Damon Lindelof's (2019) television series *Watchmen*. Transformations from comics/graphic novel to film/video necessitate creative alterations to the source text, and critical engagement with the film often centers on what is done differently between the texts (see Van Ness, 2010, pp. 172–179). Yet, as is the case with many adaptations, these critiques often imply that any alteration is unnecessary or inferior to the initial enunciation—a symptom, as Robyn McCallum (2018) writes, of the emphasis in discourses on adaptation toward venerating fidelity, excluding the possibility that a filmic adaptation might improve upon or positively enact ideas that diverge from its pre-text. Our consideration here is not to envision the film as an imperfect reimagining of the graphic novel, but rather to assess how each text represents comparable ideas of simultaneity, science, and the Capitalocene crisis.

---

[6] See https://www.philamuseum.org/collections/permanent/51449.html.

As noted above, representations of simultaneity are key aspects of the novel's narrative. Attempts to render simultaneity in the filmic text of *Watchmen* (2009) are necessarily transformed by the constraints of the medium. Following Brian McFarlane (1996, p. 28), we consider that just as spatiality is key to the filmed image, it is also essential to the graphic novel page in a manner distinct from the spatiality of the unembellished written word. Yet these media types each hold divergent relationships with linearity in terms of their engagement with the movement of time. As Scott McCloud (1994, p. 100) writes, "in learning to read comics, we all learned to perceive time spatially, for in the world of comics, time and space are one and the same." While the comics page is spatially stable and unchanging, differential access by the reader in individuated moments creates a highly subjective sense of time. By contrast, the film-image is constantly moving forward, and particularly in conventional mainstream film, presents only one filmic sequence at a time within the viewing frame.

Consider, for example, the sequence in which Dr. Manhattan's televised interview is presented in parallel with Dan Dreiberg (Nite Owl II) and Laurie Juspeczyk (Silk Spectre II) fighting a gang attacking them in an alley (Ch. III, pp. 86–91). Whereas the graphic novel uses sequential art to express the simultaneity and parallels in these scenes through repeated panel placement, or matched transitions and dialog, the film attempts to replicate this sequence through cross-editing and parallel editing. In the novel, images and speech echo and resonate between the sequences, but the film can only present the sequence as a series of cutting shots, and alters the dialog and context to better fit within the flow of the film's narrative arc. In other words, the film cannot replicate the simultaneity of spacetime because it can only depict the forward flow of time, whereas the comic presents an illusion of time in which past, present, and future exist simultaneously. As McCloud (1994) observes, the past and future always surround the present in comics, because unlike the forward motion of film, the reading eye can change direction, moving up, down or across pages. Moments of simultaneity remain embedded within the narrative of the film—for example, Dr. Manhattan's existence remains suspended in a past/present/future assemblage—but the expressive possibilities of simultaneity enacted in the graphic novel are difficult—perhaps impossible—to render on the filmic screen. As such, the sequential art of comics/graphic novels constitutes a uniquely accessible site of engagement with the possibilities of simultaneity.

What, then, of science? If, as Fishbaugh suggests, the sciences in the graphic novel are represented implicitly by character-avatars, they are explicitly enacted in the film by manifestations of science as a force of creation—and potential destruction—that are central to its narrative. This transformation is interwoven with one of the most common criticisms of the film, namely an alteration of the novel's ending amusingly referred to in social media as "Squidgate" (see Van Ness, 2010, p. 184), with the shadow of Richard Nixon's presidency informing both works. In the novel, Veidt (Ozymandias), initiates a purposefully bizarre cataclysmic event—summoning a monstrous (genetically engineered) psychic squid creature that kills millions of New Yorkers in its explosive death throes—which the film replaces with a similar plotline in which Veidt, ostensibly working with Dr. Manhattan to create a new source of energy, uses his insights into Manhattan's powers to cause explosions in major cities across the globe, framed as being attacks by Manhattan. Van Ness (2010, pp. 179–184) discusses the motivations for this change at some length, but here we are more concerned with how this change alters the relationships of each text with science and the products of scientific inquiry.

Early in the film, the shift of emphasis is signaled by Veidt/Ozymandias discussing the world's environmental crisis, which is also flagged as an issue in the novel, where it is positioned as a consequence of the Cold War and the military arms race. The film inverts this logic, portraying the Cold War conflict between East and West as less an ideological battle than a war for resource control, with Veidt emphasizing that his work with Dr. Manhattan would end the conflict by creating free, limitless, clean energy, thereby eliminating reliance on fossil fuels. The novel depicts Manhattan engaging with subatomic physics in an attempt to discover gluinos, which is reframed in the film as collaborative work with Veidt, and his cadre of scientists in Antarctica, to create a power generator derived from Manhattan's unique energy structure—science with more obvious use-value to a layperson than discovering hypothetical elementary particles. Similarly, throughout the film, Veidt is presented as a benevolent celebrity capitalist—his pursuit of clean, free energy brings him into conflict with oil barons and captains of industry, whom he silences with the reminder that he can buy and sell them at his leisure. This benevolence is undermined by the film's conclusion, in which noble aims of science and environmentalism are repurposed into a force of destruction. Science, in the film, is more directly tied to the fate of the globe than the posturing of state-ideologies, and is depicted as offering an apolitical solution which extends beyond their subjective reach.

These changes point to a broader reorientation in the film toward emphasizing the power of science as both a creative and destructive force. Whereas the novel emphasizes the hidden labor of artists and creatives in the creation of the monstrous squid, it is scientists—bedecked in the requisite white laboratory coats—who provide the labor for Veidt's scheme. Working with Manhattan to understand how his abilities may be harnessed for a greater good, their intentions are ultimately betrayed by Veidt, who poisons them all before using their work to kill millions. Their fate, more so than in the novel, reflects that of many scientists who have either been killed by, or come to regret, the fruits of their labors, and thereby offers a valuable illustrative point: that even the most noble pursuit can, and perhaps will, be coopted for purposes outside of what was initially envisioned. Much of the science depicted in *Watchmen* remains implausible in film and novel, yet the metaphor of Manhattan transforms. In the novel, Manhattan is viewed as a potential obstacle to Veidt's scheme, but in the film the global fear of Manhattan is central, because this accidental product of science is no longer under control, and must be resisted for the sake of human survival.

This alteration points to a broader change for *Watchmen*, namely, the creative context within which it is embedded. *Watchmen*, in 1986, was created in a time where the Cold War was the existential threat facing the globe, but as Tim Rayner (2009, n.p.) argues, "cinema audiences today no longer experience the mortal dread of nuclear annihilation that MAD [mutually assured destruction] left hanging above our heads." The apocalyptic posturing of nation-state superpowers in the novel is replaced with something far more pertinent to the contemporary viewer: the gradual, ongoing destruction of the global environment and biosphere through human (in)action, unconstrained by geo-political affiliations. In 2009, New York is no longer the sole site of destruction, as it is in the novel—major cities across the globe are destroyed, including Hong Kong and Moscow. The world does not unite in the wake of a single attack, but through the realization that nobody is safe if Manhattan can no longer be depended upon for safety. The idea that humanity might have to unite in a global defense against the uncontrollable products of our own scientific and technological development is now more pertinent than the shock of interdimensional invasion, and more accurately depicts the crises of the Capitalocene: as we enter the era of climate change, how do we defend ourselves against the results of our own "progress"? What will the eventual cost be? As Rayner (2009) puts it, the utilitarian argument of Veidt/Ozymandias presents the "friendly face of fascism," by suggesting that sacrificing millions of people may be worth the cost if it saves the remainder. Or perhaps it is more appropriate here to extend the metaphor to the friendly face of *scientism*—the idea that we can imagine the morally untenable as something palatable so long as we cloak it in the "klansman's sheet" of white-coated rationality. As we presently watch celebrity capitalists promote their funding of developments in rockets, electric vehicles, and artificial intelligence, we must acknowledge that beneath "scientific" surfaces there are human drives, emotions, and irrationalities that underpin these labors, including the work of the countless thousands who mine the resources required. There is always a cost, although we may not always be aware, or wish to know, of what price has been paid—until, as in Percy Shelley's (1818) sonnet "Ozymandias," we "look on [our] works and despair."

Watchmen's transformation from comic to film, and the transfigured readings made possible through such adaptations, is only one example of textual simultaneity. The intellectual property notoriously remains under the control of DC Comics despite creator Alan Moore's protestations, and has been expanded by multiple comic books, including *Before Watchmen* (Azzarello et al., 2018), an anthology series featuring backstories for the characters from the original work, *Doomsday Clock* (Johns & Frank, 2019), a sequel and crossover within the DC Comics universe. More recently, Damon Lindelof's (2019) television series *Watchmen* negotiated the comic's world and anxieties through the lens of the turmoils of race in American society.

Each of these texts provides a form of access to *Watchmen*, and all in their own way reimagine the chronologically prior novel's relevance for contemporary audiences. Regardless of fidelity-first criticism, reading *Watchmen* today is a wildly different context from reading the novel in 1986–1987, and the scientific imaginaries it communicates and critiques extend beyond the contexts in which it was created.

As a collaborative endeavor, this chapter demonstrates that *Watchmen*'s interrogations of the goals, products, and outcomes of scientific research—and of how ideologies are intertwined with "progress"—generates dialog and engagement across generations. In the opening section of this essay we affirmed our support for de Freitas and Truman's (2020, pp. 1–2) quest to foreground SF as a way to "open up scientific imaginaries... to think through the many pasts, presents, and futures of science." By opening such imaginaries, *Watchmen* demonstrates that popular media (in this case selected comics/graphic novels and films/television series) can function as generative texts (as distinct from "textbooks") for contemporary science educators.

References

ACARA [Australian Curriculum Assessment and Reporting Authority]. (2020a). *Australian Curriculum Science*. https://www.australiancurriculum.edu.au/f 10-curriculum/science/key-ideas/

ACARA [Australian Curriculum Assessment and Reporting Authority]. (2020b). *Australian Curriculum, Senior secondary curriculum, Science, Earth and Environmental Science, Structure of Earth and Environmental Science*. https://www.australiancurriculum.edu.au/seniorsecondary/science/earth-and-environmental-science/structure-of-earth-and-environmental-science.

Appelbaum, Peter. (2019). Speculative fiction, curriculum studies, and crisis: An introduction and invitation. *Journal of the American Association for the Advancement of Curriculum Studies*, *13*(2), 1–6. https://doi.org/10.14288/jaaacs.v13i2.192252.

Atkinson, Daniel. (1985, June 8). Evolutionary process from chaos to order. *Los Angeles Times*, Part II, p. 2.

Azzarello, Brian, et al. (2018). *Before Watchmen Omnibus*. DC Comics.

Bernard, Mark, & Carter, James Bucky. (2004). Alan Moore and the graphic novel: Confronting the fourth dimension. *ImageTexT: Interdisciplinary Comics Studies*, *1*(2), n.p. http://www.english.ufl.edu/imagetext/archives/v1_2/carter/.

Bleier, Ruth. (1988). Lab coat: Robe of innocence or klansman's sheet? In Teresa de Lauretis (Ed.), *Feminist studies/Critical studies* (pp. 55–66). Macmillan. https://doi.org/10.1007/978-1-349-18997-7_4.

de Freitas, Elizabeth, & Truman, Sarah E. (2020). New empiricisms in the Anthropocene: Thinking with speculative fiction about science and social inquiry. *Qualitative Inquiry*. https://doi.org/10.1177/1077800420943643.

Deleuze, Gilles, & Guattari, Félix. (1987). *A thousand plateaus: Capitalism and schizophrenia* (Brian Massumi, Trans.). University of Minnesota Press (Original work published 1980).

Demos, T. J. (2016, September 11). Anthropocene, Capitalocene, Gynocene: The many names of resistance. *Frontiers of Solitude*. http://frontiers-of-solitude.org/blo g/442.

Doll Jr., William E. (1986). Prigogine: A new sense of order, a new curriculum. *Theory into Practice*, 25(1), 10–16.

Eastman, Kevin, & Laird, Peter. (1986). *Teenage Mutant Ninja Turtles*. First Comics.

Fishbaugh, Brent. (1998). Moore and Gibbons's *Watchmen*: Exact personifications of science. *Extrapolation*, 39(3), 189–198.

Gibson, William, & Sterling, Bruce. (1991). *The difference engine*. Bantam.

Gleick, James. (1987). *Chaos: The making of a new science*. Viking.

Gough, Noel. (1991). An accidental astronaut: Learning with science fiction. In George Willis & William H. Schubert (Eds.), *Reflections from the heart of educational inquiry: Understanding curriculum and teaching through the arts* (pp. 312–320). State University of New York Press.

Gough, Noel. (1993a). Environmental education, narrative complexity and post-modern science/fiction. *International Journal of Science Education*, 15(5), 607–625. https://doi.org/10.1080/0950069930150512.

Gough, Noel. (1993b). *Laboratories in fiction: Science education and popular media*. Deakin University. https://www.researchgate.net/publication/246635016.

Gough, Noel. (2002). A postmortem on dissection. In John Wallace & William Louden (Eds.), *Dilemmas of science teaching: Perspectives on problems of practice* (pp. 118–121). RoutledgeFalmer. https://ecommerce.tandf.co.uk/catalogue/Det ailedDisplay.asp?ISBN=0415237629&ResourceCentre=ROUTLEDGEFALMER& RedirectPage=PerformSearch%2Easp&curpage=1.

Haraway, Donna J. (1989). *Primate visions: Gender, race, and nature in the world of modern science*. Routledge.

Hasse, Catherine. (2015). The material co-construction of hard science fiction and physics. *Cultural Studies of Science Education*, 10(4), 921–940. https://doi.org/ 10.1007/s11422-013-9547-y.

Hayles, N. Katherine. (1984). *The cosmic web: Scientific field models and literary strategies in the twentieth century*. Cornell University Press.

Hirsch, E. D. (1988). *Cultural literacy: What every American needs to know*. Houghton Mifflin.

Johns, Geoff, & Frank, Gary. (2019). *Doomsday Clock Part 1*. DC Comics.

Lindelof, Damon (Director). (2019). *Watchmen*. HBO.

McCallum, Robyn. (2018). Introduction: "Palimpsestuous intertextuality" and the cultural politics of childhood. *Screen adaptations and the politics of childhood: Transforming children's literature into film* (pp. 1–32). Springer.

McCloud, Scott. (1994). *Understanding comics: The invisible art*. HarperCollins. http://www.scottmccloud.com.

McFarlane, Brian. (1996). *Novel to film: An introduction to the theory of adaptation*. Clarendon Press.

Michel Duchamp. (1912). https://www.philamuseum.org/collections/permanent/ 51449.html.

Moore, Alan, & Gibbons, Dave. (1987). *Watchmen*. DC Comics.

Morrison, Grant, Truog, Chas, Hazlewood, Doug, & Grummett, Tom. (1991). *Animal Man*. DC Comics.

Percy, Shelley. (1818). https://www.poetryfoundation.org/poems/46565/ozyman dias.

Porush, David. (1991). Literature as dissipative structure: Prigogine's theory and post-modernism's roadshow. In N. Katherine Hayles (Ed.), *Chaos and order: Complex dynamics in literature and science* (pp. 54–84). University of Chicago Press.

Prigogine, Ilya, & Stengers, Isabelle. (1984). *Order out of chaos: Man's new dialogue with nature*. Bantam.

Rayner, Tim. (2009, March 8). *Watchmen* in times of change. *Philosophy for Change*. https://philosophyforchange.wordpress.com/2009/03/08/watchmen-and-change/.

Salvador, Dali. (1931). Museum of modern art. https://www.moma.org/collection/works/79018.

Schwab, Joseph J. (1964). Structure of the disciplines: Meanings and significances. In Gervaise W. Ford & Lawrence Pugno (Eds.), *The structure of knowledge and the curriculum* (pp. 1–30). Rand McNally.

Shiner, Lewis. (1988). *Deserted cities of the heart*. Doubleday.

Snyder, Zack (Director). (2009). *Watchmen*. Universal Pictures.

Stableford, Brian. (2003). Science fiction before the genre. In Edward James & Farah Mendlesohn (Eds.), *The Cambridge companion to science fiction* (pp. 15–31). Cambridge University Press. https://www.cambridge.org/core/books/cambridge-companion-to-science-fiction/science-fiction-before-the-genre/6C616545DB0452A28686806850DCBDA9.

Sterling, Bruce. (1985). *Schismatrix*. Ace.

Sterling, Bruce. (1989). *Islands in the Net*. Ace.

Vale, V., & Juno, Andreas (Eds.). (1984). *Re/Search # 8/9: J.G. Ballard*. Re/Search Publications.

Van Ness, Sara J. (2010). *Watchmen as literature: A critical study of the graphic novel*. McFarland.

Weaver, John A. (1999). Synthetically growing a post-human curriculum: Noel Gough's curriculum as a popular cultural text. *Journal of Curriculum Theorizing*, *15*(4), 161–169.

Weaver, John A. (2019). Curriculum SF (speculative fiction) Reflections on the future past of curriculum studies and science fiction. *Journal of the American Association for the Advancement of Curriculum Studies*, *13*(2), 1–12. https://doi.org/10.14288/jaaacs.v13i2.191108.

CHAPTER 18

Curricular Experiments for Peace in Colombia: Re-imagining Science Education in Post-conflict Societies

Carolina Castano Rodriguez, Molly Quinn, and Steve Alsop

When a person decides to share their stories after years and years of war and pain,... they are allowing us to pick up some "lenses" to see through their eyes at that black huge space full of tiny spots back to the past, and recover history in their hearts and souls.

—Dalia (Participating Student, Personal Conversation, 2016)

In 2016 we were given the opportunity to dream, large, of a science education that emerged from a transitional period in a country torn apart by more than five decades of internal conflict, Colombia. Invited by Universidad Nacional de Colombia, we were given freedom to reimagine science education for a 10-day course, part of the International School, which opened both to students and professionals from any background. Approaching reality as changing and emergent, the science education we were able to dream became the one we were able to construct as part of a moment still marred by bullets and yet hopefully (re)written by education: "Science Education for the Construction of Peace." As one of the students who contributed to this

C. Castano Rodriguez (✉)
Australian Catholic University, Melbourne, VIC, Australia
e-mail: Carolina.CastanoRodriguez@acu.edu.au

M. Quinn
Louisiana State University, Baton Rouge, LA, USA

S. Alsop
York University, Toronto, Ontario, Canada
e-mail: SAlsop@edu.yorku.ca

© The Author(s) 2022
M. F.G. Wallace et al. (eds.), *Reimagining Science Education in the Anthropocene*, Palgrave Studies in Education and the Environment, https://doi.org/10.1007/978-3-030-79622-8_18

dream, Layla, states: "Maybe peace building is not necessarily to forget and forgive, sometimes it has to do with remembering."

In this chapter, we present a story of dreams and hopes, of what is possible, of a changing present, of a transitional period positioned in Colombia but relevant to any corner of the world. Most of all, we present a story of a possibility to imagine a science education that contributes to rewriting history, through entering the window to the hearts and souls of those who have lived in fear and also hope for so many years. To best represent the complexities of this journey we were privileged to take, we constructed this chapter in a nonlinear way.

Firstly, we present a narrative we created drawing upon truths as articulated in the participants' comments, sometimes presented as direct quotes, and fictional realities of possible futures founded on our experience with the participants and the collective perceptions of the future we were imagining together. We worked with the narratives, reflections and chains of conversations that emerged over 10 days between the three of us, who guided the course, and the participants. The emotional tone of the story reflects the multiple subjective views and feelings regarding the future and the past that the participants shared. The narrative is placed in the near future.

We positioned the story in Escher's (1953) invention of multidimensional worlds in his art piece *Relativity*. We were privileged to come to understand diverse visions of transitional Colombia through the stories of futures yet to be created that some participants shared. Their visions of futures we found similar to the defying, creative and innovative nature of Escher's work in *Relativity*. In his work *Relativity* (1953), as viewers "we have to reconcile blatantly impossible images. While we can easily make sense of the individual components in the picture, trying to establish their relationship to each other is bewildering and forces us to see either a crazy world, or just lines, shapes and shading" (Romanin, 2018, p. 78). The transitional and future Colombia will only be possible as a multidimensional world in which diverse realities, visions and past experiences of healing and reconciling co-exist in a yet-to-be-constructed, innovative, peaceful society.

Escher's work inspired us to create a multidimensional narrative, to move the story beyond representing the diverse views of the present and future of the participants, to represent the hopeful, positive, emotional, interconnected and complex views of futurity that the participants had regarding the transition towards a peaceful Colombia. For this, the main character of the story represents multidimensional worlds, created from the diverse stories and views of futurity presented by each participant, and the personal experience of one of the authors, Carolina, of growing up in Colombia during periods of possibilities of peaceful futures that did not become materialized until 2016. The narrative is set in the future, year 2050.

To create the narrative of futurity, we also followed Feminist Standpoint Theory (Harding, 2004). Feminist Standpoint Theory allowed us to situate knowledge in context, and consider context as emergent, and part of constructed realities where all actors interact and participate, including the non-human. This theory helped us to open spaces hardly imagined in what some have called the Anthropocene; to reimagine the non-human as central to human life, and how the human and non-human share indefinite entanglements that can destroy but also construct dreams of hope. Moreover, as was observed in our course, the non-human can impact how humans generate language for expressing feelings and create context. Many of the participants used metaphors with the more-than-human created world to represent their views of futurity. As we have also seen in other studies, science education can also help to share experiences of loss and hope that otherwise would not be unlocked (Castano Rodriguez et al., 2019).

We hope that this chapter inspires others to dream and construct an education that offers opportunities to share experiences of loss, experiences of truth and reconciliation alongside dreams of futurity.

COLOMBIA, YEAR 2050...

2050, this will be my last entry in this diary. My journey to reconcile my past and reimagine my new me has finished. I am my new me. During my last 34 years, since the Peace process with FARC (Fuerzas armadas revolucionarias de Colombia) started, I worked to reconcile the loss of several of my friends who disappeared during a time when many people were disappearing and had been kidnapped. I grew up in a very difficult social situation in Colombia. A time when cars were more than transportation devices, they were also bombs that could and did kill many. A time when my parents would go to work to the Candelaria, the city centre, and I would not know whether I would see them again. I wrote what I felt about Colombia when the peace process started:

> When mass media feeds us descriptions endlessly confirming this condition these thoughts only get worse. When they lead us to believe that we have been in the longest civil war in all the history of humanity, a civil war that has claimed the most number of lives out of any other, a civil war responsible for the most of the missing persons out of any other, we are left in a state of sorrow that is difficult to overcome.

The peace process meant, among many changes, that ex-FARC members were now going to live among us and that at any time I was going to interact with one of them. I have to first accept and understand that FARC's original ideology did represent many Colombians, and that the violent Colombia in which I grew up, represented more than just one group using violence to accomplish their goals. It was hard to imagine a less violent Colombia; really it was impossible for me to imagine it.

I started this diary during the peace process. One of my first entries was about a reflection I wrote regarding how I was feeling about going through the process towards a peaceful future; water served me as metaphor:

Water is steady sometimes, quiet and apparently calm, but there is always something happening beneath; like us, steady, quiet and in apparent calm, just because it´s so painful and even scary to think what we can find if we submerge and look deep in the bottom. … But water always moves on, that movement includes rapids, vertiginous falls, knocking rocks, to at the end finish the trip in a sinuous and calm way to the immense and unknown.

We need to look down, to submerge in the pain and recover stories from the deep, we need to throw the fish nets and catch those stories, bring them back to the boats and navigate with them.

Like water, I decided to navigate with my past, to remember it and wear it to continue flowing through the present and into the future. Not moving away from it but moving with it. So, I started my life project. My journey had metaphors not only with flowing water, but also with colours. I learned to accept my feelings and accept that everyday could be felt differently. It was like a rainbow, many feelings were caused by a storm, producing many colours and creating a beautiful rainbow. In that moment I wrote: "Tell me, what is the colour of your peace?" And the answer is that it is all colours; it is a rainbow of colours, representing diverse emotions and experiences, that when together they form a beautiful rainbow full of hope.

When I accepted my rainbow of emotions, and was willing to navigate with them in a river full of rapids and also with many sections of calmness, I realised I was "ready to forgive" and be part of a peaceful Colombia. One of the many thoughts I started to reflect on during that initial time of my journey, through remembering and forgiving, was the thought about "otherness." I needed to start seeing people who had committed acts of violence, for whatever reason they had, not as others, but also as Colombians working towards peace. That process made me think beyond FARC to any member of a society, including all life forms. I asked myself, how could we start seeing and interacting with all living beings, in a way that could help to construct peace? So, I came up with a few words that I needed to incorporate in my daily life to move beyond otherness: "tolerance, respect, identity, compassion, care, integration, affect, sensibilization, LIFE."

During these first stages of my healing process, and while thinking of otherness and a peaceful future, a group of people often forgotten by society came to my mind: people in the prisons. I decided to investigate more about prisons, and came across this story in YouTube, https://www.youtube.com/watch?time_continue=183&v=WxKIt_t1wD4&feature=emb_title, of one of the prisoners who spent some time in Bogota's prison and then in a rural one. He stated that:

the people believe that the one who has been [a thief] will always be. Society does not let one change …. there is discrimination. And besides discrimination, there are obstacles. One is always classified like the bad one. And the bad one does not find work outside.

His interview made me question what role people in prisons could play in the construction of peace. How could they be better re-inserted into society if we want to construct peace? If we were serious about a peaceful Colombia, then we needed to consider all members of our country. So, I decided to focus on prisoners and what it meant to give them an opportunity to be part of a transitional period towards peace. I created a workshop called "Harvesting Forgiveness: Creation of Scenarios of Peace Futures in the Prison, through Education in Indigenous Agronomy" (translated from Spanish). The workshop addresses the needs for coexistence and reintegration of the prison population, empowering and paving restorative and reconciling routes for prisoners relative to the larger society of which they are a part and will return: in facilitating the education and participation of prisoners—during and after serving their sentences, in concert with farmers and other knowledgeable community members—in cultivating organic and native foods in networks of fair trade across the country. A focus on Indigenous agriculture and native foods, I felt, was also a way to recognize, remember, restore and revitalize the indigenous traditions of my country through the production of food in *chagras*. Chagras is a traditional Indigenous integrated system to cultivate plants, in which minerals, humans, other animals and plants all are considered to work in interdependent ways.

I first thought about this workshop during a course I took regarding peace and science education in transitional Colombia. The peace-oriented science sessions offered a subject context which enabled/supported sensitive conversations of a deeply divided and emotional nature. It was then when I felt empowered to consider how all the diverse communities that make up our beloved Colombia could co-exist and work towards peace. Now, 34 years since the peace process took place and my inner journey started, it is hard to remember how we lived among so much violence. As it was beautifully said by the former Colombian president, Juan Manuel Santos in 2016: "Las balas escribieron nuestro pasado, la educación escribirá nuestro futuro" (Bullets wrote our past, education will write our future). "Reconciliation" is how I can describe my journey towards healing, towards forgiving, towards remembering and towards co-constructing a peaceful Colombia. When I look back to that year, 2016, I remember all the diverse emotions, sometimes conflicting ones I had, and realised how all those emotions flowed like a river, sometimes moving in peaceful ways and sometimes in a storm kind of wave, but always with power; it moves forward, creating diverse possibilities like my work with prisoners and the endless possibilities it brought for me and for their sense of a future in co-creating peace.

The Beginnings of Our Journey

Colombia, 2016, the year when the peace treaty with FARC was signed by the Colombian government to end one of the longest civil conflicts in history. Two local academics from a Colombian University and a Colombian residing in Australia dared to dream about a science education for such a transitional period towards the construction of peace. We reflect on how peace-oriented citizenship considers the human and the non-human as central to harmonious living (Carter, 2015). We asked: What will collectivity and solidarity look like when most citizens, humans and non-humans, have endured a violent past but dream of a peaceful, harmonious future? What could and should be the role of science education in such a context?

The university's international school, which opens to students and the general public every year during June/July, provided the opportunity to explore these questions and allowed us to dream of a course in science education that responds to the social, political, and environmental context of contemporary Colombia. Over a week, these three academics explored the local context and envisioned how science education could respond to this transitional period. Two more academics, Steve Alsop and Molly Quinn, joined this team of dreamers. Wellbeing and science had already been explored by Carolina in diverse contexts (Carter et al., 2019; Castano, 2012a, 2012b; Castano Rodriguez, 2019). Steve brought his expertise in environmental justice and activism, and Molly provided the bridge to consider critical peace education. We discussed possible curricular approaches and disruptions needed to construct a course that was anchored in the needs and priorities of the Colombian society and which positioned educational spaces as sites of possibility and transformation. The course was open to everyone, students and professionals. Forty-five participants from diverse sociocultural and professional backgrounds, and at diverse stages of their careers, joined us. They were all interested in the space this course offered to reflect upon and to re-think society and the construction of peace for the human and non-human.

This journey of re-thinking science education towards the construction of peace in a transitional society entailed envisioning education and peace anew together, subjectively informed via context, culture and lived experience. Engaging participants with us in this transformational work, we had to inquire deeply into peace itself as well; the way of cultivating social and ecological justice, and cultures of peace, sustainability and activism. The inspections and interrogations we undertook to create this course were supplemented by the insights and offerings of curriculum studies— curriculum theory specifically, critical peace studies/education, transformative theory, critical pedagogy and feminist theories. These diverse theories informed the process of creation of this course.

RE-THINKING EDUCATION IN TRANSITIONAL COLOMBIA: CURRICULUM STUDIES, CRITICAL PEACE STUDIES/EDUCATION AND CRITICAL PEDAGOGY

Via curriculum theory—foregrounding the question of what knowledge is of most worth (Schubert, 2009; Spencer, 1884/2015), we were able to initiate "the interdisciplinary study of educational experience," examine curriculum as the intellectual framing of such experience, embracing subject matter, society and self, and engage autobiographical (*currere*) processes towards illuminating our own meaning-makings respecting such things as teaching, learning and justice—as lived (Pinar, 2012). Through curriculum theory we created spaces for sharing with each other our own stories of war, pain and possibility as well as our own avenues therein for agency, healing, and transformation— *experiences of loss, of truth and reconciliation alongside dreams of futurity.*

Critical peace education positions education as a site for transformation towards peaceful societies from the lenses of social justice and pedagogies of resistance (Quinn, 2014). Insights gleaned from the field of peace education—offering lenses for re-viewing, remembering and reimagining herein as well—we felt were also particularly apt for our shared consideration. With its ties to international and comparative education and to modern human rights movements, peace education—through its focus upon understanding the problem of violence writ large (Harris & Morrison, 2012), particularly structural violence, and its myriad forms (e.g., cultural, domestic, interpersonal, etc.)—enabled us to give studied attention to it (and in relation to identity, culture, context, education, epistemology, peace, science and such, as well).

Additionally, through the work of critical peace studies/education (Bajaj, 2008; Bajaj & Brantmeier, 2011; Verma, 2017), we drew upon analyses of colonialism, class, race, gender and other personal-political sites of difference making for complex, conflicting, confounded and contested expressions and experiences of violence and responses to them. Such theoretical grounding connected well, as well, with that of critical pedagogy, both traditions of thought rooted in the work of Paulo Freire (1970/1993)—also of Latin America—on oppression and liberation and the role of thoughtful awareness and reflective action (e.g., *conscientization*, problem-posing and *praxis*) in undertaking the "vocation of humanization" in a world of injustice, violence, and suffering—and this, via education.

We also learned from work aimed at critically understanding and attending barriers to peace—a term often negatively defined as freedom from various kinds of marginalization, domination, dehumanization and hostility (Oxford English Dictionary, 1989; Quinn, 2014)—and creatively considering ways to positively promote it: reimagining, *futuring*, education, science, subjectivity and peace. The role of education in the production and perpetuation of *epistemicide*, epistemic violence (Paraskeva, 2016; Paraskeva & Steinberg, 2016), specifically through the ongoing control and force of Western knowledge systems, and the abiding curricular legacy and legitimization of colonial and imperial ways of knowing and being could be interrogated (Noddings, 2012). The way and wisdom of other, indigenous, home-grown and -generated onto-epistemologies also could be explored, as counternarratives, decanonizing and interrupting the normative storyline(s), potentially offering opportunities for transformational restorying rooted in the context, culture, subject matter and subjectivity at play among us in coming together and collaborating in such study. For instance, the peace park in Japan, inspired by the brave young girl Sadako and her thousand paper cranes (Coerr, 1979) in the radiation aftermath of the bombing of Hiroshima, compelled us to build our own peace park memorial in class, expressive of our dreams of and for peace, as issuing from what photographic and reflective writing concerning its presence in our own lives.

Such, while not explicitly so at the time, certainly involved "staying with the trouble" that Haraway (2016) endorses, through which we creatively composed new visions and vistas for peace for ourselves and others—seeking to "make kin" of each other and beyond, of the worlds of our own makings via our part and participation with, and posture towards, the natural world too—and via science education, from the place of as well as commitment to our entangled relationality and responsibility before the living and dying on a damaged earth, the yet to be born and fragilely emerging... especially here, intended reconciliation efforts in Colombia. As she so eloquently describes such work:

> The task is to become capable, with each other in all of our bumptious kinds, of response. .... The task is to make kin in lines of inventive connection as a practice of learning to live and die well with each other in a thick present..., to make trouble, to stir up potent response to devastating events, as well as to settle troubled waters and rebuild quiet places … , requires learning to be truly present … as mortal critters entwined in myriad unfinished configurations of places, times, matters, meanings. (p. 1)

We knew that it was vital here, too, then to examine the forms of intelligibility made possible/impossible by our normative ways of seeing, being, thinking and doing—e.g. as epistemologically framed via capitalism, colonialism, Western imperialism and universalism. We knew it was crucial to consider whose/what lives are made unintelligible, and most vulnerable, therein as well (Butler, 2009)—in striving to advance peace as a critical (educational, curricular) demand and collective responsibility, to recognize the ways in which we not only fail but continue to further precarity in so many and to reach for an ethics of relationality that would enrich us—on/and the earth—all (Smits & Naqvi, 2015).

TRANSFORMATIVE LEARNING AND CARE-ORIENTED PRACTICES IN SCIENCE EDUCATION

How could science education contribute to empower marginalized societies and, particularly, how could it contribute to create lasting peace in Colombia? *Different figurations for worlds that can barely be imagined*, as the editors of this book encouraged us to consider, were possible to be imagined and constructed through the inspiring stories that the participants shared with us, their views of the future and the experiences of the past. Curriculum studies and Critical Peace Education engaged us in reflecting and creating spaces for the participants to rethink peace and the many possibilities for reaching a more peaceful society, including reaching an ethics of relationality that includes the non-humans. Transformative learning encouraged us to consider transformation as central to access rational and emotional aspects of experiences of loss, truth, and dreams of peaceful future worlds.

Described originally by Jack Mezirow's work, transformative learning requires a process of emotional and rational confrontation (Dirkx et al., 2006). For transformation to occur, disorienting dilemmas, described as confrontational situations that will increase awareness of the diverse causes of an issue need to be present. The goal of this process is not only to increase awareness, but also to empower people to act and engage in change. In our journey of working with the possibilities of a peaceful future, the experiences the participants had with loss and their considerations of reconciliation as the main force to envision a peaceful future, provided the central triggering events for transformation.

Although transformation has not been central to science education, some authors have argued that science could promote agency and activism (Alsop & Bencze, 2015). When science education's aims are expanded to reconsider an ethics of relationality for creating a more peaceful world, transformation occurs and new possibilities for engaging with humans and the non-humans that enrich all are possible (Carter et al., 2014).

Finally, an ethics of care, initially described by Carol Gillian in 1982, helped us connect critically, transformation and peace education, bringing the local to the forefront of our consideration of education in a transitional period towards the construction of peace. Ethics of care rejects the notion of universalisation and includes not only considerations of underrepresented groups in societies, but also of non-human species and natural systems (Gilligan, 1982; Noddings, 1992). Particularly, Noddings' (1992) centres of care provided reconsideration of science education towards the wellbeing of all life forms in the context of this transitional period in Colombia. Themes of care, now known as "centers of care"—caring for the self; caring for the inner circle; caring for strangers and distant others; for animals, plants, and the Earth; for the human-made world; and caring for ideas (Noddings, 1992) require expansion of our compassionate circle beyond those who are close to us, to include other species, nature, and diverse thoughts and ideas. Through the centres of care, a new science education can be created to foster consideration of all life forms and expressions such as diverse systems, cultures and social groups (Carter et al., 2019). Transformative learning and ethics of care theory allowed us envisioning a future of a post-conflict society and contributed to create a bridge to reconsider science education's role in this process.

Re-imagining Science Education in Post-conflict Societies: Transformation and Reconciliation

We were given the opportunity to dream of a science yet to exist and a science education that is central to the construction of peace. We envisioned peace as a harmonious coexistence between humans, and with the non-human. Particularly, we were privileged with access to create a space of hope and solidarity with locals to rethink science education in transitional Colombia. Together, with the participants of this program, we re-considered how success in science education could look and what role it could and should play in a society that has endured more than five decades of internal conflict. Theories such as ethics of care, transformative learning and critical peace education provided the foundation for disrupting imported curricula to attend local needs and validate local experiences. We positioned participants as co-constructors of the course and agents of change and hope in their society.

Escher's art, and particularly the work *Relativity* (1953), also reminded us that imagining a science for a peaceful Colombia requires believing in science education's capacities to reformulate, share and experience loss, truth and reconciliation. A peace-oriented science education that contributes towards transitioning to peaceful societies could be perceived almost as impossible to be imagined in the Anthropocene. Yet, when we were given the space to reflect on the impossible, and to share it with those who have experienced a violent Colombia and were transitioning towards a peaceful future, we were able to co-create a space for this underexplored dimension of science. As Escher's work demonstrates, "Believing in the impossible allows you to invent and innovate. When the normal rules no longer apply, anything and everything is possible" (Doughty, 2018, p. 88).

Feminist standpoint theory served as an analytical and methodological resource for our curricular explorations, bringing local experiences as central to the consideration of futurity of science education in a context of peace building. Specifically, the concepts of "horizontality, reciprocity and solidarity" (Jaramillo & Carreon, 2014, p. 395) were brought by the participants as central to re-thinking science education towards peace and harmony. Central to this envisioning were pedagogies of resistance which view education as sites for practising more democratic participation, with horizontal decision-making structures (Bajaj, 2015). In this view of a new science, rural education and local knowledge and experience were prioritized by the participants. The concept of "reconciliation" was identified by the participants as fundamental to the construction of lasting peace in Colombia. Reconciliation was envisioned as a new form of solidarity and care for others, where others could be any actor of the conflict endured in the past.

This new vision of education sees understanding, remembering and also forgiveness as central to solidarity. It includes diverse experiences of loss and hope to construct a peaceful society. For science education it means to prioritize the local: local understanding, local practices, local personal histories. This strongly localized knowledge and envisioning of the future requires education to be re-positioned as a site for resistance and transformation towards hope, care and peace. Within this framework, success is envisioned as the active construction of peaceful interactions between humans and with nature, with individuals equally valued, capable and empowered to become agents of change and transformation towards a peaceful Colombia.

In post-conflict scenarios, science and education could offer an important, powerful and very necessary way (standpoint) to reach out across radical differences and inequalities in acts of wishful dreaming together for a brighter future. And yet, education and science in this regard are so conflicted because they enable these dreams through familiar imaginaries that simultaneously obfuscate controversies and associated histories of inequalities and ongoing structural violence. Thus, following Escher's work *Relativity* (1953), we gleaned something of the "gravitational forces" and "impossible realities" constitutive of this world within which we live, and of the "impossible objects" with which we wrestle: a different science, an alternative education, new relations among us upon earth, peace restored or realized for the first time—and as an enduring verb, a way of being and becoming together here—perhaps, too, impossible. Such means, as Harding (1996) reminds, opening up to ways of knowing, alternative epistemologies, that counter the present dominant paradigm wherein social advantage advances epistemic disadvantage and vice-versa, and through such a future science that impacts life on earth-human and non-human—in more life-affirmative, sustainable and peace-producing ways. As ourselves, as justice, peace too, we pondered, is ever incomplete, ever becoming, and yet to come.

REFERENCES

Abbot, E. (2019). *Flatland: A romance of many dimensions*. Warbler Press (Original work published 1884).

Alsop, S., & Bencze, L. (2015). Activism! Toward a more radical science and technology education. In L. Bencze & S. Alsop (Eds.) *Activist science and technology education*. Springer Press.

Bajaj, M. (2008). "Critical" peace education. *Encyclopedia of peace education* (pp. 135–146). Information Age.

Bajaj, M. (2015). Pedagogies of resistance and critical peace education praxis. *Journal of Peace and Education, 12*(2), 154–166. 1080/17400201.2014.991914.

Bajaj, M., & Brantmeier, E. (2011, November). The politics, praxis, and possibilities of critical peace education. *Journal of Peace Education, 8*(3), 221–224.

Bowell, T. (n.d.). Feminist standpoint theory. In *Internet Encyclopedia of Philosophy*. http://www.iep.utm.edu/fem-stan/.

Butler, J. (2009). *Precarious life: The powers of mourning and violence*. Verso.

Carter, C. C. (2015). *Social education for peace. Foundations, curriculum, and instruction for visionary learning*. Palgrave Macmillan. https://doi.org/10.1057/978113 7534057.

Carter, L., Castano Rodriguez, C., & Jones, M. (2014). Transformative learning in science education: Investigating pedagogy for action. In J. L. Bencze & S. Alsop. (Eds.), *Sociopolitical activism and science & technology education* (pp. 531–545). Springer.

Carter, L., Castano Rodriguez, C., & Martin J. (2019). Embedding ethics of care into primary science pedagogy: Reflections on our criticality. In J. Bazzul & C. Siry (Eds.), *Critical voices in science education*. Springer.

Castano, C. (2012a), Fostering compassionate attitudes and the amelioration of aggression through a science class. *Journal of Research in Science Teaching, 49,* 961–986. https://doi.org/10.1002/tea.21023.

Castano, C. (2012b). Extending the purposes of science education: Addressing violence within socio economic disadvantaged communities. *Cultural Studies of Science Education, 7*(3), 703–718. https://doi.org/10.1007/s11422 012 94124

Castano Rodriguez, C. (2019). Pushing the political, social and disciplinary boundaries of science education: Science education as a site for resistance and transformation. In J. Bazzul & C. Siry (Eds.), *Critical voices in science education research: Narratives of academic journeys*. Springer.

Castano Rodriguez, C., Barraza, L., & Martin, J. (2019). Rethinking equity: Standpoints emerging from a community project with victims of violence and abuse in Argentina. *Cultural Studies of Science Education, 14,* 393. https://doi.org/10.1007/s11422019099203.

Coerr, E. (1979). *Sadako and the thousand paper cranes*. BDD Books.

Corso, L. M. (2018). Biography. In *M. C. Escher: More than meets the eye*. Council of Trustees of the National Gallery of Victoria.

Derrida, J. (1990). Force of law: "The mystical foundation of authority" (On deconstruction and the possibility of justice). *Cardozo Law Review, 11*(5–6), 919–1044.

Dirkx, J. M., Mezirow, J., & Cranton, P. (2006). Musings and reflections on the meaning, context, and process of transformative learning. *Journal of Transformative Education, 4*(2), 123–139.

Doughty, M. (2018). Design. In *M. C. Escher: More than meets the eye*. Council of Trustees of the National Gallery of Victoria.

Emmer, M., Schattschneider, D., & Ernst, B. (2007). *M.C. Escher's legacy: A centennial celebration*. Springer.

Escher, M. C. (1953). *Relativity*. Lithograph, Cornelius Van S. Roosevelt Collection All M. C. Escher works © Cordon Art-Baarn-the Netherlands.

Escher, M. C. (1989). *Escher on Escher: Exploring the infinite*. Harry N. Abrams.

Fien, J. (2003). *Learning to care: Education and compassion*. Lecture given 15 May to the Australian School of Environmental Studies Director, Griffith University EcoCentre.

Freire, P. (1993). *Pedagogy of the oppressed* (M. Ramos, Trans.). Continuum (Original work published 1970).

Freire, P., & Freire, A. M. A. (1994). *Pedagogy of hope: Reliving pedagogy of the oppressed*. Continuum.

Gilligan, C. (1982). *In a different voice: Psychological theory and women's development*. Harvard University Press.

Haraway, D. (2016). *Staying with the trouble: Making kin in the Chthulucene*. Duke University Press.

Harding, S. (1996). Multicultural and global feminist philosophies of science: Resources and challenges. In L. H. Nelson & J. Nelson (Eds.), *Feminism, science, and the philosophy of science*. Synthese Library (Studies in Epistemology, Logic, Methodology, and Philosophy of Science, vol 256). Springer, Dordrecht. https://doi.org/10.1007/978-94-009-1742-2_12.

Harding, S. (2004). Introduction: Standpoint Theory as a site of political, philosophic, and scientific debate. In S. Harding (Ed.), *The feminist standpoint reader* (pp. 1–16). Routledge.

Harris, I. M., & Morrison, M. A. (2012). *Peace education* (3rd ed.). MacFarland.

Jaramillo, R., & Carreon, M. (2014). Pedagogies of resistance and solidarity: Towards revolutionary and decolonial praxis. *Interface: A journal for and about social movements, 6*(1), 392–411.

Magendzo, A. (2005). Pedagogy of human rights education: A Latin American perspective. *Intercultural Education, 16*(2), 137–143.

Noddings, N. (1992). *The challenge to care in schools: An alternative approach to education*. Teachers College Press.

Noddings, N. (2012). *Peace education: How we come to love and hate war*. Cambridge University Press.

Oxford English Dictionary. (1989). Oxford University Press.

Paraskeva, J. (2016). *Curriculum epistemicide: Towards an itinerant curriculum theory*. Routledge.

Paraskeva, J., & Steinberg, S. (2016). *Curriculum: Decanonizing the field*. Peter Lang.

Pinar, W. (2012). *What is curriculum theory?* (2nd ed.). Lawrence Erlbaum.

Poole, S. (2015, June 20). The impossible world of MC Escher. *The Guardian*.

Quinn, M. (2014). *Peace and pedagogy* (Primer Series). Peter Lang.

Romanin, D. (2018). Mathematics. In *M. C. Escher: More than meets the eye*. Council of Trustees of the National Gallery of Victoria.

Schubert, W. H. (2009). What's worthwhile: From knowing and needing to being and sharing. *Journal of Curriculum and Pedagogy, 6*(1), 22–40.

Schwarz, M. (2018). Psychology. In *M. C. Escher: More than meets the eye*. Council of Trustees of the National Gallery of Victoria.

Smits, H., & Naqvi, R. (Eds.). (2015). *Framing peace: Thinking about and enacting curriculum as "radical hope."* Peter Lang.

Spencer, H. (2015). *What knowledge is of most worth?* Palala Press/Alibris (Original work published 1884).

Taras, N. (2018). Philosophy. In *M. C. Escher: More than meets the eye*. Council of Trustees of the National Gallery of Victoria.

Verma, R. (2017). *Critical peace education and global citizenship: Narratives from the unofficial curriculum*. Routledge.

Part V

Complicated Conversations

CHAPTER 19

A Feral Atlas for the Anthropocene: An Interview with Anna L. Tsing

Anna L. Tsing and Jesse Bazzul

Jesse Bazzul: Thanks for agreeing to do this interview. One of our goals for the book is to get educators thinking differently about this peculiar and urgent time many are calling the Anthropocene—which for us means thinking differently about science, pedagogy, kinship, nonhumans, and politics. Your work has been inspiring for many people across education because it engages diverse forms of collective life in the hopes of reimagining the future. I wonder if you could talk about how you see this moment of the Anthropocene and perhaps how it changes how researchers and educators engage their work.

Anna Tsing: For me, the biggest challenge of trying to tell students about the Anthropocene, and tell the public in general, is that we have to learn new ways to tell stories that are simultaneously about human histories, and also about histories of the natural world. Because of the way particular structures of knowledge have dominated the last several centuries, people have managed to separate these histories. The result is that there are particular ways of storytelling about humans, and then there are wholly different ways of storytelling about plants and animals, or rocks and climate, and we don't know how to mix these up very well. They have different genres, different expectations, and so most of the time we tell stories about humans as if we humans lived in a

A. L. Tsing (✉)
University of California Santa Cruz, Santa Cruz, CA, USA
e-mail: atsing@ucsc.edu

J. Bazzul
University of Regina, Regina, SK, Canada
e-mail: jesse.bazzul@uregina.ca

© The Author(s) 2022
M. F.G. Wallace et al. (eds.), *Reimagining Science Education in the Anthropocene*, Palgrave Studies in Education and the Environment, https://doi.org/10.1007/978-3-030-79622-8_19

vacuum. And, in the same way, when we tell stories about plants and animals, or rocks and climate, it's as if they lived without humans. So somehow, the challenge of the Anthropocene is to figure out how to bring these ways of understanding the world together.

Jesse: In your book, *The Mushroom at the End of the World: On the Possibility of Life in Capitalist Ruins*, you talk a lot about different ways of noticing. And now you're currently working on another transdisciplinary project with colleagues at Aarhus University called *Feral Atlas*, where you're trying to illuminate different ways of telling stories and new storytelling techniques. I wonder if you can tell us a little bit about what *Feral Atlas* is about, and why such a project is urgent?

Anna: It's great to talk about this project in the context of science education because it's meant as a project of science education through the Anthropocene. That's the purpose of it.

Feral Atlas is an online digital platform for stories about the Anthropocene, which I'm putting together with many colleagues, like Jennifer Deger, Alder Keleman, Feifei Zhou and others, who are co-editors or curators of this project. In addition, we have something like 75 natural scientists, social scientists, artists, writers, and many other kinds of people contributing to it, in order to tell particular stories about the Anthropocene. None of these people or stories, however, tell the story of the Anthropocene as a whole. Instead, we have argued that one way to get a better grip on what's going on around us is to tell *granular stories* of the feral effects of the Anthropocene, understood as a spatial and temporal phenomenon. We felt like this would better attend to the kinds of questions of social justice on one hand—the uneven distribution of resources and forms of violence around us—and on the other, the natural phenomena relevant to the natural sciences emerging around us. So, for example, we have an entry on carbon dioxide, but not on a planetary scale all at once, but in terms of its particular effects. Coral reefs, for example, have been very sensitive to temporary changes, and so a key story of the Anthropocene is about the bleaching of coral reefs and their decreasing ability to support vast ecosystems because of the warming of the oceans, etc. In other words, parsing out the Anthropocene to the places where substantial differences can now be sensed, observed, and imagined, we argue, necessitates the skills that both the humanities on one hand, and natural scientists on the other, bring to us in order to think about the Anthropocene.

Jesse: And perhaps you could also clarify what you mean by feral... or ferality. In this case, it is in relation to various human infrastructures, yes?

Anna: Yes, let me backup for a minute to explain what both "feral" and "atlas" mean here. Designing the project, we decided to begin with the kind of human infrastructures that have been built for imperial conquest on one hand, and industrial development on the other, and how these infrastructures have created Anthropocene effects. Which were not the effects they were necessarily supposed to have! The word unintentional, however, would not be entirely accurate—because in many cases infrastructural designers knew perfectly well that they would have effects that weren't the ones that they designed for! In many cases, designers didn't seem to care about the fact that certain infrastructure projects were going to have these "extra" effects. Feral is a term that we have employed and stretched to talk about the ways that living beings, as well as non-living beings, react to the kinds of infrastructural projects humans come up with. So ordinary things like roads, nuclear power plants, and the burning of fossil fuels have effects beyond those that they were designed and promoted as having, and that's the feral role that we want to tell people about. And we call it an *atlas* because we're trying to pay attention to the particular places that matter. So, for example, carbon dioxide produced by fossil fuel burning happens in northern industrial centers, and only after a few days drifts around the world. We always tend to think of carbon dioxide as a planetary phenomenon because of how it's often measured, but it comes from particular locations and is chemically reduced in particular forests and other places. So, *Feral Atlas* is arguing that we need to think about the Anthropocene in patchy ways... an Anthropocene that's socially and environmentally uneven.

Jesse: I wanted to ask you about this, because I find the mapping part of the project very interesting. It seems to coincide with an increasing focus on, and a need for, topologies of environmental problems, political issues, and material realities. One of the things I came across in reading about *Feral Atlas* was that you utilize a map that disrupts scale and temporality and size. So, it's not a map like a google map of the planet, made by satellite imaging. Rather, it's a sort of disparate, forever ranging kind of map. I wonder what your thoughts are around the use of these maps for transdisciplinary work: how does this mapping project bring scientists and artists and humanities scholars together?

Anna: I think everyone expects a digital project that involves mapping to use GIS systems and to make everything commensurate so that it all builds up to one planetary map, and we've refused that from the very beginning. So, while there will be some completely recognizable maps in *Feral Atlas*, there'll be many that are at scales that are just incompatible with GIS systems. You see, some of the maps are just at a very small scale. For example, we have an entry on the scaly side of a salmon fish; which is about salmon farming and the subsequent death of a lot of wild salmon. And the scale here is incompatible on purpose. Another example of what we're calling a map, is a painting by an aboriginal artist from northern Australia, who has shown his people's relation to land and country through a painting of goanna lizards, or what we in the United States call monitor lizards, that have disappeared because of the introduction of the cane toad, which killed off many animals in northern Australia. And so, for him, showing that his people are celebrating the goanna lizards, even though they've disappeared from the area, along with the menace of the cane toads, is a map. We want to include this kind of first-hand empirical knowledge of a place, and in a different set of conventions than might otherwise be included in an atlas. So, we're purposely pushing the meaning of map in an atlas.

Jesse: I like this idea of patchiness, because it introduces a different way of viewing the world ontologically speaking. So, on an ethical level what does patchiness afford us?

Anna: An anthropologist named Eduardo Vivieros De Castro has made an argument for what he calls ontological anarchy, that is, imagining the world as having many kinds of ontological gaps that all work across each other at the same time. And, in a sense, I deem that to be quite useful as an approach to understanding the kinds of empirical knowledge we're going to need to understand the Anthropocene. *Feral Atlas* sometimes juxtaposes completely different ways of knowing and living in a world with Anthropocene creatures.

For example, we have a woman's memoir about how important elm trees were to her growing up in the UK. They defined the boundaries and contours of her family's farmland, as well as created a sense of locality. Then Dutch Elm Disease arrived in the early twentieth century, and they had to kill and burn all of their elm trees. And this really changed her sense of self. And so, we juxtapose that narrative with a scientific article by a natural scientist, who is a plant pathologist studying how that incredibly virulent strain of Dutch Elm Disease, which again came and killed all the trees in the UK, was formed in part through hybridization practices of the commercial nursery trade at the time. So, the industrial infrastructure here is this concentration of the nursery trade in a few places, which has created and has allowed the emergence of so many new plant diseases that have then spread to new places around the world. So, we put these two perspectives on Dutch Elm Disease together side by side, while not forcing them to talk to each other directly. In doing so, we can see how perhaps, different kinds of immersion, knowledge, and ways of being might come up against each other in helping us with the Anthropocene. That is, in knowing what the Anthropocene is all about.

Jesse: One of the things that's impressive about the project is the bringing together of natural scientists, humanities scholars, and artists, etc. I am sure you understand the tensions that come up with transdisciplinary work especially. For example, on one hand scientists will say, "Ah, this isn't scientific enough." and a humanities scholar might say, "Well, your scientificity misses much of the point. " Your introduction to *Feral Atlas* encourages scientists, and all participants, to stay true to their training. So you encourage people to talk across boundaries, while allowing them to be true to what they know best. I think this is a really important point for educators trying to bring different kinds of people, ways of knowing, and disciplines together. I wonder if you could talk a little about your experience in bringing diverse people together.

Anna: Yes, well maybe a place to begin is by saying that the first question some contributors to *Feral Atlas* had was, "Please send me a model, a template." We've been very cautious, because we want the natural scientists to write like natural scientists, and we want the artists not to be illustrating science but rather bringing their vision and the way that they particularly understand things to the project. The last thing we want is to homogenize all the entries, because epistemological and ontological juxtapositions are at the heart of *Feral Atlas*... and I think science education for the Anthropocene.

Another way of putting it is that there is a need to enlarge what the term natural history has meant. Too often it is understood as a European-Colonial kind of practice that involved conquering the rest of the world, such that the term has connotations only of conquest. Yet, if you imagine natural history as everybody's vernacular engagement with the natural world around them, along with all the kinds of things they could observe and pay attention to, then there are many, many kinds of natural history. And although these kinds of natural histories may involve different conventions of knowledge, that doesn't mean they are not empirical. So, I believe that to get better understandings of the Anthropocene, we're going to need all different kinds of natural history. So, starting out with a set of regulations, and saying we will only accept work that meets these kinds of standards, means we lose the possibility of learning many different kinds of knowledge that are important for understanding the world around us. I would be really delighted if new kinds of research projects come out of the imagination that *Feral Atlas* is trying to put forward! Projects that start to blend different ways of thinking, the methods and skills of natural scientists on the one hand and social scientists, humanists, artists, etc. on the other. And I am very much hoping that segregation does not have to be the fate of these really contingent disciplinary silos, but instead, by beginning with their integrity, we might be able to see diverse projects going forward that make use of the talents on all sides.

Jesse: In our email exchanges you mention different kinds of media, and we may have touched on that a little bit at the beginning. What kinds of media and communication does *Feral Atlas* draw attention to, and how does it attempt to innovate on that level?

Anna: In some sense the timing is wrong for this question, because by the time the interview comes out, we might have designed everything differently. This is what's wrong with conversations in general. They are always meant for a particular moment!

Jesse: Haha, that's true! Whatever you say in this interview is going to be read after *Feral Atlas* is officially launched, and the final decisions already made about media formats, and things like that.

Anna: Well, *Feral Atlas* is trying very hard to make use of what digital media can do for us in terms of having a system for working across the different reports and essays. This digital media system involves all of the pleasures of the visual, as well as provoking people's curiosity. For example, Feifei Zhou is an architect who has drawn four magnificent landscapes that show what we call Anthropocene Detonators; which involve examination of the historical contingencies from which infrastructure projects have come into being. So, a user on *Feral Atlas* will begin by focussing their attention on a feral entity on these landscapes that depict Anthropocenic Detonators. They can then explore the landscape by looking around, or they can follow the little thing that they picked at the beginning and use the digital functions of being able to zoom in, zoom out on these landscapes. This, in turn, takes them to a set of musings about what human infrastructures are, and how they change the world. Also, a set of short videos being made for *Feral Atlas* will try to orient viewers to the way that infrastructures change the world by immersing viewers in a particular context and location, and emphasizing the gap between what we knew before and what we know now. And only then do they get to go and read documentation about this phenomenon. This documentation will draw attention to what we call feral qualities; which are the ways that nonhumans somehow gain purchase with these infrastructures, transform themselves, and become something different. Furthermore, at each of these analytic axes users have a chance to think about other entries in *Feral Atlas* that they might want to look at. Thereby creating comparisons or contrasts across entries. We are aiming to engage users, so that it's not just a reading project. In fact, *Feral Atlas* bears a notable resemblance to games, art projects, and research archives due to the possibilities afforded by new digital formats.

Jesse: I think it's very curious that you've specifically chosen to talk about infrastructure from a human perspective, as infrastructure can mean many different things to different people. So instead of a media studies, or nonhuman perspective, of infrastructure you specifically look at infrastructures designed and built by humans. It seems that you and your colleagues are doing this in order to inject a much needed political context for the project. I am reminded of your earlier work where you focus on feral outcroppings created by political–ecological disturbances. One of the things that seems to be true of our current moment in time, is that it is impossible to live outside of things like capitalism and human infrastructures. I feel that, just like with *Mushroom at the End of the World*, *Feral Atlas* demonstrates the need for finding new forms of collectivity in the ruins of human infrastructures and capitalism. So I am wondering if you could speak a little bit about the political urgency in which this project unfolds.

Anna: There's a lot bundled together there, so let's begin with the infrastructure question. For me, one of the reasons to focus on these human-kind of infrastructures is that, rather than the Anthropocene being a kind of planetary or abstract-nothing situation, we would like to ground our understanding of what's going on with particular entities. So, for example, if it's a railroad, and it stretches quite a long way, and we can see that the railroad has created certain effects, then it should give people pause the next time someone wants to build a railroad in that particular way!

Part of the conceit of modernity, since at least the middle of the nineteenth century, is that there would be no regrets. That you would build all of this stuff and it would only lead to a brighter world, and that you just didn't have to pay attention to what else was going on around you, what else might emerge from these things that were being built. So, infrastructure was designed, built, and installed without a care. It's not that people didn't know that if you put in a nuclear reactor, radiation was going to get out and become part of the metabolism of plants and animals, including humans. Everybody knew that. But they thought, who cares, it's a great idea! The engineers are going to make it happen, it's a wonderful thing! Look, wow, you can put corn chips in bags! Nobody thinks ok, where's that plastic going to go and how many marine mammals are going to subsequently die? So it's these kinds of issues that have not been on the radar. To me, one really important part of science education in the Anthropocene is that people think about these kinds of infrastructural projects that all of us bring ourselves into every day. Many of which really are unnecessary. I can't think of any reason why styrofoam is allowed to be manufactured at all, but it lasts pretty much forever. So I'm hoping that the students that science educators influence, that we all influence as educators, might be able to think about these infrastructural projects and say: "You know, there are some that we really could do more thoughtfully and carefully... perhaps not at all."

Jesse: I want to dwell a little on the concept of disturbance in your work overall. In *Mushroom at the End of the World* you note that ecological succession patterns of human-designed environments sometimes don't follow the typical categories of succession in the biological sciences. You also show similarities between psychic disturbance and ecological disturbance. Namely that... once disturbed life patterns change. Some disturbances, as you show, can be ambiguous—a volcano can be a disturbance and so could a traumatic event. If the Anthropocene involves learning to live with disturbance, what might an ethics of disturbance look like?

Anna: Ah, you've just identified, I think, a big ethical question, which is: which kinds of human disturbances are we willing to live with? And which kinds are we not willing to live with? And I think this is where the term Anthropocene can help us. One of my colleagues, Zachary Caple, has introduced the term *Holocene fragments*. I mean, he's really spacialized what constitutes the Holocene and what constitutes the Anthropocene. He uses the term Holocene, which you know was only less than 12,000 years ago, being cognizant that humans were very much a part of the Holocene. Farming was part of the Holocene, the trade and travel of humans was also part of the Holocene, and so what is left of the Holocene still exists in fragments. I'm sorry, I'm sort of mixing up the order this should be introduced...

A Holocene fragment is basically an ecological space in which mutualisms and complementarities of various sorts, often having ecologically foundational moralities or moral terms, continue to exist. Caple's example is the Florida scrubland, where a particular kind of bird feeds on a particular kind of acorn from a particular kind of oak. And, that it's these kinds of specific long-term arrangements between birds and trees that actually allows the scrubland to exist in the first place! And so he compares these Holocene fragments to Anthropocene landscapes like the ecological disturbance caused by a Florida phosphorus mine, along with the attempts to cover up this disturbance. For example, just simply rolling grass out on top of the barren clay landscape created by the mine! This is what Caple calls an Anthropocene landscape, and in his work he richly describes an area where you have a long stretch of fencing, where on the one side you have this scrub forest, and on the other you have this rolled out turf! And in a sense, they're both forms of nature. Since you asked the ethical question, we should be able to distinguish between those two kinds of natural environments around us. Perhaps holding on to the Holocene fragments that we already have is a major ethical commitment that we might want to make as more and more landscapes get transformed into these uninhabitable landscapes; not just for humans, but for all of our nonhuman living companions too.

So, I think what you're saying is right, that there's an ethical challenge in distinguishing the kind of disturbances that work more toward different kinds of flourishing. Perhaps then it is a good time to turn to the feral, in that we need to attune to the nonhuman response to the way humans are living. Again, that's when a field is allowed to go back to forest; the trees growing and replacing a highly specific human-created environment. That's the feral—and it's absolutely necessary to the worlds we live in. We can't do without it. Yet, at the same time, we have to learn that some kinds of ferality are not okay.

Jesse: So looking to ferality could also be where we somehow begin to explore an ethics. It is a very exciting question, because ethics would include a differentiation between ecologies, what might have been more consistent to the Holocene without destructive human interventions versus these Anthropocene landscapes. It might also then mean looking to flourishing, or non-flourishing, as a kind of ethical substance. For me, this seems like a really important ethical project.

Anna: I have also started to think of these things affectively in terms of wandering mists of dread. It's really important to acknowledge the severity of the environmental challenges in front of us, while remaining appreciative of all that feral action around us. To try and access it with all our skills and curiosity and passion, in order to have a sense of what we can and want to live with.

Jesse: Education and anti-capitalist movements haven't been the best partners historically—and what I get from your work is that we are going to have to face or live with capitalism as a massive disturbance going forward. How must attention to capitalism be a part of the context for education, especially in and around the sciences, going forward?

Anna: Well, to begin with capitalism has brought us this really bad idea that very distanced investors, who don't care about places, can have full control over them and wreck them completely without suffering any consequences. That is a really bad idea! But, rather than starting at that abstract level, we could also talk about aspects of capitalism that are very new and very dangerous, but are already treated as if they're set in stone and you could never change them! Take the example of drug derivatives, and the destruction of many different communities by things like opioid addiction. However, again, this phenomenon relies on these distant pharmaceutical investors—who don't care about places—and yet the prevalence of these drugs has risen enormously in the last twenty to thirty years. And over and over again, a prevailing belief amongst scientists, scholars, journalists, and the general public is that we can't mess with free trade... as if it were some kind of sacred establishment! But what free trade often means is a complete lack of regulation of incredibly dangerous things, even when many kinds of precautions and procedures exist for controlling these dangerous things. For instance, I myself am really horrified by the fact that we allow soils to be sold and shipped large distances at an incredibly large rate and scale. The whole world already has well distributed soils! You don't need to move them around the world, yet we allow massive corporations to ship them everywhere, spreading plant diseases and endangering plants without ever really thinking about how dependent we are on those plants.

But, it has been one expansion of capitalism after another, and no-one really has, or has been allowed to, speak up about regulating things that are completely unnecessary for human health and destructive for the wellbeing of many other kinds of organisms. Just because of the so-called sacrosanct status of free trade, no one even discusses it!

Jesse: And so *Feral Atlas* becomes a challenge to destructive ideas of economic and ecological scalability. In your past work you examine specific places where capitalism is operating through detailed supply chains and ask what other forms of life are still possible despite these operations. Do you see resistance to the commodification of life involving a confrontation with capitalism's own sacred pillars like scalability and private property as the basis of law? Or is the historical point that things went drastically wrong much earlier? Say with the advent of colonialism. Are there also temporal distortions, as well as the spatial ones, you focus on in *Feral Atlas*?

Anna: Well, to be clear, *Feral Atlas* has some temporal questions woven into it. So maybe I can answer by describing the four Anthropocene Detonators, or historical conjunctures, that have given shape to human infrastructure and give depth to the *Feral Atlas* project. The first detonator is *Invasion*. So we really begin with the invasion of the New World by Europeans. And the kinds of murder and displacement that happened, and the displacement and desecration of native ecologies and native peoples went together. And we don't say that just happened 500 years ago, because it is still happening in the present. Not only because indigenous people are still being displaced, like for example, in Australia, but also because of the continuing ways that colonialism as a set of institutions, world-making practices, and ideas continues to rule. For example, one of the *Feral Atlas* pieces is about the kinds of pasture grasses that were imported to the Amazon in order to purposely disallow the forest to grow back. Again, the feral at work! The forest can't grow back when these very aggressive pasture grasses are introduced and so the contributor talks about how people viewed them as a form of invasion. Certain people wanted this invasion, the conquest of the Amazon through grass, so that both the people who lived there and specific ecologies could no longer exist. So, the first Anthropocene detonator is Invasion.

Our second detonator is *Empire*, and we begin with European, and move into Asia, and deal with things like the triangular trade in which captive, kidnapped Africans were brought to do plantation labor in the New World. We emphasize that there are many forms of imperial conquest. So not just invasion, but the governance of people from afar created huge infrastructural programs involving water management, plantation agriculture, and all kinds of ways of channeling resources to imperial cities. And those things are all still with us. Part of what has happened since World War Two, and the partial decolonization of the Global South, is that these imperial infrastructure projects have been used by national elites all around the world, to continue the same kinds of water management and plantation agriculture systems—all of which were started some time ago. Thereby making that imperial mode of governance even more entrenched.

Our third detonator is *Capital*. Here, we begin in the nineteenth century with the interconnection between things like plantation agriculture, in which slavery was such an important part, and the burgeoning industrial development in the metropole. So altogether, how these things together create that whole world of capital, in which the rationalization of resources and commodities is spread around the world.

Our fourth detonator is called *Acceleration*, which relates to the great acceleration after World War Two: where Invasion, Empire, and Capital are brought to new heights. In part, this is because they were distributed around the world in a new way, such that all kinds of infrastructure building programs, however destructive, were marketed as the way to a better life. They were also marketed on the presupposition, which is entirely false, that all their waste products can go to some other place. And, of course, because of imperial programs writ large, there were those other places for waste to go unseen. But now, those other places are everywhere. Furthermore, the kinds of waste that got invented in the first half of the twentieth century, from radioactivity to plastics, have proved to be much more long-lasting, and much less able to be recycled, than what came before. We have a new set of landscapes created by these non-disposable wastes, and each of these create a landscape of injustice and mistreatment that just gets ever more dense.

I think my biggest fears about the Anthropocene is that the gap between the super elite and disadvantaged will keep growing. As the elite try to find ways to shelter themselves from the toxins, waste, and death they've unleashed on the rest of the world, there is also a kind of densification of poverty, inequality, and deprivation.

Jesse: I wonder if you could share some of your experiences in trying to motivate people to take up the call to action and revisioning. As science educators, we talk to young people and students about environmental destruction a lot, but how might we bring people on board?

Anna: Well, as far as *Feral Atlas* is an educational project, some people have compared it a little bit with Rachel Carson's *Silent Spring*. The reason I find this comparison interesting is that when I was last reading about the history of DDT, and those terrible insecticides, what I learned was that DDT was one of our most well-studied toxins in the history of science. Even before Rachel Carson, there were hundreds of studies of DDT, and each one established levels of safety, rules of proper use, and concluded over and over again that if people used it safely enough DDT was just fine. What Rachel Carson did to blow that out of the water was start from a different place. Rather than think, well, could we simply adjust the levels a little bit? She asked, "What happened to the songbirds? Why did they all die?" So, sometimes a change in angle can get one to a new place, and I would like *Feral Atlas* to be a small part of a way of saying, "Okay, here's some things that can be addressed." Even specific things that can be addressed. So, for example, a scientist in *Feral Atlas* has written about the extinction of many kinds of Pacific tree snails, in part because of the purposeful distribution of a carnivorous snail, called the Rosy Wolf snail, that eats all the other tree snails. And on top of this, the Rosy Wolf snails are still being distributed to new Pacific islands where they're not yet part of any problems. At the very least we could stop that; and things like the distribution of live soils around the earth, as well as the continued use of styrofoam. You know, enough is enough!

Jesse: It seems that some things in our world are so "siloed" they really don't make sense anymore. Your work really tries to expose some of these vicious and strange aspects of the Anthropocene, so they can be seen in different ways. Teachers really cherish examples of how we might try doing things differently, because that's what we are trying to do in educational institutions every day.

Anna: Well, I hope you bring students outside of the classroom, to look around them. I think it's very easy to see many kinds of Anthropocene phenomena such as human infrastructure and feral action. Even sidewalks. The feral action part is when we notice the weeds that are coming out of the sidewalk, as well as the sidewalk itself. And students who have a chance to be observing these phenomena might become concerned and informed about how to think differently about these issues. I'm going to say one more thing because, you know, what school kids are doing right now is so exciting to me, around the world. I just got back from a conference in Ireland a little while ago and everybody was talking about the Swedish girl Greta [Thunberg] and the schools movement that she has stimulated globally. I hope science education for Anthropocene doesn't dampen the spirit of young people, but instead encourages them to think of creative interventions before it's too late.

Jesse: Right. So, seeing human infrastructures and ferality as a kind of unraveled complex.

Anna: I also find this idea of ferality in young children. This difference between not noticing the sidewalk, but the grass growing in between the cracks of the sidewalk—it seems to me that children might be the best observers of this. Perhaps college students aren't much different. It is my dream to have students go out and identify, discuss, and perhaps intervene in feral entities in their communities and neighborhoods. Put a bit of themselves alongside these feral entities!

Jesse: Great idea! I think this might be a good place to end. Well, thank you so much for sharing today. I really appreciate it.

Anna: Yes, and thank you.

CHAPTER 20

In Conversation with Fikile Nxumalo: Refiguring Onto-Epistemic Attunements for Im/possible Science Pedagogies

Fikile Nxumalo and Maria F.G. Wallace

Maria Wallace: One of the big threads that we were talking about that we wanted to hear your perspective on is related to the complicated context of science. So in science as both a discipline of knowledge, but also a context of study and so forth. There are so many problematic kinds of assumptions that get kind of taken up in that, whether it's independent, historical, or ontological considerations of how we might romanticize our understanding of what constitutes nature, as well as destructive or hegemonic socio-ecological relationships. So, I know that a lot of your work deals with children, but also, just thinking about how we inherit these dialogues or narratives of nature and culture come through. If you could talk a little bit about the nature–culture divide either from whatever perspective you want or if you think about children as a particular starting point.

Fikile Nxumalo: Yes. My engagement with the idea of inheritances and their unevenness is about trying to bring to the context of early childhood education critiques of the universalizing effects of discourses of the Anthropocene as the so-called age of Man, where humans are the primary drivers of planetary changes. There have been several critiques of the Anthropocene from

F. Nxumalo (✉)
University of Toronto, Toronto, ON, Canada
e-mail: f.nxumalo@u.toronto.ca

M. F.G. Wallace
Center for Science and Mathematics Education, University of Southern Mississippi, Hattiesburg, MS, USA
e-mail: maria.wallace@usm.edu

© The Author(s) 2022   321
M. F.G. Wallace et al. (eds.), *Reimagining Science Education in the Anthropocene*, Palgrave Studies in Education and the Environment,
https://doi.org/10.1007/978-3-030-79622-8_20

many different perspectives, including feminist geographies, Black studies, and Indigenous studies that have pointed out that dominant discourses of the Anthropocene don't bring sufficient attention to the ways in which marginalized populations are not only differently responsible for anthropogenic damage; they're also disproportionately vulnerable. So my ethical commitments in my work include centering the ways in which early education, including science education, can resist and subvert rather than reinscribe the universalisms and erasures of the Anthropocene. This includes asking how early childhood pedagogy might include attention to the specific geographies of children's inheritances of environmental precarity.

This attention to uneven inheritances is also part of my thinking on possibilities for unsettling nature/culture divides in early childhood education. So asking, for example, when it is said that children are needing reconnection with nature, which children are we talking about? What assumptions of childhood underpin pedagogies of reconnecting to nature? What is meant by nature—this is an important question to ask in relation to current environmentally damaged places that children are unevenly inheriting. What is left out about the more-than-human connections that are always already there? This is not to suggest that there are universal "right" answers to these questions but rather to think about how place-based pedagogies might disrupt colonial and universalizing assumptions about nature and children or childhoods. Megan Bang and Affrica Taylor are two people whose work has been particularly inspirational for me thinking about these questions and their pedagogical and curricular implications in the actual places and spaces in which I work with young people and educators.

Maria: Yeah. And as you were talking, I was starting to think about how you were pointing out a couple of different kinds of inheritances. So there's the human dimension, but also the nonhuman inheritance, whether that's the physical material or the ideological. And so maybe if you could expand a little bit on that, the ways in which there's this intergenerational, whether it's human or nonhuman, exchange of our conceptions of what constitutes nature or even culture and how that's taken up personally in different bodies, geographical contexts, and so forth. So there are these different, multifaceted ways in which we can think about the inheritance of science or nature or ways of knowing and being, I think. And so some of what I heard you talking about too is that that gets mapped onto human bodies in particular ways, but also the nonhuman dimension. So like the ideas about nature and how human bodies carry those through wherever they're at. But also then the inheritance of the land or the water and how stories get kind of brought into that assemblage.

Fikile: Yes. I also want to add that part of this work with educators is unpacking our assumptions about nature, and children's relations with nature and how they may influence what we do with children and what we pay attention to as we engage in place-based learning. So with regards to inheritances I would say they show up in the everyday in different ways. For example, I have written about the discourse of discovery as it relates to the ways in which children's encounters with nature might be narrated. So one could think of discovery as something that might seem somewhat benign but can actually be a reproduction of colonial inheritances of nature as something separate from humans, awaiting children's meaning-making rather than, for instance, considering how nature is always already pedagogical, agentic, and in-relation with humans in different ways.

Something else in relation to inheritance is that for me it is important to attend to the ways in which inheritance is fraught, political, and filled with inequities. A lot of my work has been in contexts where the children are for the most part middle-class, white settler children and I think in these settings it is easy to ignore these political aspects of inheritance. So I have found it important to be curious about which inheritances are easily covered over and might need to made visible in our pedagogies. Most often for me this has meant thinking through and finding ways to unsettle the colonial and anti-Black erasures in which our pedagogies are implicated.

Maria: I think that's very helpful and an important point. Especially, thinking about the role of the teacher and the educator of exchange, even if it's not within a school, about reproducing the colonial ways in which our ideas about nature get marked or inscribed on the space as we navigate, or who has access to the spaces. So I think some of the work that you've done, when you talk about working with the children, was in Austin, Texas, as part of the camp or there was an organization? Right?

Fikile: Yes, so when I was referring to working with predominantly white settler children this was in the context of my work in preschool settings in British Columbia and more recently in a kindergarten classroom in Austin, Texas. The summer camp work you are referring to is led by Coahuiltecan elders at the Indigenous Cultures Institute in San Marcos, Texas, where I have been privileged to witness and document a place-based summer camp for predominantly Indigenous and Latinx children and youth. So yes, in relation to the question of access, there continues to be a racial and class divide in terms of which children have access to place-based inquiry learning. At the same time, however, the inspiring pedagogies enacted at the summer camp in San Marcos also show what is possible beyond formal schooling spaces.

Maria: And then wrestling with who has access to these different encounters with nature, there's huge dimension to it, but an inheritance problem as well. The last just question was related to how we document children's experiences with nature. And so one of the things that we're really inspired by too in your work is thinking about what it means to research on these types of questions and these experiences, whether that's on or to have a particular idea about science or education or nature or culture. We were wondering about how you find yourself casting this kind of wide methodological and theoretical perspective net to inform your work with children or unsettle the ways in which we study the child or their experience with nature or culture.

Fikile: Well, perhaps I can talk a little about how I came to that. I want to acknowledge one of my mentors, Dr. Veronica Pacini-Ketchabaw, who has helped me to think through the ways in which the complexities of working with young children require shifting the authority of child development to find ways to challenge normative assumptions about children. The challenging and also most generative part of these ongoing shifts has been working through them in everyday encounters with places, children, and educators. In other words, working through what these shifts might look like as more than theoretical or conceptual shifts but as pedagogy that impacts what we do with children, the questions we ask when we are outside with children, the stories we bring to children, the ways in which we write about the places we encounter and children's relations with them... and so on. In terms of working with broad perspectives, I don't think I have a singular answer in how that has come about since, for example, part of that has arisen from unsettling moments that emerge in practice, where it becomes evident that the usual learning theories I might turn to are not enough to, for instance, consider how racialization and colonialism come to matter in the everyday places and spaces of my work with children and educators. I think another important factor has been that in my work I am often trying to grapple with the question of how place-attuned education can simultaneously pay attention to issues of human inequity (for instance, anti-Black assumptions of nature education) and also attend to the more-than-human world in ways that unsettle anthropocentrism. In doing this I have found that I need multiple tools. So recently, in writing about an inquiry on water pedagogies, I found that theories of affect, Black, and Indigenous feminist concepts were all necessary to help me attend to how the planned curriculum that we engaged with as well as what emerged spontaneously, together mattered in relation to decolonizing water pedagogies.

Maria: Yeah, definitely. That's very helpful. Just as a more personal note, that's one of the challenges within science too, we find is there. Are these very linear-bounded ways of knowing the scientific experience or encountering what constitutes science? And so really embracing the complexity of all of these kinds of compounding things that get mashed up into a human subject and also beyond the human subject. It requires a much more diverse set of skills to draw on. And your work is really one great example of that.

Fikile: Thank you. I would like to add to the example that I just mentioned in relation to trying out different water pedagogies with the kindergarten children. In this work we have also engaged with Western science, so, for example, doing all kinds of water testing with the children and learning about aspects of the ecologies of the creek where we spent our time together. At the same time, I am always thinking about the limits of Western science as ways of knowing, relating, and becoming with environmentally damaged places. Dr. Megan Bang's work is an important inspiration to me in relation to place-based science learning that is socio-culturally situated and also takes seriously the agency of the more-than-human world. So alongside Western scientific ways of knowing and their affordances, I'm interested in intentional relational practices that have anti-colonial resonances. From my perspective, this anti-coloniality can take multiple forms, including the development of an ethic of caring that is less about individual responsibility or stewardship and more about recognizing intrinsic human/more-than-human relationality. I also want to add that joy, play, and playfulness are an important part of this work even as we are working to complicate what Affrica Taylor refers to as the romantic seductive appeal of Nature's child in settler colonial places. Borrowing from Haraway, I am interested in what it might look like to stay with the trouble of learning-with and practicing reciprocity with ecologically damaged landscapes, like the polluted creek we spent time within Austin.

Maria: Right. I think that's a really helpful example because so often folks who end up being drawn to these more critical and complex perspectives want to totally disregard the dominant paradigm. But also what I heard you also talking about, which was really interesting, is the way in which as a researcher you draw on these different theoretical and methodological perspectives to come to understand how we research a child's experience or sense making in complicated ways. But the example you also gave that was really helpful was how could we engage children to do that with their encounters with the world as well? So even the pointing out that they engage the testing and the Western science, but also there you're giving them the space in the room to complexify that one account of their water testing with non-Western perspectives at the exact same time.

Fikile: Yes and this work has been immensely enriched by being able to work with Marleen Villaneuva in the first year of the research project. Marleen is a member of the Miakan/Garza Band of Coahuiltecan peoples in Central Texas. She was able to bring to the children situated Indigenous place and water stories. For example, we have written about her sharing with the children a Coahuiltecan creation story that has important lessons about water as human relative, water as agentic, water as life, and more (Nxumalo & Villanueva, 2020a; Nxumalo & Villanueva, 2020b). We have then, together with the children and educators, enacted different practices, some planned and some emergent, of being with water in ways that embody and materialize the lessons of this story and other water stories that we have encountered together.

Maria: Right. That's amazing. That sounds like a beautiful practice. We talk about in science ed, the common dominant narrative often is like, bring in a scientist to talk to your class. But again, that then continues to privilege a particular way of knowing science. And so the example you just gave starts to privilege a very different way of knowing nature and culture combined and the complexity of its materializations. So that's really helpful.

Fikile: I also want to acknowledge that this kind of work is difficult within the current structures of many public schooling contexts, particularly as this was a sustained long-term inquiry over the school year, not a "one-time" activity.

Maria: Right. That's a very good point, that it's not just a drop-in moment, but there's a continued relationship between different ways of knowing, being, and communicating with alternative perspectives over an extended period of time.

Fikile: Yes.

Maria: Okay, great. So we've started to talk about it a little bit, but how have children informed your multifaceted perspectives? Whether that's theory or method or your work writ large?

Fikile: Yes. I think this is difficult for me to articulate, because I find it easy for me to slip into the kinds of romantic ways of viewing children and children in nature that I'm wanting to unsettle. I would say the majority of my perspectives have been informed by working with children in particular places and spaces. I am always inspired in observing countless times the ways in which young children can shift their perspectives to attend to the liveliness of the more-than-human world, I think often, though not always, with more of an ease than adults. So in the research that I've been talking about with children and water, myself and the educators tried to be conscious of how we speak of water and engage with water. We found that we also at times slipped into ways of speaking that we were trying to unsettle and at the same times noticing that for the children it did not seem as difficult to speak of and with water in ways that noticed its liveliness and agency, including asking for permission, which is one of the teachings that Marleen gifted to the children and educators. So again I am wary of romanticizing and reinforcing a "naturalized" relationship between children and the more-than-human world but also want to emphasize that I learn a lot from children all the time, including their affective, embodied expressive ways of learning—which that can come about when these ways of being are nourished and encouraged to flourish in educational settings.

Maria: Right. That's so cool. You've already started to exemplify this so much in all of your other responses, but you're very helpful and reminding us about the pedagogical encounter—whether it's in a classroom or beyond—and it's dynamic and it's not sometimes intentional and sometimes just always already emerging. And so I'd like to spend some time going more specifically in that direction a little bit of talking at teachers or educators in a broad sense. And so can you discuss a little bit how you enact or invite educators and, however, you conceptualize that to provoke something new or unfamiliar or like you referred to earlier, the unsettling. So how do you invite somebody to start enacting or imagining that in their work?

Fikile: Yes I think I would preface that by saying that I don't have a how to and that this has been possible within the specific contexts where I have worked. In these contexts there have been I think two things that have been vitally important in working with teachers: pedagogical documentation and learning circles (Pacini-Ketchabaw et al., 2014). So, for instance, in the kindergarten classroom, I would spend once a week at the creek with children, teachers, as well as often with Marleen or another graduate research assistant. Pedagogical documentation was a really important way for us to collaboratively and critically reflect on what happened, to ask each other challenging questions, and to collaboratively think together about ways to extend or build on what had emerged the previous week. As part of our documentation for that week I might include a quote for the educators to think with and to collectively push our thinking further about the pedagogy and curriculum that we were trying out. Alongside the pedagogical documentation we also worked with what we called learning circles, where on a regular basis I met with educators to discuss more in depth our documentation alongside readings that I would have shared beforehand to stretch our thinking and bring multiple perspectives to make meaning of children's learning and our pedagogies. I have found these two things together, pedagogical documentation and learning circles, to be really helpful in inviting educators to collaborate in thinking and doing early childhood pedagogies differently. Also not to say that that's necessarily always a comfortable space because it's also about bringing difficult questions about our practices and the taken-for-granted assumptions that inform them—so I am asking for educators to be vulnerable about their practices. For instance, in my book *Decolonizing Place in Early Childhood Education* I've written about how, particularly when it comes to engaging with issues of settler colonialism, that can be really a difficult, unsettling space.

Maria: That's helpful. And thinking about the unsettling as a productive or generative moment rather than the achieving of the outcome. In education—and I appreciate it that you also mentioned this—we don't have a "how to."

So often teachers and educators are so interested in the how to, but sometimes the unsettling is the generative exchange or the pedagogical encounter.

So you kind of started to speak at this too, but considering the colonial legacy spread throughout place and practice and ways of knowing the educational experience, can you discuss how you witness or refigure these tensions in your work? And by doing so, or is there anything that is rendered thinkable or actionable for you?

Fikile: I think I would come back to the pedagogical documentation as a starting point; an artifact that materializes, through video, photos, children's and educators' words, our everyday encounters, and then also becomes a site to witness and question what it is that is made visible and what is erased or marginalized—so perhaps the pedagogical documentation helps to make "otherwise" ways of thinking possible. In terms of what is then actionable, that is more difficult. Some questions come up and are taken up, and others less so. I would say though that even those provocations or questions that are not taken easily up directly in practice or perhaps bring some resistance are also still doing something, if that makes sense. That interruptive "doing" can take many forms, some of which is when I'm with children and teachers, and some of which is in returning to those moments by myself. So, for example, I have written about how returning to unsettling moments was inspiration to think with what it might mean to refigure what was actually present in those everyday moments, and for me that meant (re)storying places by foregrounding multiple marginalized stories of particular places.

Maria: There's this tension too in the temporal as a teacher—like the moment when you're with the children or with the students and then the moment after that you come to these different kind of insights to how you might have troubled what we did as a teacher before, or the things related to the colonial impacts of our practice in that moment after the fact. And so the temporal for me and also in the context of the Anthropocene, which is bounded by a different dimensions of time, but feels extra complicated for teachers too and so I appreciate you sharing that. And in the example we see the tension between the moment with and the moment after of having those emerge and be visible either to you or to folks that are in the classroom too.

Much of your scholarship highlights unique partnerships with communities beyond academia, and you've given some of those examples so far in our conversation. So in times like the Anthropocene or Anthropocenes, how do you view your work within or becoming with these multispatial or transdisciplinary contexts, or the partners that you choose to involve or invite, or do they come to you in your work?

Fikile: Thank you. I am really excited by the community-based work I have been involved in, in San Marcos with the elders at the Indigenous Cultures Institute that I mentioned earlier. Importantly for the opportunity to witness and document the important land and climate change education that they are doing. While there are specific teachings that belong to that particular place, I do think there are important lessons on relational, reciprocal, and anti-colonial science that can have broader resonance, including in public school contexts. I should also mention that the work with kindergarten children and teachers in Austin was in an alternative school where there were already openings to do place-based pedagogies. So I'm really interested in what might be possible in formal school settings.

Maria: Right. Yeah. I think that's a very helpful reminder because there are so many constraints that get put on public school teachers and formalized school settings that places like a camp or informal space or even walking down the street in some regards have more mobility to work and to de-colonialize particular pedagogies or ways of knowing the world in schools presents a particular kind of challenge for this work. That's beyond many of the settings and where the work is currently happening too.

Do you have any insights into how to navigate that? I work close with teacher education, and I am very attuned to the complexities of the responsibility of a teacher and the tension and the dynamics. And I think many of our potential readers probably will be interested in that as well. And so I'm curious how you imagine navigating that space too with de-colonial pedagogies or decolonizing science in the classroom or nature.

Fikile: Yes. I think that are always possibilities to try something that makes a difference, however, small. So in the Texas school context where I worked, the teachers still had to make connections between what we were doing and the Texas Essential Knowledge and Skills (TEKS). I learned from the teachers in that respect in how they are able to powerfully express our work in connection with and also surpassing those learning requirements. I don't have specific advice in navigating these complexities; I just want to highlight that I learn from teachers with respect to the ways they navigate working within, subverting, and going beyond what the public school system requires with respect to what it means to teach and learn science. The place-based aspect of this of course comes with particular challenges and I again want to emphasize the uniqueness of being able to, as we were, spend an hour outside by a creek each week.

Maria: Right. That's a really good point. And I think an important reminder, especially when we think about working with teachers, that they have the particular skillset when merged in these collaborative ways that can start to render new ideas about decolonizing pedagogy thinkable that weren't maybe before.

So the last question I have is related to the stories and storying and how you find that helpful or in ways that might trigger something different or reconfigure how we move our responsibilities, whether they're temporal or not. How do you see storying and storytelling coming into play for your work or how you conceptualize the nature–culture kind of dichotomy?

Fikile: Yes, I don't think I can emphasize enough how stories are central to everything to me—and this comes from storied teachings I have grown up with in eSwatini as well as what has emerged in my teaching and research in the North American contexts—so yes stories as pedagogy, curriculum, methodology, theory, as ethics and more. I could say a lot about stories and nature/culture divides but perhaps here I would say that old stories are important and at the same time I'm always thinking about ways to tell stories that are not freezing Indigenous people and knowledges to the past. So, for example, asking what are the stories, including science stories, that we can tell that are for Black and Indigenous futures. I wrote an article with my dear friend and colleague kihana ross on thinking with Black speculative fiction storytelling and what those might do in imagining what else is possible for Black children's futures outside of current anti-Black formations of schooling, where this also includes resisting deficit constructions of Black children's learning relationships with the more-than-human world (Nxumalo & Ross, 2019).

Maria: And so I know that when we can talk about or present an alternative story of the future, it renders a new way of being into existence, even in the present moment or the present day. When we talk about futurities in a ways that are unfamiliar or inspiring in regards to the present moment it helps us complexify what it means, the power of time and the conditions in which we move within this moment, whether it's physical or virtual. The stories then allow us to navigate a different plane of being, so it's helpful to hear you talk about the specificity also in that work around Black or indigenous communities and why it's important to render a new image beyond the present moment to escape or to de-colonialize the present moment.

Fikile: Yes and in the article kihana and I are the ones who create the story but I'm really interested in what this could look like working with young people to create situated and anti-colonial speculative science fiction stories, and I am sure there all kinds of examples of this already. One person that comes to mind is Dr. Stephanie Toliver, whose work with Black girls and speculative fiction storytelling is amazing and inspiring.

Maria: And what stories they might tell for themselves that exceed their current moment. Yeah. That's a powerful idea too. I think just the idea that educators might start with stories. So often teachers or educators might bound their pedagogy to a "lock step" lesson plan, but when we just tell a story of a moment, whether it's past or future or non-existent, it produces a new way of coming to know or be.

Fikile: I think you put that really well.

Maria: Now I'm thinking back to the constraints of the classroom experience versus the space that a camp affords or an informal space. And that when or if you invite students to write these speculative futures for themselves, that they might be bound within a school wall or at the institution of a school, but can they write another way of being or beyond that space too?

Fikile: Yes. I think this is really important, particularly in thinking about Black and other historically marginalized children and how to complicate and tell different stories of what it means to belong to certain places, including so-called natural places. So I want to think with what are the possibilities, whether it's through speculative storytelling or through some other pedagogies, about how to interrupt colonial, capitalist, and racialized nature/culture dualisms.

Maria: Right, yeah, that's helpful I think. In science education or nature studies, there's always the leaning and the role of capital. And so I think that's a tension in the desire to capitalize financially, economically, or socially on non-white bodies in ways that continue to reproduce the oppressive conditions of the colonial narrative that we might be trying to escape or destabilize or unsettle.

Fikile: Yes. Thank you.

References

Nxumalo, F., & Ross, K. M. (2019). Envisioning Black space in environmental education for young children. *Race, Ethnicity & Education, 22*(4), 502–524.

Nxumalo, F., & Villanueva, M. (2019). Decolonial water stories: Affective pedagogies with young children. *International Journal of Early Childhood Environmental Education, 7*(1), 40–56.

Nxumalo, F., & Villanueva, M. (2020a). (Re)storying water: Decolonial pedagogies of relational affect with young children. In B. Dernikos, N. Lesko, S. D. McCall, & A. Niccolini (Eds.), *Mapping the affective turn in education: Theory, research, and pedagogy*. Routledge (Invited peer reviewed book chapter adapted from article in *International Journal of Early Childhood Environmental Education*). https://doi.org/10.4324/9781003004219.

Nxumalo, F., & Villanueva, M. (2020b). Listening to water: Situated dialogues between Black, Indigenous & Black-Indigenous feminisms. In C. Taylor, J. Ulmer, & C. Hughes (Eds.), *Transdisciplinary feminist research practices: Innovations in theory, method and practice*. Routledge (Invited peer reviewed book chapter).

Pacini-Ketchabaw, V., Nxumalo, F., Kocher, L., Elliot, E., & Sanchez, A. (2014). *Journeys: Reconceptualizing early childhood practices through pedagogical narration*. University of Toronto Press.

CHAPTER 21

In Conversation with Vicki Kirby: Deconstruction, Critique, and Human Exceptionalism in the Anthropocene

Vicki Kirby and Marc Higgins

Using and Troubling the Anthropocene

Marc: We find ourselves in strange and unprecedented times. As you state, the "most pressing questions about the achievements of science or about environmental dramas that threaten species diversity and human survival require stories that are heavily reliant on scientific evidence for their political credibility and *gravitas*" (Kirby, 2017, p. 7, emphasis in original). Science, and in turn science education, present themselves as both poison(s) and panacea(s) in the ways in which we can both imagine and enact our response-ability to this contemporary moment that is often referred to as the Anthropocene.

Within your 2018 piece, "Un/limited Ecologies", you make this argument both explicitly and carefully. Particularly, through a beautiful reversal and (re)opening you ask us, in this moment of urgency in which we are politicizing taken-for-granted understandings of ecologies, what it might mean to understand politics as ecologies: finding fissures in the statement that the current moment is caused by humans and the ways in which this admission of culpability might mask more than it reveals.

V. Kirby (✉)
School of Social Sciences, The University of New South Wales, Sydney, NSW, Australia
e-mail: v.kirby@unsw.edu.au

M. Higgins
Department of Secondary Education, University of Alberta, Edmonton, AB, Canada
e-mail: marc1@ualberta.ca

© The Author(s) 2022
M. F.G. Wallace et al. (eds.), *Reimagining Science Education in the Anthropocene*, Palgrave Studies in Education and the Environment,
https://doi.org/10.1007/978-3-030-79622-8_21

Could you speak to the ways you are witnessing irruptions of anthropocentrism in the ways it is resisted and why these particular slippages are significant in how we conceptualize and mobilize our responses to the Anthropocene?

Vicki: My answer risks being a bit long-winded for two main reasons. I'd like to say something about how I arrived here, because these concerns take me back to my early studies in anthropology which in some ways inform my current position. And not unrelated, the knot of investments that preserve the identity of *Anthropos* as an analytical departure point in most of the literature involve investments and assumptions that I want to interrogate further. So, to begin with the discipline of anthropology, it seems fair to say that it strives to document and understand the myriad behaviours, mores, and beliefs that reflect what it is to be human. There is surely an irony in the perception that the unitary essence of human exceptionalism is secured in the sheer diversity of societal and cultural expressions; indeed, our ability to interpret the world in comparatively idiosyncratic and inventive ways is said to be remarkable among other animals, as culture effectively generates our world as a meaningful place. When I began my studies I was aware of the diversity of sexuality and gender roles across different societies and histories, and perhaps more compellingly, I had always been exercised by the vagaries of moral and ethical belief across cultures and even historical moments within the "same" culture. And yet the appeal of this inclusion by exclusion, namely, the belief that we fabricate a world rather than respond to its enduring and universal truths, became something of an obstacle as my research progressed. Was there really no access to a reality outside or before human mediation? This is how the question is conventionally configured, as if two quite separate systems, or entities— nature and culture—pre-exist their possible encounter and interaction. As I describe it above, we tend to think of nature as a steady and stable sameness that endures, its truths unchanging, whereas cultural insights are mutable and always shifting. But what are the implications of refusing this division and its attending logic?

I was fascinated by the riddles that anthropology generated but also taken aback by the defensive rhetoric that surrounded them. Yes, there was an implicit commitment to cultural constructionism well before the mantra took hold in emerging disciplines such as cultural studies. However, the appeal of fieldwork's immersive cultural experience—you can't presume to understand a society from the outside—assumed quasi-scientific credentials through its detailed evidence gathering "on the inside". It was as if cultural relativism was both true and false, as if the inherent integrity and enclosure of a particular culture prevented access even as it revealed its mysteries to a specialist few. If we concede that we inherit an invisible legacy of accumulated interpretations and subjective prejudice that misidentifies our particular cultural/historical inheritance as a universal, shared, and self-evident reality, then where are we standing when we interpret *other* cultures? In anthropology, the threat of cultural relativism seemed grafted into the discipline's every assumption, and

yet the practice of ethnography erases the question of how these purportedly separate systems of "making sense" are translated by, and into, something
apparently "other", foreign and external.

I think these questions about the unifying term "culture" and its capacity to
accommodate myriad modes of being that are considered separate yet strangely
inseparable and entangled (hence, their translatability), rehearse larger conun-
drums regarding the special status of human identity versus its non-human
others. To explain this, we tend to align being human with the unique ability
to "world" a world through language and representation. Consequently, to be
human is to be cultural, to reinvent our world to reflect specific human needs
and understandings. However, the elephant in the room is nature; not the
"nature" generated within the hermeneutic circle of cultural interpretation—
nature as a cultural artefact that we misrecognise as what is not culture—but
a nature whose identity, at least according to cultural construction, pre-exists
human arrival and escapes representation. This escapee nature surely complicates the truth claims of the sciences because a history of political (cultural)
prejudice inevitably informs all of human endeavour, albeit to different effects.
And if scientific inquiry can't remain immune from the cultural life that
produces it, is the belief that nature and culture are two quite separate systems
of operation still plausible? After all, we board planes that provide us with rapid
transport, take medications that cure myriad ailments, and rely on algorithms
to deduce forensic and scientific evidence. If we invest in the hermeticism of
cultural construction, then how can the models and representations forged in
the sciences presume any working purchase whatsoever?

One way into the conundrum is to query any commitment to the in-
itself of culture, whether defined against nature, or whether conceded internal
variation in the myriad cultures whose different locations and expressions a
more traditional anthropology tried to capture and define. The implication
here is that an escapee nature, one which I previously described in terms of
its "enduring and universal truths", will also lose its referential status as a
foundational "something" against which change can be measured. No longer
outside and other to culture, it reappears as a force and energy intrinsic to
being anything, including being human. And this is where things get interesting. Instead of assuming that to be human is to be radically different and
separate from the world, a belief that understands the activities of science in
terms of a tripartite division between human subject, intervening model, and
object of analysis, the ability to scrutinise how the world works, to invent
technologies and languages as instruments of inquiry, are no longer unique to
being human. The inquiring subject now manifests in worldly form, and all its
possible individuations (of itself) appear as specific manifestations of a larger
field of agentic forces. This way of thinking finds the subject enfolded with/in
the object; however, this is an auto-affection that is inherently heterogeneous
and diffracted. Just as anthropology divides culture into myriad individuations,
assuming their differences are somehow both separable and yet inseparable
(such that the opposition no longer holds), we could say that the world

individuates itself infinitely, and yet these differences from and within itself (technologies) are not as autonomous and separate as they appear, but rather, articulations of what Jacques Derrida calls "systematicity". Although various authors in critical theory have emphasised this sense of a shared world, a world in which relationality gives us our respective identities, I want to nuance this assertion to better explain a more counterintuitive sense of relationality. "We are part of the world" is now a standard reminder that we are not innocent or separate by-standers in regard to the world's well-being. And yet this sense of "the part" fails to capture the ontological complicity of individuation. For me, we are not so much a component, or individual member of a larger world picture, one that might be likened to a jig-saw puzzle where each piece is local and situated[1] while still necessary to the whole picture's possibility. In other words, I am not trying to evoke an assemblage of parts, but rather, a sense that any "one" part is always/already an articulation of and by the whole. Thus, the limits that secure "one" identity are diffracted through and by the whole, both local and non-local at the same time.

If we generalise such formative complicities then there are only intra-active ecologies, congealings of spacetime whose apparent differences are transversally implicated. When I liken political positions and ideologies to ecologies it is within such a framework of profound intra-dependence. This means that culpability is a fraught notion because causal forces don't arise in a single origin or author: if identity (of whatever sort) is internally diffracted, then we begin with structural entanglement rather than clean and isolated identities. This is not to endorse the rejection of, say, anthropogenic climate change or to celebrate political quietism, as if one position is as good as another. But it does question the identity of *Anthropos* as an autonomous agent, able to realise his intention to destroy or remediate, as if it is his decision to make. For example, the profligate and violent behaviour of humans in our squandering of resources and indifference to the consequences is a comparatively easy criticism to make. This is a confession, a mea culpa. And yet there are questions to be asked about such a gendered and racialised understanding of power that identifies the other, nature in this case, as inherently passive, incapable, the dumb and helpless victim of human action. To admit that we are subject to forces that far exceed our ability to control them, and that our individual identity and agency is a fractured ecology of seemingly alien chemistries and "creatures", such that "nature speaks us", leaves us feeling uneasy. ~~I'm going to suggest here that life/nature is always out of step with itself and in constant argument about~~

---

[1] Much cultural analysis has countered the pretentions of a universal, god's-eye-view perspective with the modesty of "situated knowledges" that are experientially grounded and therefore, more politically and ethically pertinent. Indeed, science has often been seen as the culprit that denies its embodied location in the rush to secure its status as objective, repeatable, universal. Although these have been important interventions, the separation of the universal from the situated and local, the objective from the subjective, or the notion of what is rational and abstract from what we claim as embodied experience, is considerably more confounded than a choice of sides can explain.

~~what its myriad identifications imply, demand, secure or prevent.~~ If there is no stable integrity or final reference that can still this dynamic, neither God nor a transcendental truth, then our arguments will need to be alert to the flows and vagaries of our shared situations: how we engage specificity can matter profoundly.

Situating and (Re)Committing to Deconstruction at the Ontological Turn: "What if Culture Was Nature All Along?" (Kirby, 2017)

Marc: As Latour (1993) frames in *We Have Never Been Modern*, the ecological problems that we face and attempts to account for and be accountable to are at once cultural, natural, and deconstructing; in turn the ways in which we formerly relied on bracketing out nature to make strong cultural claims (e.g., sociology), bracketing out culture to make strong naturalistic claims (e.g., science), and rupturing the stability through which truth claims are made (e.g., deconstruction) might no longer be sufficient.

However, because deconstruction subverts often taken-for-granted assumptions about mediated access to an external natural *or* cultural reality (and, more recently, subverts the binary distinction between the two, as you do in your work) by subverting their stability, it is often and always already unwelcome within the fields of science and science education. Or, as Barad (2011) states, Derrida is "the science warriors' darling stand-in for all that is wrong with the humanities" (p. 448).

What might it mean to commit to deconstruction in fields wherein it might already be a precarious position from the get-go? Or, how do you leverage that ambiguity towards productive ends? Further, in this contemporary moment, what can deconstruction (after Derrida) offer, perhaps not as corrective, but as productive orientation towards responding otherwise?

Vicki: If we set up an opposition between what science claims to achieve versus the value of insights from the humanities and social sciences—and I agree that such divisions are routine, as the pragmatic concern in your question makes plain—then we inevitably find ourselves in a structural impasse. How to reconcile entirely different and even contradictory truth claims, modes of valuation and methodology in these different endeavours? Is the only response to insist that "the two cultures" are irreconcilable, and leave it at that? I suppose my own strategy is to question the ways in which we quarantine these different practices and explore what might be at stake in maintaining their respective autonomy. Feminism, for example, was especially interested in the phenomenon of hysteria in the late eighties because it appeared that political discrimination could be somatised (Gatens, 1983; Irigaray, 1985). However, Elizabeth Wilson made the important point that much of this commentary, despite its explicit assault on the nature/culture division (which carries myriad discriminatory associations), failed to ask how biology could "perform" these politically charged symptoms (stigmata, hysterical blindness, dermographism, etc.) (Wilson, 1999). Today, scientific evidence in epigenetics tells us that the effects of difficult social and psychological circumstances are inheritable across generations, so perhaps ironically, it is science that questions the definition of culture as the other of biology. Brain plasticity is another example that illuminates how cultural and social activities are cerebrally registered. And there are arguments coming from the sciences that "the knowing subject" is an impossible fiction. Indeed, myriad examples support what we might call post-structural insights into the complexity of reality.[2]

However, when you get down to the finer descriptive details in many of these examples there remains an almost automatic need to locate agential efficacy in culture, as if that's the only way to understand the direction of these dynamics. Even the term "epigenetics" implies it, something I hadn't realised until I was brought up sharply by a conversation with Astrid Schrader, a scholar whose work in deconstruction and science studies is full of intriguing challenges and insights.[3] I recall her saying rather matter-of-factly, which only underlined the point, that the "epi" in epigenesis that signifies supplement, addition, something nearby (which is how we think of culture's relationship with nature) was redundant and misleading. "It's just genetics" she said, reminding me that, because the gene isn't closed, there is no addition. Although the causal logic from culture to nature tends to remain intact—and the sciences and humanities are both invested in preserving it—I've always found that students are especially intrigued by this particular riddle if you guide them through its surreal logic. How, for example, does biology—meat—*read* the variety of culture's messages and enact its corporeal theatre through tissue, blood, nerves, chemical arrangements, and so on, if meat is deaf to the literacies of culture? Put simply, if it isn't in the nature of biology to speak, to read, to think and metamorphose, then how do we achieve these behaviours? And why do we assume that this "we" that acts is not a biological dynamic?

I guess I'm trying to say that an intervention that strives to bring the humanities and sciences into some kind of dialogue or disruptive displacement has myriad sites of potential engagement because there is cross-over, dispute, and ambiguity on both sides of the ledger. Yes, there may be opposition from the sciences about the value of something like deconstruction just as there is in the humanities, but that's because "deconstruction" is so misunderstood. Why brandish that particular banner if your audience has already turned the page? For example, you cite Bruno Latour, a thinker whose ethnographic work has significantly reconfigured how a scientific truth can emerge from a cacophony of forces. The scientific apparatus for Latour is arguably a *field*

---

[2] As references are myriad, a quick scan of the cover stories from *New Scientist* surely underlines the point. I mention just two. "Reality: The Greatest Illusion of All" (August 2019); "Memory: The Exquisite Illusion That Creates Our Sense of Self" (October 2018).

[3] For a fascinating argument about the ontological shapeshifting of life forms that is representative of this thinker's innovative perspective, see Schrader (2010).

of involvements whose distributed agencies and networks include the non-human. However, despite enabling a more comprehensive understanding of truth production and what constitutes evidence, Latour remains aggressively dismissive of deconstruction, which he interprets as a destructive enterprise, smugly returning us to the prison house of language—"a brain in a vat"—that is hermetically sealed against the material world that he wants to investigate (1999a, 2003).

Although this is an all-too-common misreading I doubt that a corrective will achieve very much because Latour's appeal is his apparent ability to resolve a problem. So perhaps a more effective strategy that Derrida himself recommends (1994–1995) might be to read Latour's achievement grammatologically, acknowledging his insights while magnifying and extending their implications further. Let's take "Circulating Reference: Sampling the Soil in the Amazon Forest" (1999b) as our example. Latour provides us with a moving and evocative image of the intrication, or sociality, of different people, languages, plants, soils, make-shift technologies, and idiosyncratic behaviours, all of which are nevertheless "reduced" into words on a page, letters that translate the Amazon forest into what Latour holds in his hands at the end of the exercise. Questions about translation, indeed, transubstantiation, are alive in this scene, and the notion of reduction as "less than" is overturned and made strange, for the words evoke a vastness, a worlding, whose reference has evidential leverage. Latour brings genuine wonder to this scene of radical metamorphosis: how did all these different ontologies prove so intimate in outcome? And how could the world be here, on a page? I think we can redeploy the terms and investments in Latour's argument to offer a more deconstructive reading, shifting his rigid (albeit disavowed) commitment to anthropocentrism and the politics of aggregation as we go. Remaining faithful to deconstruction's canonical format, vocabulary, and disciplinary commitments can elide the grammatological mysteries and fascinations within texts that we would otherwise criticise and reject too quickly.

Marc: A recurring theme in your work is the question of *What if Culture Was Nature All Along?* through which you address how deconstruction might differently respond to and disrupt the nature/culture binary. Of particular importance, your work explicitly takes up what it means to consider as co-constitutive nature/culture without reverting nature to a pre-critical status; or, along Spivak's (1976) line of thinking, that "to make a new word is to run the risk of forgetting the problem or believing it solved" (p. xv) through reproducing the problem elsewhere, albeit differently; displacing rather than disrupting. Significantly, and repeatedly across your scholarship, you call us to come-to-know the philosophical history which has come to inform this recent turn so that we do not differentially repeat ourselves: a trajectory from humanist, to anti-humanist, to post-humanist approaches.

Could you carefully take us through, albeit with broad strokes, the trajectory through which you understand the relations between nature, culture, and representation as they are articulated within humanist, anti-humanist, and post-humanist approaches?

Vicki: Again, I think I'll have to answer your question in a rather round-about way because if I'm being true to the detail of what motivates my own practice I'd have to say that this notion of "co-constitution" has always bothered me. In fact, I've never really liked the corollary expression, "nature–culture", either. My reservations might sound pernickety given the routine use of these notions in cultural and critical theory circles. But as we are getting to the pointy end of what is at stake in these arguments I should take the opportunity to clarify my position. This will also allow me to better explain why I've glossed Derrida's "il n'y a aucun hors-texte" ["there is no outside (of) text"] as "there is no outside (of) Nature". First of all, you rightly note that my aim is to leverage nature out of its pre-critical position as "the before" or "primordial", that material "stuff" that lacks the more evolved capacities of language, agency, and intelligence. Such assumptions rest on a circumscribed concept of identity, because even if we acknowledge the genesis or emergence of something we tend to assume that this processual dynamic is in the past, before identity "set" into something unique and locatable. However, I'm not trying to give nature a Latourian dynamic whose extent, by definition, will nevertheless remain circumscribed (because nature for Latour isn't culture—he remains committed to the human as master of the dance). Instead, I want this sense of one plus one to collide and fracture, not into ever tinier aggregate parts, but something whose mysterious involvements defy simple division. In my teaching, I used to evoke this sense of an originary energy that remains alive and ubiquitous "throughout" a field, or system, by using "magic-eye" images that were popular some decades ago. You look at a mess of electric colours and lines, you relax your focus, and suddenly you see a 3-D vase of flowers, a heart inside a box, or perhaps three dolphins. My aim was to under-line that these images are generated from *within* the same frame, that what appears to come second is already alive with/in "the original". This differ-ence without addition, a discrimination that requires no distance, complicates the concept of difference as something straightforwardly other and elsewhere. For me, "nature–culture" looks too much like a solution, an amalgam whose respective differences remain identifiable. And just as Spivak warns, we see this recuperative tendency at work in the appeal of the cyborg and in earlier versions of feminism's critique of the Cartesian mind/body split. Such inter-ventions can seem to resolve a problem when they underline inseparability and interdependence and yet they remain committed to difference as an amalgam. The popularity of supplementary logic is that the difficulty we are engaging is perceived as a problem to be solved rather than an enduring and constitutive dynamic. Making these differences adjacent might feel more satisfying because hierarchies are deemed to be bad, avoidable, wrong. However, to read is to hierarchise, to learn, evaluate, and adjudicate involves hierarchies, to reject a binary for an apparent non-binary is to hierarchise (and inadvertently bina-rise). If we just say yay or nay to a binary, as if we are confronted with a moral dilemma, we remain blind to the incestuous structures, the perversity and impurity that discovers another perspective *within* the one that seems to refuse it.

With these considerations in mind, my aim is to reframe what is conventional in our thinking by acknowledging how "systematicity as such"—a heterogeneous "unity" wherein "firstness", or "originary *différance*", remains ubiquitously at work—can accommodate myriad perspectives within one perspective, a bit like a nesting of Russian dolls but without distinct, internal borders that neatly separate one perspective from the next. To refer to this transubstantiating energy as "nature" challenges what is routine in humanist, anti-humanist, and even many post-humanist assumptions regarding human exceptionalism; developmental narratives of progress and increasing complexity; as well as language and cultural representation as relatively recent and unique technologies. Without an absolute referent against which to define, locate, and identify anything in absolute terms, the historical framing of humanism against anti-humanism and post-humanist approaches is as useful as it is misleading. Where, for example, should we place Spinoza, or Leibniz, because their writings are in many ways exemplary of deconstructive criticism, and yet deconstruction had yet to arrive ... or was it already at work?

Marc: This last move is not a small pivot and shift. As we collectively move towards considerations of ontology you invite us to use and trouble Butler's criticism of particular cultural uses of "natural facts" as they still stand today: "'natural facts' are always informed by cultural bias" (Kirby, 2011, p. 94). Further, this is "one of the most important contributions that scholars such as Butler have made, ... any return to the question of Nature will need to accommodate or reconfigure such insights rather than put them aside" (p. 94). Yet, to consider the ways in which matter comes to matter, you state that this is "not to suggest that we need to 'get real' and add Nature's authorship to this strange text, as if Culture's inadequacies might be healed with a natural supplement" (Kirby, 2011, p. 13).

This question bears coming at it once more, differently: particularly because, as you mention elsewhere (Kirby, 2017), the insight that "*relationality is not mediation; it is not an in-between entities*" (p. 11, emphasis in original) often gets lost. Could you elaborate upon this?

Vicki: This is a wonderful question that allows me to dilate on what must seem like a contradiction in my commitments. Why would I want to respect the oppositional logic that underpins Judith Butler's understanding of nature, the body, and matter if elsewhere my aim is to undermine this way of thinking? Can we get out of this conundrum and resolve its stickier inconsistencies, or does this sense of being caught up in the very thing we want to question and refute set the scene for a different heuristic, a different understanding of what this new buzz word, "entanglement", might involve?

Let's begin with Butler. I was approached by Continuum to write *Judith Butler: Live Theory* (2006), and as I knew Butler's work well I was confident that I could despatch the task quickly. However, as I surrendered myself to the intimate detail and pulse of her arguments I realised how closely her commitments tracked with my own, even though her departure point and its associated assumptions—namely, there are no outside *cultural* (human) representations of a world whose extra-linguistic reality remains inaccessible to us—inadvertently recuperates the very binary divisions and circumscribed identities she strives to complicate. However, this is not a simple mistake that can be put aside. I share Butler's appreciation that "structure" and "system" are organisational, allowing culture to constitute entities "from within itself", that is, that "the how" of identity formation is inherently systemic and entangled. Butler offers a compelling argument that illuminates why nature is not the other of culture; why nature is not a passive, feminised, and racialised "primitive" whose difference provides the reference point against which masculinism, racism, and today, anthropocentrism, might justify their superiority and developmental complexity. However, to name and enclose that systematicity as *properly* human/cultural commits Butler to notions of identity and integrity that undermine and contradict the overall direction of her argument: power becomes "power over", its productivity largely negative. The sign becomes an entity *within* a context of different/other signs rather than an articulation *of* that context with no outside "itself". Nature, now under erasure, reappears as radical alterity, "something" that pre-exists human arrival, language, and mediated representation. To make the point succinctly, Butler has no way to even consider that the nature she excludes from the chattering activity of cultural production has the capacity to re-present itself *as* cultural production.

As my position vis-a-vis Butler involves an acknowledged intimacy my strategy was to follow her logic to the letter as best I could and ask questions when she seemed to depart from her own insights: what justified the integrity of the boundaries and limits she presumed when her entire argument called integrity of any sort into question? How could she justify difference as a gap *between* entities as if these entities pre-exist their ongoing manufacture through *différance*; why does she assume that the body that speaks and reads is itself illiterate, returning us to the Cartesian subject whose residence is necessarily outside or other than the body/nature? I tried to ventriloquise her arguments as if I were inhabiting them, wearing them like a piece of clothing but to very different effect.

The reason why arguments that privilege "cultural construction" are, to my mind, not simply mistaken—at least, not in the conventional sense that might hope to repair an omission or correct an error by adding something presumed to be absent—is that "systematicity" complicates this logic. For example, when I displace the conventional reading of "no outside textuality" with "no outside nature", I haven't really added anything. And yet the systemic implication of what we understand as "productivity" is radically transformed, and all those entities or capacities regarded as primordial, "the before", begin to manifest and resonate throughout the system, and vice versa.

These are really difficult and subtle concepts to negotiate, and Derrida's insistence that identity is not secured by an "in-between"—because there are no entities that would anchor that sense of spacing/timing—is a challenge that continues to exercise my attentions. How to represent this extraordinary insight when conventional understandings of representation and perception "work", at least on the surface, by denying these entangled onto-epistemologies? I think it's a question of timing. If I think of my students who are naïve empiricists when I meet them—the world is what it seems, questions of phenomenology and subject formation are entirely unknown to them, the "rational subject" is, indeed, "the one who knows", language is a thing that we acquire, a thing that separates us from everything that is not human—I can only disrupt that worldview and allow a more generous and complex appreciation if I work *with* their convictions and interrogate their hidden reasoning. A bit like my strategy with Butler, I need to challenge their commitments by using the logic that makes sense to them, because on closer inspection, that same logic will manifest errancies and slippages whose disavowals and defenses can prove insightful. I'm not really introducing something new that isn't already at work within their very own worldview. To this end, what might make the difference in providing those students their aha! moment could be an excerpt from a science journal, Butler's work, or even Descartes', something apparently conservative and wrong. Because surely, all these texts are already alive to each other, open to different interpretations and perspectives. Different worldings can emerge, but from *within* one perspective rather than as separate alternatives.

Critical Consequences: Critique After the Critique and Subject of Critique

Marc: The consequences for the earlier (and above) are multiplicitous: one which you continue revisiting in your scholarship is what it means to engage in critique *after* (or perhaps more within) the critique of critique. Or, to be more precise, that criticality stemming from the realization that critical negation has "run out of steam" (Latour, 2004). As you state, "critique is a messy business that can surreptitiously recuperate and affirm what it claims to reject" (Kirby, 2018, p. 122).

Can you speak to the importance of attending to snags and irruptions not as *flaws* to point out in ways that dismiss the entirety of the argument, but rather as grammatological opportunities: openings to deconstruct the textuality of that which lay before you to pursue meaning-full-ness? Also, why is it significant to engage an affirmative critique that emerges out of proximity rather than distance?

Vicki: My previous answer illustrates something of the manoeuvres that might promote a more affirmative practice, but it's worth emphasizing this issue because negative critique carries many traps and pitfalls that can ambush the wary as well as the unsuspecting. The perverse antics, errancies, and unpredictabilities of intratextuality (systematicity) aren't problems that fall away and disappear once diagnosed, nor are they resolved with a change in approach that reads "generous" as a call to add what was previously excluded. The spatial logic that underpins this apparent choice between an inside or an outside—whether to include or exclude what is absent or missing before we "get generous"—turns an ontologizing, structural entanglement into a problem which a sovereign subject, presumably situated outside this same entanglement, could resolve. A more affirmative reading works with an understanding, however, fragile (because this is not a knowledge that is separate from what constitutes our own being-in-the-world), that the implications of systemic complexity are already at work in all arguments. Given this structural complicity a grammatological practice isn't better because it includes more than another practice, or because it affirms rather than rejects, or because it doesn't narrowly restrict its focus. The point here is that even a very focused argument, or one with a perspective that we might deem risible and patently wrong, is nevertheless an expression of this same, shared, involvement. An important clarification, however, is that "shared" doesn't commit us to a soup of sameness, for what is "common" is an ever-changing field of cross-referential forces whose internal differences are "in touch" even when they appear separate. An illustration of how this difference within apparent sameness might work can be seen in language use. Every individual inherits a mother tongue, or *langue*. And yet we know that every individual speaks that same *langue* in a unique way (*parole*), deploying a particular vocabulary, relying on certain rhetorical structures, slang expressions, and local idioms whose repeated patterns constitute a very specific signature of language use. Given this, we could liken a society to the collective aggregation of these independent voices which together express their individual intentions and personal perspectives. Such a view conceives of society as a federation of individuals where each person is an autonomous agent, the sovereign source of their actions and choices. However, the analytical division between society and the individual can leave us wondering about the relation between the whole and its part, or how an individual becomes social if they are not socially primed to begin with.

If we begin our analysis with a sense of an entangled "whole", an implicated, self-reflexive force field, then any "part" is not so much an entity *in* a context—in this case, an individual in a society—but rather, an individual whose specificity is a particular expression of that "same" society. Returning to our language example, we could say that *langue* individuates *itself*, or that *langue paroles*. But why should this matter at all? The relevance of this latter description is that it allows a robust understanding of specificity, not by circumscribing identity as an island surrounded by a context (an outside identity), but by an appreciation that the system reconfigures itself in specific (individual) ways. In other words, the specificity of *parole* isn't isolated and autonomous but deeply embedded in and through those entities that appear other and elsewhere.

In sum, despite renewed awareness of the consequences of negative critique, the critique of critique can operate as yet another opportunity to leverage (while disavowing) judgmental self-righteousness. It's as if the left hand doesn't know what the right hand is doing, and we are all persuaded by this naïve belief that we can truly take our distance. I recommend Ash Barnwell's (2017) wonderful engagement with this dilemma in "Method Matters: The Ethics of Exclusion". To take one simple illustration from her argument, she returns us to Latour's irritation with critique as "essentially limited and destructive" (p. 30), citing his comparison of critique with the destruction of a hammer. According to Latour, a hammer can undo, but it can't compose. However, Barnwell's response is, as she describes it, "prosaic"— "hammers create and build homes, repair cars after panel beating ... the list is endless: destruction from one perspective is hope and restoration from another" (p. 30). It is this paradox that sees one identity, position, or methodology within another—such that "their" respective differences are strangely confounded, that to my mind is more politically and ethically provocative and dare I say true, than those arguments that seek to make a virtue out of distancing themselves from the very errors which they unwittingly recuperate and promote.

Marc: Another of the recurrent themes within your work is the notion that Cartesianism is not so easily remedied. Not unlike Butler's (2005) line of questioning "does the postulation of a subject who is not self-grounding, that is, whose conditions of emergence can never be fully accounted for, undermine the possibility of responsibility and, in particular, of giving an account of oneself?" (p. 19), you invite the question of whether a decision and its associated notions of responsibility and forethought can exist without Cartesian notions of self-possession. Particularly, after the epistemological uncertainty of the subject as posited by Butler, you offer an ontologically indeterminate subject as the location from which critique emanates.

What might it mean to take seriously the notion that if the subject is not separate from its object, the task of criticality too becomes less self-evident?

Vicki: Let's set the scene in order to appreciate the question's difficulty. Responsibility and accountability presume a sovereign subject who has full command over his actions and is "at one with himself" such that he can realise his intentions.[4] In other words, how he acts is his choice, his decision, and this allows us to apportion blame and determine culpability. Although this is surely a pragmatic requirement for a society's survival—and I think your question is attentive to this necessity—this doesn't change the fact that the sovereign subject as we conventionally understand him is more fiction than fact. We know that poverty, racial and sexual discrimination, and other forms

---

[4] I've chosen to use the masculine pronoun to acknowledge the masculinism that structures these seemingly straightforward and universally applicable logics.

of social denigration and suffering have deleterious effects on an individual's life chances and life choices. Indeed, the very notion of choice seems thin and rather shabby when considered against social inequality and deprivation, and this, against a backdrop of exploitation both domestic and global. However, the fiction of the sovereign subject isn't explained by social difficulty alone. The point isn't that some people aren't in a position to exercise choice, enterprise, and initiative whereas others are (although this is surely an important consideration for how a society explores the significance of opportunity and agency).

In respect to the larger question of subject formation you don't need Freud or more contemporary psychoanalytic insights to tell you that we are all of us motored by unexamined and uncontrolled fears and desires that drive and direct our behaviours. How often, for example, do we narrate a story and a friend who shared the experience counters with an entirely different interpretation; or we are forced to the painful conclusion that our positive experience of a relationship was sadly misconstrued or made foolish by the unexpected outcome of events? The evidence suggests that we remember in interested ways and that to a significant degree we live our lives in disavowal. And we do this in good faith, not knowing that the archive of our memories can prove marvellously creative. Unfortunately, the awkward apprehension that the truth can't be determined once and for all, or that the subject is duplicitous, even self-deceiving, feels like a cruel insight that leaves us with nowhere to turn.

I've tried to elaborate why the decision to say yes or no to something via the routine manoeuvres of critique can refuse to acknowledge the myriad involvements that inform the very "thing" that is in question. It's like asking, "Do you believe in God?" and being satisfied when your interlocutor responds with a yes or a no. What both answers leave intact and uninterrogated, despite their apparent difference, is the object of the question—not just the fact of God's being and possibility, but the more general mystery of being and existence that confronts us all, regardless of religious belief. To my mind, the Butlerian perspective shares something of this avoidance even as it seems to target the question of the subject by underlining its impossibility, its deficiency, incapacity, and self-deception. However, it is a sovereign subject, an "I, me" who is called upon to admit their failings and know their deficiencies.

I'm more challenged by the way Derrida broaches this "self-capture" in *The Beast and the Sovereign* (2011). He is clearly impatient with confessional declarations about peccability and insufficiency because they actively affirm a subject whose sovereign identity is made manifest in guilt, acknowledgement, and a recuperated (if surreptitious) appeal to enlightened responsibility. Although Derrida doesn't reject this approach outright, indeed, he refuses to refuse its motivational direction entirely, he nevertheless mounts a counterintuitive tactic that complicates its conceptual investments while not dismissing them out of hand. Importantly, the enigma of the subject or "who" remains open for Derrida, and in such a way that it appears displaced, a seemingly ubiquitous authorial presence whose enigmatic identity haunts his entire argument. I've tried to explain how this sense of displacement and breadth (systematicity) might be alive in/as the unique singularity of individuation by referring to the frontispiece of Thomas Hobbes' *Leviathan*, a lithograph of the sovereign, sword and sceptre in hand, by Abraham Bosse. On close inspection we find that this unified subject, this body of supreme and sovereign power who appears outside the law (inasmuch as he pronounces it), is already hostage to a populace, broken up and into by myriad vociferous divisions, ventriloquised and strung through like a dummy or marionette, the agency of this "I, me" invaded, peopled. Through such a reading, *l'état c'est moi* takes on a very different complexion that complicates individual agency, blame, and causal explanation. The point isn't that there is no subject, no sovereign agent simpliciter. As Derrida explains,

> In a certain sense, there is no contrary of sovereignty, even if there are things other than sovereignty... even in politics, the choice is not between sovereignty and nonsovereignty, but among several forms of partings, partitions, divisions, conditions that come along to broach a sovereignty that is always supposed to be indivisible and unconditional. (2011, Vol. I, pp. 76–77)

What I take from this is that we need to think again about what we mean by control, and whether either denying or acknowledging the possibility of sovereign responsibility, the authority of an "I, me", can actually do justice to the intricate political complicities that compromise such adjudications. I don't see this as an excuse for political quietism, but rather a call to think again in ways that aren't complacently routine. Although we still have to decide and take a position in regard to specific political and ethical concerns, a more robust acknowledgement of the forces that "author" a decision might allow us to better appreciate why positions aren't as stable and fixed as they might seem and why any one response might be replete with myriad perspectives and even contradictory interests.

Response-Ability and/at the Anthropocene

Marc: Not that the logics of containment or closure could come to contain the arguments made here, revisiting the guiding question of this interview, how might some of the arguments made earlier (re)open the ability to respond to this particular moment that we call the Anthropocene?

Vicki: The concept of "human exceptionalism" describes our species' unique intelligence and creative capacity, whether for calculating purposes and self-interested goals—in which case we are described as perpetrators and misfits, destroying the natural order—or inventive change agents who can remediate previous mistakes and even improve and augment our natural inheritance. By reconfiguring Derrida's "originary *différance*" *as* "originary humanicity" I wanted to destabilise the automatic belief that "the human" is supremely powerful and intelligent because no longer subject to the capricious rhythms of nature. It is common to explain human exceptionalism in terms of a break, or transcendence of our natural origins. However, as I tried to explain above, I don't think "the human" has a circumscribed identity that is easily individuated from what we perceive as outside and other than human. As the markers of what make the difference have fallen—opposable thumb, language use, brain size, forethought, upright posture and forward gaze, complex social organisation—we learn that identity is ecological through and through, a mangle of variables that inhabit and make possible (or impossible) every unique being. As I write this response I'm in self-isolation, the coronavirus my newest and most gregarious neighbour. However, the virus's ability to use my body to reproduce itself speaks of life more generally, even in this case when we are told that a virus is pure techne, lacking life. If its RNA already knows how to utilise my DNA, and if my response to the virus is already forged in cultural beliefs and lifestyle, social affluence, diet and exercise, reading habits—indeed, there isn't anything we could leave out here—then the "meeting" is already underway and the virus has a well-developed suite of literacies and strategies as a result.

In short, I think any engagement with anthropocentrism would do well to acknowledge the how of being, the ontological complicity that gives all being, all identity, even something as tiny and apparently life-less as a virus, an ecological dimension that confounds the local with/in the global. This is not to dismiss ethical and political concerns, but it is a plea to understand why restricting our analyses into yay or nay responses might inadvertently encourage the very outcomes that we most fear, and ironically, deny the intricacies of a general ecology.

References

Barad, K. (2011). Erasers and erasures: Pinch's unfortunate "uncertainty principle." *Social Studies of Science*, *41*(3), 443–454.

Barnwell, A. (2017). Method matters: The ethics of exclusion. In V. Kirby (Ed.), *What if culture was nature all along?* (pp. 26–41). Edinburgh University Press.

Butler, J. (2005). *On giving an account of oneself*. Fordham University Press.

Derrida, J. (1995). For the love of Lacan (B. Edwards & A. Lecercle, Trans.). *Cardozo Law Review*, *16*(3–4), 699–728.

Derrida, J. (2011). *The beast and the sovereign* (Vols. I & II, G. Bennington, Trans.). The University of Chicago Press.

Gatens, M. (1983). A critique of the sex/gender distinction. In J. Allen & P. Patton (Eds.), *Beyond Marxism: Interventions after Marx* (pp. 143–160). Interventions Publication.

Irigaray, L. (1985). *Speculum of the other woman* (G. C. Gill, Trans.). Cornell University Press. (Original work published 1974).

Kirby, V. (2006). *Judith Butler: Live theory*. Continuum.

Kirby, V. (2011). *Quantum anthropologies: Life at large*. Duke University Press.

Kirby, V. (Ed.). (2017). *What if culture was nature all along?* Edinburgh University Press.

Kirby, V. (2018). "Un/limited Ecologies." In D. Wood, M. Fritsch, & P. Lynes (Eds.), *Eco-deconstruction: Derrida and environmental ethics* (pp. 121–140). Fordham University Press.

Latour, B. (1993). *We have never been modern* (C. Porter, Trans.). Harvard University Press. (Original work published 1991).

Latour, B. (1999a). "Do you believe in reality?": News from the trenches of the science wars. In *Pandora's hope: Essays on the reality of science studies* (pp. 126–137). Harvard University Press.

Latour, B. (1999b). Circulating reference in the Amazon forest. In *Pandora's hope: Essays on the reality of science studies.* Harvard University Press.

Latour, B. (2003). The promises of constructivism. In D. Ihde & E. Selinger (Eds.), *Chasing technoscience: Matrix of materiality* (pp. 27–46). Indiana University Press.

Latour, B. (2004). Why has critique run out of steam? From matters of fact to matters of concern. *Critical Inquiry, 30*(2), 225–248.

Schrader, A. (2010). Responding to *Pfiesteria piscicida* (the fish killer): Phantomatic ontologies, indeterminacy and responsibility in toxic microbiology. *Social Studies of Science, 40*(2), 275–306.

Sokal, A. D. (1996). Transgressing the boundaries: Toward a transformative hermeneutics of quantum gravity. *Social Text, 46–47*, 217–252.

Spivak, G. (1976). Translator's preface. In J. Derrida (Ed.), *Of grammatology* (pp. ix–lxxxvii). The Johns Hopkins University Press (Original work published 1967).

Wilson, E. (1999). Introduction: Somatic compliance—Feminism, biology and science. *Australian Feminist Studies, 14*(29), 7–18.

CHAPTER 22

Conversations on Citizenship, Critical Hope, and Climate Change: An Interview with Bronwyn Hayward

Bronwyn Hayward and Sara Tolbert

Sara Tolbert: Thank you so much for agreeing to do this interview, Bronwyn. I know I've told you a bit about the book, and how we are bringing together scholars, activists, and educators who are thinking about and interested in education in and for the Anthropocene. I've been editing a section on vectors of power and am very interested in your work as a political scientist, particularly the work that you've done with social agency and trying to bring more attention to youth voice and other areas of ecological citizenship.

Bronwyn Hayward: Ah, fantastic. I just had to write a blurb for Mike Hulme, who wrote the book, *Why We Disagree about Climate Change*. He's a climate scientist. He's been around for a long time. He's just bringing out a new book, *Climate Change*. His argument is that we have to move beyond the science, but he structures the science in a really interesting way that makes you think about "thin environmentalism" and modernism. But then he looks beyond that to Indigenous writing and feminist writing and artists to join the push back on the [prevailing narratives about how] "there is no time left" and "it's all going to end" to "we're actually going to live in a changing climate"... we always have—it's going to be difficult and different, but it's going to be diverse and the scientists don't get "The Say" in this. It's very interesting.

B. Hayward (✉) · S. Tolbert
University of Canterbury, Christchurch, New Zealand
e-mail: bronwyn.hayward@canterbury.ac.nz

S. Tolbert
e-mail: sara.tolbert@canterbury.ac.nz

© The Author(s) 2022
M. F.G. Wallace et al. (eds.), *Reimagining Science Education in the Anthropocene*, Palgrave Studies in Education and the Environment, https://doi.org/10.1007/978-3-030-79622-8_22

Sara: It is, I mean, that's a really interesting point. I think my co-editors and I have all been really situated in that space of understanding, but also moving with and beyond, the science. Kind of like expanding the boundaries of what you know, what we even call science, and who gets to be seen as qualified to have a perspective. Not only science, but even in science education, we do that a little bit, too, to each other. And I wondered, what you see, as a political scientist, as the relationship between political science and science because you do a lot of work on the IPCC with scientists and you have to try to have those transdisciplinary conversations. What do you see, given your role on IPCC and given your positionality as a political scientist—how do you see those relationships? And what do you think is working well and in those collaborations in that space? And what needs to change?

Bronwyn: I can say for me, I've found working on the Intergovernmental Panel for Climate Change [IPCC] has been really empowering for several reasons. First, you're working with whole teams, huge teams of people, who are concerned and interested in the issues, and that really creates a sense of purpose. And we talk about how empowering it is for children and young citizens, to have a sense of collective agency, when there are many of you working together to effect change on complex issues, but it's also been really true for my experience on the IPCC. It's not that we expect to save the world one citation at a time. But you do feel like you're a small part of a much wider project to tackle the underlying drivers of climate change, but also to document and bear witness to the changes that are happening. So that's the first part.

The second part, I think, is that it is quite confronting as a political scientist to work with other scientists on a complex wicked problem. Because I think in the social sciences, we often assume that the problem lies with the sciences, if only they [scientists] would be more open to our reasoning, if only they would accept wider knowledge and other perspectives. I think what has been important for me to understand is that some of the biggest battles that I have with the discipline boundaries that police who's allowed to talk about what and how an issue is framed are actually with my social science colleagues. They have more astute language to obscure the power plays, but there's a definite policing of what happens, and who's allowed to say what. And in many ways, working with natural and physical scientists is quite refreshing because they just ask questions like, "why do you think that is?" because they are based on observation, often, and trial and error. They're often just interested in other ways of understanding problems. And I think that's a bit salutary to think that the social sciences are part of the problem of why we can't collaborate very well, that we're very defensive often and we're quite determined to frame the problem, that we [also think we] have "The Way" of framing the problem. So bringing a bit of humility into research is quite interesting. And it's quite important, I think, especially on these big projects and big problems.

I think the third feature is that, while the IPCC works very hard to be policy relevant but not policy prescriptive, meaning they don't try to give governments "advice" because the governments will sign off [i.e., approve, or not] the final reports, politics are imbued in all processes. And I think as a political scientist, one of the things I am conscious of is, as are many colleagues, is to avoid self-censoring because an option may not be accepted by governments, but where there are deep value applications in the options that we're offering—and to notice sometimes if you're not trained or not thinking about how power affects people and communities and non-human nature, you just don't notice it, or you're not aware of it.

Sara: It makes me think about some of the more recent work that we've been talking about in terms of working across these diverse socio-climatic nexuses, looking at working across interest groups or different stakeholders that might not necessarily see themselves as working together. But part of this idea of, as you said, living as well as we can together in the context of climate change and whatever language you want to use, "on a damaged planet," let's say, does kind of require this sort of forging of new collaborations… and new solidarities—like with the Federation of Farmers, for example, that we've been talking about, so across scientists and iwi [tribes] and bringing communities in conversation together. And that's been an aspect of some of your recent work. Can you say a little bit about that?

Bronwyn: Well, I do think that Michael Hulme (2021) in his book on climate change puts it really well. When he says that there are a multitude of experiences of climate change and climate change is a complex series of problems. It's not a single problem and it's not a single experience—that it's *not* something that we are all in together. I think one of the problems about why it's been hard to get collective solidarity to address climate change is that our changing climate is experienced as weather in different ways in local communities, over time, so it's not like a global pandemic where everybody is experiencing an acute risk at the same time. Instead you're having some communities exposed to extensive storms and others to extended droughts and a lot of the suffering and a lot of the problem solving is at local community level and requires the nuance and understanding of local issues, and is also subjected to power imbalances and ways of framing the problem and how the story is told to others. So I think there is a real role for researchers and storytellers and community advocates to do the connecting so that people can understand how their experiences in one time and place are connected to and influence the experiences of others. And Occupy and those kinds of movements, and the School Strike movement actually, have created that sense of a shared experience, even though our local experiences of climate change are really different. We share some key concerns, and those can be articulated as a larger narrative, even though the local experiences are going to be really diverse.

Sara: So you mentioned School Strike 4 Climate as a recent example. I think there's a lot of interest in what's going on there because it's youth-led and has some parallels to youth-led movements in the sixties. How did you become involved in that movement because I know you were the only adult [speaker] at the school strike for climate demonstration last March (2019)?

Bronwyn: I was very honored to be the token adult. Actually, I don't know how I was invited. They just emailed me. Some of their mums and some of the students knew about the *Children, Citizenship, & Environment* book. And some of them have been my students, some of the organizers were former students. And I think that's how it came about, really, from teaching. So it was a lovely connection. And the fact that it happened on the day of the mosque terror attacks [in Christchurch, 15 March 2019], and all of those issues collided, is a reminder that this generation and the ones that follow are dealing with such complexity. And so many layers of challenge of social and economic and inequality and racial and identity challenge all at the same time. I find it really important that we support their capabilities to cope in what will be almost a sped up hyper-reality of these colliding issues because our global connections are so much closer, our ways of communicating so much more rapid and far-reaching, that many of the problems that we have built up over the centuries, like colonization, racism, gender inequality, our changing climate, are all kind of coalescing in… moments in time. Now much faster even than they were in the Sixties in the civil rights movements because our communication is just that much faster and so much more personalized into people's homes and their phones and their Facebook messaging. So I think having the skills and capacity to cope with that is really important.

Sara: Could you talk a little bit about more about what you mean about their abilities to cope? Because you've written a lot about that and I think it's very useful for educators.

Bronwyn: I've been really interested in how we maintain democracy through disaster and change, over time, as a political scientist. Originally, my work on *Children, Citizenship, & Environment: Nurturing a Democratic Imagination in a Changing World*, the reason that that started was because I was asked by the New Zealand Electoral Commission to do a small study on why kids aren't voting. And why young people aren't voting. And I had wanted to do some interviews with eight to twelve year olds, sort of that time of a rising sense of citizenship and community. What really struck me is fascinating because I was also interested in environment and geography as well, is that when we talk to people about what do they do around here, what do they like to do with other people, actually their physical environment and their experiences of their social and natural world collided with their understandings of what it was to be a citizen in very holistic and rich and deep ways for every child that we interviewed. And so what began as a study of just "what do young people think about voting?" rapidly became "how do young people learn to feel that they are part of a community and can effect change?" And that communities have a right to be heard because we know that a lot of the reasons that youth aren't voting really is not the problem of a deficit of information or a deficit, period.

It's the series of suppression effects that we do in different societies. Even repeating the argument that young people are apathetic or aren't interested or that they can't vote then reinforces a kind of a disjuncture with voting. But when they're in contexts in which they're talking about issues that matter for the community and the world around them, then they're very strongly motivated to have a say when they're supported. And I was also very struck by the really practical, physical, emotional, and social factors in whānau [extended family] and family in neighborhoods and schools that restrict and reduce children's voice. And understanding whether it's fear of physical violence at home, whether it's lack of money and feeling shamed to speak out, whether it's feeling not worthy. So that's what I was interested in, I sort of rapidly found a way to draw together the two things that I love, you know, how is it that we maintain democracy, and how is it that we experience and understand our natural world and support it. Because actually in real life, we do those things together all the time. We just separate them when we're researching them. So that got me thinking about the fact that a lot of our teaching goes in two directions. And this was before Trump. But back in 2012. Because of the barriers that children were experiencing and reporting just here and in my local city and the hugely diverse experiences of participation [children were experiencing], I started thinking about what suppresses their voice and their opportunities to engage. So what political power do kids experience that silences them? And that's what I described as the "FEARS" of citizenship [(Hayward, 2012, 2020)]. It's almost a pathology of experiences and where you feel frustrated about your lack of agency and not able to express yourself because of fear of being hurt. It's a very basic fear of speaking out or lack of money. And when you are excluded from the environment around you. We're building walls, whether it's just fences to constrain where children can play, and very strict limits on what they can do, where they can be—seen and not heard. But also, not even physically in spaces that are increasingly being securitized and kind of semi-public private spaces and the way that we encourage a very basic sense of retribution of justice, the eye for the eye, and find someone to blame. And it's all very simple. You direct your anger and feelings of helplessness to another group or organization or individual that you can blame. And its effect is we see very low participation in formal voting and we have a really kind of silenced sort of political imagination.

When we're teaching environment, we risk the same kind of participation because we reinforce this kind of "SMART" way of thinking about the environment [(Hayward, 2012, 2020)], which I would argue is quite a thin approach to environmentalism, where there's a lot of emphasis on the self-help and self-responsibility of individuals, which is great, but it leaves that anxious and overwhelmed child unsupported. There's enormous influence on participating in the market. So you express your sense of citizenship through doing things like buying eco-products or creating sustainable new innovative products that you can sell as an entrepreneur. And if you are a good, smart environmental citizen, then your ideas about justice are very contractual. Like we encourage a lot of a priori reasoning, because we want the market to work well and we want what's fair in a market sense, so we encourage young people to think about, "I will do my bit if you do yours." There is no sense of "we're doing the right thing in the current situation," but it's more contractual justice. We teach a lot of that in classrooms. We set up what our classroom justice is going to be, and we expect children to do it. As teachers, what we're also setting them up for is to understand how the market will work in the future because we're expecting people to behave in particular ways. In one sense, their political imagination is very transformative. So we're looking at developing new technologies that are solar powered, that are low carbon, that are innovative. And we want them to vote for and be good citizens, voting for representatives that are going to speak for us.

And there's nothing wrong with all of that, but it leaves unquestioned and unchallenged the underlying drivers of our climate problems. For example, the social inequalities, the economic growth models, the consumption models are not really questioned—what's driving it [climate change]. It is terribly important that we have individuals who care and take responsibility and take action, but in these conflicts, wicked huge global problems, it is overwhelming. And that's where we see very anxious children. People who feel that they just can't cope or that they could just pay and that would solve the problem. So, in thinking about that, I was thinking, well, that is the classic neoliberal citizen, but it doesn't explain everything that we were seeing when we were interviewing children just in the city.

And we did also interview children up in the north of New Zealand as well because we just worked with a research team who went home to their own primary schools, but what we also saw was this kind of resistance where children were not just thinking as individuals, they were learning social agency, and they were learning that through participation in school camps, in choirs, through their iwi, through kapa haka, they were learning to be part of a collective. And those skills of the collective come out in things like the Student Volunteer Army [a collective of student volunteers that formed a disaster response team in post-earthquake Christchurch, 2011, and then became institutionalized, https://sva.org.nz/our-story/]. I mean, it just doesn't happen that everyone collaborates if they haven't been nurtured and encouraged and socialized to collaborate as a norm in response to shared problems or risks. And we also saw a strong sense of embedded justice, the ability to be able to reason about what's fair in place. A lot of children could tell you that it was unfair that something [e.g., such as a disciplinary action] had happened to one of the kids in their class because they knew the conditions or issues that were facing that child. So children and young people can be very reasonable if they're encouraged to think about—not just if you fail to meet expectations— but actually we live in a messy world, how can we do the right thing over time, understanding the impacts of colonization and gender violence and all the things that shape the trauma and the lives that we live.

For me, environmental education is more than just the science. The science is deeply important but the children who had a strong sense of being able to participate in their environment and effect change had some connection to it as well. So often, they had a sense of tūrangawaewae ["a place to stand" or empowerment]. That might be their local marae [communal or sacred place] or it might be their local neighborhood that they had grown up with, where they enjoyed and found a restful and emotional connection with that space. And also when there's an opportunity to decenter deliberation so that children are not just overwhelmed with their own local experiences but are able to put it in context through storytelling is really basic, and, for example, you just look at common children's stories like *Winnie the Pooh* (Milne & Shepard, 1994). When you actually look at that as an English narrative, there's a lot embedded in there that's talking about how children deal with large social environmental change that's beyond their immediate capacity to control, whether it's a big flood or the big wind or somebody leaving home that they weren't expecting. And then that sense of self-transcendence, so how young citizens are supported to do and be in the context of a history of what's gone before and a feeling of what might be to come. And sometimes that's achieved through strong spiritual values, but often it's achieved, we noticed, by children who feel that they are part of something that grandparents have worked for. For instance, Ngāi Tahu young people, many who are taught within bilingual schools [i.e., kura kaupapa Māori], are very aware of generations working for Te Tiriti o Waitangi [The Treaty of Waitangi] settlement. Others are aware of their parents and grandparents, having made changes and that sense that you're able to defer not only immediate gratification, but to understand that change takes time and happens across generations. It's not just all down to you.

Sara: I think that's really important, I mean for my perspective, working with preservice teachers and helping them unpack some of the ways in which they've learned about the environment. And taking a deconstructive approach to that. It's really interesting seeing how they peel back some of those layers around how some have come to understand their ability to act as very individualistic, I would say, even now and so maybe you've found that as well as undergraduates that you teach, but definitely the conception of how we impact on our communities; many of them, who are mostly in their early 20s, they have this sense of engaging individually. But if you can make links to the ways in which they have experienced social agency, it's really powerful.

Bronwyn: And I think it's really essential, because I don't know if you notice in teaching, but when we talk about how you effect political change with students, I noticed a lot that students will know things about the suffragettes or they'll have learned about the Civil Rights Movement in America, and they know about Rosa Parks sitting on the bus, but what they don't know is the huge amount of training and support that happened to enable Rosa Parks to sit on that bus. They think it was an individual who simply sat on that bus. But then, actually, ironically, there's that great story about the two young girls that sat on the bus, and did the same protest just a couple of days earlier and were thrown off and nearly killed. And the difference when students understand that Rosa was part of the Highlander School's network of citizenship teaching, that she worked with Martin Luther King, Jr., that she'd been auditioned for the role, and that it was all thought about quite carefully. That another young woman of color who was pregnant, who already was a married mum, was not selected because they decided that there would be too many distracting issues for the media. So it doesn't detract from what Rosa did but there was enormous training and a group of people around her and several attempts at doing this and an infrastructure that would move people who [were boycotting] the buses because of the protest to and from work safely for the next three months. There was a massive network of organization around it.

And that's the thing that's extremely important for students who want to effect change, to understand that you can't do it alone. It is a network of change and I think that in many ways, that's what some of the student leaders are learning in the student school strike. But I'm also anxious that they're doing it at a time when there is such focus on the individual through social media that reaches to the individual in an unmediated way that young people are exposed, as they're learning to conduct themselves politically, to enormous scrutiny and challenge. Several of them have said, you know, "it's very scary to make a statement." You're only 16, you make a political statement, 180,000 followers have liked it and then, you know, what happens in a couple of years when you want to add more nuance and change the view. So it's a very public political learning where there are not many avenues for mistakes, and I think that brings its own burden on young protesters and young activists.

Sara: And we've talked a little bit about that, too, as well in terms of having the importance of recentering youth voice and making sure that young people's voices are part of the public dialogue in meaningful ways, but then also that, you know, are there some tensions around Greta Thunberg, for example, being in a position where, I don't think it's exploitation, but it's sort of like there's some kind of fragile boundaries around that, isn't it?

Bronwyn: So yes and I think tying a movement too closely to one individual and to one concept exposes the movement itself to a lot of risk, because Greta is only human. She's only 18, enormous things will change in her life and in the lives of all those millions of protesting students. And actually, one of the interesting things is in this survey we are doing of children growing up in seven cities around the world (Nissen et al., 2017), is that not a lot of children, outside of a small group of urban students, are even aware of the school strikes, really. So even though they're very visible, they're very digital. They're very photogenic. And they have captured mass media. It's interesting that that depth of penetration is not as great in the developing world, or even regional communities as you'd expect. And the focus on one individual makes it very difficult for students who perhaps don't agree with her or don't identify with her. So it's both a strength of the movement, her articulation, her ability to be so visual and so present. And it's a huge weakness and vulnerability of the movement as well.

Sara: You mentioned the CYCLES project. Can you say more about the work that you're doing there and what you're finding?

Bronwyn: That's funded by the UK's Economic and Social Research Council and it's with a larger consortium that I'm a co-investigator with, which is called CUSP, which is the Centre for Understanding Sustainable Prosperity, an interdisciplinary research group of 12 universities led by the ecological economist Tim Jackson. He's based at Surrey, and they are looking at new ways of thinking about ecology and wellbeing in ways that encourage a more sustainable view of prosperity and good living. And so it's interesting because it has science and economics streams, but it also has a very explicit religious philosophies theme that's led by religious leaders like the Archbishop of Canterbury. And then our project is following children growing up in seven world cities. And we came together as a team of researchers, because we had previously worked for a UN study on sustainability, where we looked at older 18 to 24 year olds in 18 countries, and we'd really enjoyed collaborating together. And we were all interested in what the younger adolescent [e.g., 12 to 18 years old] is thinking about their lives and how they are experiencing wellbeing, or not, in their community and what it would take to support more sustainable lives. How many of those families are locked into unsustainable ways of living because they lack city infrastructure, they lack financial options? What are the children's expectations of their lives? What do they like about cities they live in and what do they want to change?

So we did some focus groups and image making with children and young people aged 12 to 24. Each city has a team of locally based researchers either in a charity or a university that's leading their own local consultation and there's only one city where that hasn't happened. It really doesn't work, I think, when you're not embedded in the communities that the children you are interviewing are growing up, and I think it's really difficult to do nuanced, thoughtful research. You can get high level, but when children and young people are talking to you about the tree that fell down or the debate about the local swimming pool, unless you know that that's been a housing development that was seized by the mayor, it's hard to understand the politics and significance of what they're saying because a lot of children and young people's political realities are embedded in their personal relationships and their family relationships. And so you need local knowledge to understand the wider political and economic significance of something that can seem quite little. Like, for instance, a good example that I learned about was a local primary school that we interviewed in the first edition of the book (Hayward, 2012), the children said that they had protested in the playground about not being able to climb trees. They wanted to be able to climb the really high trees which they've always been able to, but Health and Safety of the school board said they couldn't. So they made placards and they protested and what was very interesting about that was that—I wouldn't have known, except that one of our interviewers lives in that particular neighborhood—is that there was a big strike affecting their local supermarket and a lot of the local parents. So many of those children that were striking about not being able to climb trees were observing strikes at home in parents' workplaces, and I wouldn't have been able to make that connection, unless somebody local was doing the research.

Sara: Yeah, that's true. It does matter a lot, having that nuanced local knowledge in this work.

Bronwyn: Yes, because what we sort of forget is that children are learning about their environment from multiple ways. So when I interviewed a small group of school strike leaders, you know, some of them had watched their grandparents striking and protesting over health reforms locally. They were using those lessons directly for how they might effect change on the issue that mattered to them, which was climate.

Sara: It makes me think about how the most activist of my friends grew up in very activist or, you know, highly unionized families and were just immersed in it.

Bronwyn: Louise Chawla said that youth activists and their understanding of power and their interest in protecting the environment grows up usually from their experience with the environment, but I think the bit that's missing that we need to look at is where did they see their models for effecting social change. And because we don't often put those two discussions together— the political change and the environmental change—we forget how much our civics and citizenship teaching influences our environmental understanding.

Sara: I feel like that's such a huge theme in the book, what you just said: where did they see their models for effecting social change and how that influences their understanding of environmental ecological citizenship or engagement. And you have the updated version out now.

Bronwyn: Yes, in the significantly revised second edition, we look at how students with teachers, parents, and other activists can learn to take effective action to confront the complex drivers of our climate crisis. And that includes economic and social injustice, colonialism, and racism. Though we haven't done much on racism, and I should have done something more [in the book]. But what I wanted to do was make sure that it wasn't an environmental book at the beginning. This is very frustrating for people to just say, all we have to do is reduce carbon, as if all these other issues are slowing us down, like dealing with colonization. It's not a direct route. The thing is if it were just about reducing carbon, we would have fixed climate change. Because it's not that easy, it's complex layers of issues that have affected out climate change. Climate change is the symptom of multiple social and economic injustices and struggles. Mike Hulme has a depressing take which is probably fair enough, that as we deal with these wicked problems, we're going to shift the problem in new ways. We are always going to have a changing climate. We're just going to get some bits of it more fairly done. But I don't know, I think Covid, just suddenly stopping everything… but anyway.

Sara: I mean it is interesting to think about how these new global challenges highlight—I don't want to say opportunities but highlight—it's not an opportunity, it's a crisis, but a crisis means a turning point. Right. So yeah, who knows. But it does make me think about what you're saying in your co-authored article on learning from Pacific Small Island Developing States (Hayward et al., 2020). Can you just briefly highlight some of those lessons from the Pacific Small Island climate responses?

Bronwyn: And that was with colleagues and students because they were concerned that there is a narrative that it's too late to take action and that itself becomes a self-fulfilling prophecy, because it's not too late for the Pacific. Every climate change action that happens will make the impact on those communities less. We may not be able to stop rising sea levels now, but we will be able to mitigate some of the worst effects on communities. And also writing that article about thinking about what wellbeing means and how communities support each other through big environmental and social change in the Pacific arose out of my concern listening to teachers that were working in schools in the Pacific expressing frustration that charities and international educators were coming in with programs focused around climate which were very narrow, and increased student anxiety. They were not really related to the issues that the students wanted to deal with or the community. And they were externally donor- or grant-driven so that they weren't accountable to the local community. And so I was interested in documenting, how is it that communities have coped with past existential crises, whether it's slavery or genocide in the Pacific, and how are they addressing the complex changes of climate, given that there have been really major pressures on these communities. HIV, for instance, has had a huge effect on the demographic profile of many Pacific communities. So there are lots and lots of young people and fewer adults now in the population, and that's not necessarily just because of migration for work, but because of previous losses of population. But I also think that that comment that was made by the reviewer for our recent grant proposal that alerted us to the dangers of the assumption that Indigenous communities are all so endlessly resilient that they will be able to cope is important, too. But I think this "beyond science" idea is important here because the scientists who think, you know, the carbon is the problem, don't see the social networks and the religious and the community networks that sustain change. So it's not too late for that. But it needs different kinds of support.

Sara: You mentioned the comment on our recent grant proposal about critical hope and climate change. That there has been some critique of this concept of hope, as you know, is hope what we need? What would you say about that? It seems like it builds on what you're saying here [related to the aforementioned article, Hayward et al., 2020].

Bronwyn: Yeah, I think it is what sustains action and change. And the challenge is agency, you know, the neoliberal effective politics, so long as we've given a lot of emphasis to individual action for change, or even collective action. But you have to find a way to sustain resistance in movements and whether you call it hope or whether you call it faith or whether you call that just collaborative solidarity, the concept of the critical hope is not acting in a Pollyanna away believing that good things will happen. It's acting in the face of the knowledge of political inequality, the extreme difficulties that you face, and taking action, anyway. If we don't have that ability to take action, anyway, we are lost.

Sara: A good friend of mine, Amanda Holmes, who is Haudenosaunee Mohawk, talks about how you just keep on keeping on because you're doing it for the generations to come. And you're doing it to honor the ancestors who came before you who did that as well. So you just don't think about whether it's going to be "effective" or not. You just do, because it's part of this genealogy.

Bronwyn: And duty.

Sara: And duty, and long-term sustainability. Like you just—it's part of the ethic of being alive, right, and this is something I think about, too, I don't know, when we were talking before about individual action. I sometimes wonder if we're talking about ethical versus political engagement. That's kind of an aside, but when you were talking about individual actions we take like to make sure that—I compost, and recycle, but I don't necessarily see those as meaningful political forms of engagement, but I see them as ethical forms of engagement.

Bronwyn: And it matters that we do have ethically responsible individuals, but I have been surprised and I haven't done research about it. I'd like to, it's just anecdotal, but I've been surprised how many students say to me that they feel that they can't take political action because they're not recycling or they just don't have keep-cups [reusable cups] for the coffee. But it's just so irrelevant to the big picture whether they use a plastic straw or not. And when our ethical actions become such that it inhibits people from feeling that they can support social movements for change, that's a problem. Because it's too much of a burden. The difficult balance is to recognize that ethical actions have to take place in a social and political context. And that many individuals and whole communities are locked into behaviors that are unsustainable and might seem unethical. But it's unaffordable to do anything else, or simply impossible, or they haven't got the time, or they haven't got the social support. And they shouldn't be beating themselves up, we shouldn't be eroding people for being locked into impossible situations.

Sara: Right, exactly.

Bronwyn: Tim Jackson's done a lot of work on "lock-in." Lots of people have, but particularly in England, low-income communities and women have got really high environmental footprints, because they have to use lots of heating in poorly insulated houses. They've got to take unsustainable forms of transport, because there aren't local bus routes, all those kind of things add up. You know, it's like why would we eat bad food? Because it's nearby and affordable.

Sara: It's sort of related to the issue of awareness versus infrastructure. Like ethics being something that's contextualized and almost inherited in a way and then also you know, different points of engagement where infrastructure is a different point of engagement, like do I create a compost in my own backyard or do we lobby the city to create these organic waste bin for us—it becomes systemic change where people are much more likely to actually do something if there's the infrastructure in place.

Bronwyn: That's where I think Donella Meadows, a lot of people quote her paper on leverage points [http://donellameadows.org/archives/leverage-points-places-to-intervene-in-a-system/]. I mean, what actions would get the most traction? Sometimes you can burn up enormous amounts of energy over something that perhaps gets only a little traction, has little change in the system, whereas you can get systemic change by asking the right questions.

Sara: Do we teach that to young people?

Bronwyn: Well, that's what we should be doing in politics. It's very hard to work out. And that's the thing about the local experience of climate is, where do you put your energy? But if it's too localized you're trapped. So you do need national leadership, you do need some international leadership. And it's so interesting. At the moment, because the politics of, you know, who should take leadership over what is not clear at all. It's not clear how we're going to affect political change. We're not even clear what change would entail.

Sara: Yes, there is a lot of uncertainty in the system. I love the way you describe the history of neoliberalism and New Zealand in your book because I don't think a lot of people know that. I mean abroad our international perception of New Zealand, many of us don't realize that New Zealand also was caught up in the neoliberal movement as well, both from, you know, the left and the right, from Labor and National. Some people are saying that we're in a post-neoliberal moment now, and especially in New Zealand. Do you think that's true?

Bronwyn: No, I think we are doubling down [laughs]. I think the beginning of the change is to change the language of aspiration. So the Prime Minister (Jacinda Ardern) talking about being kind is a new set of values on top of being efficient and fair. But being kind is not a political action and has a strong individual responsibility, so being kind alone is not yet a vehicle for social change, but in the pandemic itself, watching the community just stop and prioritize public health over the economy, was amazing, and it will be interesting to see what effect that has on children and young people's political learning. Because it's been an imaginative moment which has changed the rules. For 30 years we prioritized the market and had to maintain it at all costs, and in a matter of three months we prioritized public health and stopped everything. And that's a very remarkable action. And I think having watched how just small-scale political actions have big reverberations for children's learning of change—I'm really interested in what effect that will have. So I don't think we are out of the neoliberal moment, but I think that it's expanded people's imagination about doing something different; what will happen next matters. Do we double down on deeper nationalization, stronger fear of the outside, do we increase private sector spending and reduce debt at the cost of public services? If we do that, I think we will be going backwards. We know that we've gone backwards in most countries on almost all the sustainable development goals. And I think that income inequality gaps will have grown hugely in New Zealand, but there is a whole generation of children coming into the political process, who have just seen something quite remarkable happen.

Sara: I love that language—"an imaginative moment which changed the rules." If not, at least for a short time. Hopefully, it has a long lasting impact, especially on young people.... I don't know how much of this you can speak to, but just, one of the things you and I have talked about is the notion of how to best communicate the impact of climate change to the public. And I think, also to young people in the context of growing eco-anxiety.

Bronwyn: It's very hard, I'm very anxious about the way in which we talk about urgency and emergency. Because of the democratic implications of declaring a state of emergency, meaning that you suspend normal decision-making and normal inclusion. And this idea that these decisions are urgent and have to be made is quite monolithic and yes, we have to take some urgent action, but we're not going to have one action that's going to fit everything anyway. And it's very hard to think about how to convey that. We need to have far reaching changes without disempowering or creating anti-democratic responses that then risk stripping kids of a democratic future as well as a right to a sustainable one. And that's the part that worries me always.

Sara: Such a good point. I've talked with you before about my own daughter who is learning about climate change and ecological sustainability in primary school. And the strategies that she's come up with for fighting climate change is "stop driving," which, you know, she has no control over, yes, yes. Okay. "Stop poaching" was another one her list of notes. It's something I think about a lot, how to teach it in ways that I think honor the kind of realities that we're facing, that are undeniable realities, but at the same time, don't make people feel completely hopeless, or helpless.

Bronwyn: And not make the situation worse.

Sara: But it's such a good point about, well, if we say it's an emergency, then what does that mean for democracy?

Bronwyn: Yeah, but I'm fighting a lone battle on this one. Now it's increasingly harder and harder. Now I see the UN's been using the term, emergency, I just think, oh bugger it, we really shouldn't be using that language, but anyway.

Sara: Well, Bronwyn, that's all I have. And I jumped around but we got through actually all of these questions.

Bronwyn: I feel honored to know you, Sara, it's been lovely. It's like an extension of my life having somebody I can rave with about this stuff.

Sara: Thanks so much for doing this interview because like I said, these are questions that I've always wanted to ask you more about. I'm so grateful to know you, Bronwyn. Thank you so much for your time and generosity.

REFERENCES

Hayward, B. (2012). *Children, citizenship and environment: Nurturing a democratic imagination in a changing world*. Routledge.

Hayward, B. (2020). *Children, citizenship and environment:# SchoolStrike Edition*. Routledge.

Hayward, B., Salili, D. H., Tupuana'i, L. L., & Tualamali'i', J. (2020). It's not "too late": Learning from Pacific Small Island Developing States in a warming world. *Wiley Interdisciplinary Reviews: Climate Change, 11*(1), e612.

Hulme, M. (2009). *Why we disagree about climate change: Understanding controversy, inaction and opportunity*. Cambridge University Press.

Hulme, M. (2021). *Climate change*. Routledge.

Milne, A. A., & Shepard, E. H. (1994). *The complete tales of Winnie-the-Pooh*. Penguin.

Nissen, S., Aoyagi, M., Burningham, K., Hasan, M., Hayward, B., Jackson, T., Jha, V., Lattin, K., Mattar, H., Musiyiwa, L., Oliveira, M., Schudel, I., Venn, S., & Yoshida, A. (2017). *Young lives in seven cities—A scoping study for the CYCLES project* (CUSP Working Paper No. 6). Guildford: Centre for the Understanding of Sustainable Prosperity.

CHAPTER 23

Conclusion: Another Complicated Conversation

Maria F.G. Wallace, Jesse Bazzul, Marc Higgins, and Sara Tolbert

Maria: This edited collection is not just a conventional "body of work," but a living-breathing community of more-than-human potential. Writing, researching, questioning, and yes, even, feeling; perhaps another science education is actually possible?

Jesse: I really think so Maria! When we conceived this collection some time in late 2018, we had no idea how the COVID-19 pandemic would make the realities of the Anthropocene more visible and more urgent. I think there's a few things to note as we sit here at the end of this collection.

M. F.G. Wallace (✉)
Center for Science and Mathematics Education, University of Southern Mississippi, Hattiesburg, MS, USA
e-mail: Maria.Wallace@usm.edu

J. Bazzul
Science and Environmental Education, University of Regina, Regina, SK, Canada
e-mail: Jesse.Bazzul@uregina.ca

M. Higgins
Secondary Education, University of Alberta, Edmonton, AB, Canada
e-mail: marc.higgins@ualberta.ca

S. Tolbert
Science and Environmental Education, University of Canterbury, Christchurch, New Zealand
e-mail: sara.tolbert@canterbury.ac.nz

© The Author(s) 2022
M. F.G. Wallace et al. (eds.), *Reimagining Science Education in the Anthropocene*, Palgrave Studies in Education and the Environment, https://doi.org/10.1007/978-3-030-79622-8_23

First, the variety of different concepts and new ethical–political considerations for science education that have been introduced or elaborated in this collection such as pedagogies of fire, black holes, magical realism, and Anthropocentric detonators. All of these have the potential to recast how educators engage the Anthropocene with students. The introduction of such concepts is not through any special virtue of the authors, but rather these realities and new ways of thinking are unavoidable. Second, we are writing at a time when COVID-19 has made the interconnected nature of science even more visible. It's clear that science and technology (in all their manifestations) must somehow play a positive and nurturing role in collective life going forward. Justice absolutely belongs in our pedagogies. And third, that things are even more strange than when we began. This is one of the lessons of Feral Atlas, a resource discussed by Anna Tsing in the interview section of this book. Science educators can engage the uncanny nature of the Anthropocene with different stories of kinship and just futures. How is science a welcomed element among many in these exciting, strange, and sometimes sacred stories? Maybe we're on the right track with this volume. Maybe our work as science educators will become even more unrecognizable from the field we've inherited (a word Marc has used to describe our predicament of being caught between worlds and fields).

Marc: Sharing your enthusiasm Jesse, I absolutely agree that another science education is possible (and necessary!) (see Stengers, 2018; Higgins, et al., 2019). At the same time, I also believe that Jane Gilbert's (2016) question of whether science can be transformed to respond to the Anthropocene, articulated half a decade ago, continues to lurk and linger. Despite the pluralizing forms of science education towards what would be unintelligible *as* science education a decade or two ago, there continues to be forms of science education whose commitments continue to be (re)produced by modernity, the age of extractivism; "the 'subjects' of the modern school curriculum, including science, were developed to support the growth of modern economies/societies" (Gilbert, 2016, p. 192). Above and beyond the problematic ethical, social, and political commitments that this double(d) subject engenders, as both learner and curricular content, Gilbert brings significant attention to the pedagogical modes presented in science-education-as-usual: critical, social, and ecological issues pertaining to science often remain inert as they are *added* to the curriculum and learned *about*. As she states:

> Why is it that, while science educators *say* they are committed to meeting the needs of learners in their socio-cultural context/s, they default to "aboutism"? Is there something in the way science educators are socialised that predisposes them to think like this? Or does science education attract people who think like this? Or does this have something to do with how science education is structured, with how it has developed as a discrete field of enquiry? I do not think we know the answers to these questions, and I think this is part of the problem. (Gilbert, 2016, p. 190, emphasis in original)

However, rather than present a dichotomy between progressive and regressive science education, I think that these latter forms continue to haunt our attempts to move beyond them. There is no being without inheritance, and no inheritance without responsibility (Derrida, 1994/2006). As we move toward a science education which pedagogically produces modes of being *implicated* and *called upon* (den Heyer, 2009) that bring into question the double(d) "subjects" of science education, I believe Gilbert's (2016) advice continues to bear relevance and should be heeded: "We need to look at ourselves to dig up some of our assumptions about science education—what it *is* and what it is *for*—and our assumptions about science, education, society and the future" (p. 190, emphasis in original).

Sara: Comrades, these are such critical and visionary points. In regard to Gilbert's call above—taking a serious look at our assumptions about science and education—I think the authors in this edited collection have taken up that challenge in beautiful and imaginative ways. I feel one way we/they do this in the book is by engaging unique local and global perspectives that inevitably lure us as readers, thinkers, educators, activists to look (dream) outside of the way(s) our own systems and structures of "science education" (particularly in formal schooling) are organized. These visions prompt us to see that science/education doesn't really have to be the way our arborescent "structures" demand it to be (or make us think it should be)... Imagine if the engaging and radical ideas of authors in this book were taken up and dispersed with the same gusto and enthusiasm as "NGSS three-dimensional learning © ®" ! I believe fervently in this possibility. Recently a group of educators reimagining mathematics education in Aotearoa New Zealand shared this video with me (https://www.youtube.com/watch?v=fW8amMCVAJQ), which I've adopted as a bit of a mantra: "A leader needs the guts to stand alone and look ridiculous...". We may feel, at times, like "lone nuts" but collectively we are all igniting a movement of nuts that can revolutionize science and education in ways that the Anthropocene so desperately desires and demands.

Maria: Yes, I love this idea of "lone nuts" igniting a movement. It reminds me of the hidden ways that trees communicate sending energy and nutrients to each other just below the surface (Wholleben, 2016). In many ways the chapters in this book represent some of the hidden work and ideas that have lingered in the margins of science education—quietly sending nutrients to the field. The fact that this book was even thinkable a few years ago depicts the impact of this *slow* work. In research on science education, there is an abundance of work that has sustained the factory of science education (e.g., acquisition of proper science identities, efficacy of NOS implementation, production of "good" argumentation techniques) for *quickly* and *efficiently* meeting the needs of federal mandates (e.g., NGSS, workforce development). Rather than continue to reproduce this capitalistic model of research on science education, I see this book as an invitation for science educators to *stay with the trouble* (Haraway, 2016). Across each section, we see authors providing non-formulaic examples of *slow science, minor inquiry, and disruption*—three features of what it might look like to stay with the trouble in science education (Higgins et al., 2019*)*.

WHAT REMAINS TO BE DONE?

Jesse: One of the authors of our book (Aswathy Raveendran) recently used the adjective "derived" to describe a science education that works solely for interests already outline by institutions and apparatuses of power. Perhaps, there's slightly more room after this year to explore this foreboding, yet oddly pleasurable, trouble that comes with this Anthropocene moment. What has struck me about this past year is how much easier it's been to discuss ecological–social entanglements, transdisciplinarity and matters of concerns with undergraduate and graduate students due to the pandemic. Some of the work that needs to be done involves embracing the pleasure of ecologically connected existence in general, something that well exceeds what capitalism, colonialism, and rigid hierarchies have to offer. As Sara and Maria point out, there must be overt elements of collectivity and solidarity with both nonhumans and humans in this work. One problem I see going forward is how science education might simultaneously make room for a vibrant and multiplicitous ethics, while not shying away from matters of urgent political concern. For me, the tensions between ethics and politics run through this book. Questions of how beings might live brought into relation with diverse principles of equality. There's no reason why science education can't engage the most pressing questions of collective existence—however, students, communities, and teachers might formulate them!

Marc: Thank you Maria and Jesse for the reminder about the temporality of the work that remains to be done: there is at once great urgency (as the wicked problems of the Anthropocene are pressing) and a necessary engagement with slowing down (as to not too quickly reproduce the same problematics elsewhere, if but only slightly differently). Part of this work, as mentioned above, involves *slow science*, "facing up to the challenge of developing a collective awareness of the particularly selective character of [science education's] own thought-style" (Stengers, 2018, ṗ. 100). I also think that there are chapters within this collection that aptly bring to life the notion of *slow activism* as well. After Métis scholar Max Liboiron and her colleagues Manuel Tironi and Nerea Calvillo (2018),

> If a permanently polluted world is characterized by chronic slow disasters – incremental and attritional violence that stars no one, and fails to manifest in an event or clear-edged representation... – then a complementary form of politics is slow activism, which is also incremental and attritional, stars no one, and is not premised on nor produces events or clear-edged representation. Slow activism describes some of the political and representational tropes that eschew immediate visible and measurable outputs... that the effects of action are slow to

appear or to trace... Slow activism does not have to be immediately affective or effective, premised on an anticipated result. It can just be good. (pp. 340–341)

Such a call for slow activism is of particular relevance to teaching and learning practices which, as Maria pointed out, are busy by design and caught up in logics of measurability and effectiveness. While we need clear, cracking, and concise political action at various levels of representation (including the individual), the tension proposed is productive: "ethics, rather than an anticipated result, is at the core of slow activism" (Liboiron et al., 2018, p. 342). How might we act, in relation of ethical obligation to one another (humans, other-than-humans, and more-than-humans), if we did not and could not (wholly) know what our actions would produce (recognizing that the need to act is more pressing than ever)? Certainly, we might come to be perceived as "lone nuts" as Sara pointed out (even if our activism "stars no one")! Echoing Jesse, the need for (and emergence of) critical ecological collectivism without sameness in science education feels greater than a decade ago.

But there are also larger questions here that bears attending to: while science (and science education) cannot easily be disentangled from production of the Anthropocene, what does it mean to address the myth that science can and will ultimately "solve" the problem that is toxicity and pollution? What if we treat a teleology of progress as its own form of seeping and spreading toxicity? What if the world is to be permanently polluted? What then of science education? Surely, "a permanently polluted world is one that, because of its deep alteration, reclaims the need to incite new forms of response-ability" (Liboiron et al., 2018, p. 332).

Sara: Important questions, Marc! I think this idea of slowness—as not only research and activism but also more generally as being/becoming—*is* the "arts of living on a [rapidly changing] damaged planet" (Tsing et al., 2017) or a "permanently polluted world" (Liboiron et al., 2018)—and clearly one that the authors in this edited collection, and others elsewhere, have taken up. Certainly "Science" needs to slow down. (We can look to recent and ongoing attempts at solar engineering as one poignant example of the importance of such slow ethics.) And as Maria has argued, "Education" needs to be derailed from impossible factory-like timelines and mandates. Also, in the context of growing concerns about eco-anxiety (especially among children), I am empathetic to fierce rejections of overly simplified and perhaps unproductive narratives of Despair (Mann, 2021). At the same time, this moment does necessitate both mourning *and* resurgence (Haraway, 2016). Dahr Jamail's (2020) work, *The End of Ice: Bearing Witness and Finding Meaning in the Path of Climate Disruption*, encourages us to ponder deeply: What does it mean to (slow down and) bear witness as the Earth is in hospice? "Not shying away from urgent political concerns" as Jesse states resonates as another form of bearing witness. It seems to me some activisms, which may often be the subject of critique (like Extinction Rebellion or SchoolStrike4Climate), are not necessarily "narratives of Despair" but forms of simultaneous mourning and resurgence. Meanwhile, people who are coming together to reimagine (or resuscitate) forms of life and education on a damaged planet, such as those described in this book, are similarly bearing witness for all that is lost, but also possible, in the Anthropocene. Paulo Freire's (1992) *Pedagogy of Hope* comes to mind as I reflect on what this means for science education: "What can we do now in order to be able to do tomorrow what we are unable to do today"? (p. 215).

Authors in *Reimagining Science Education in the Anthropocene* take on these nuances and complexities of science and education, looking to (and disrupting) the past(s), present(s), and future(s)—asking what science education has been, what/how should it be and do? And, what are the disruptions in/to science education that can hold space for mourning and resurgence in the Anthropocene?

References

den Heyer, K. (2009). Implicated and called upon: Challenging an educated position of self, others, knowledge and knowing as things to acquire. *Critical Literacy: Theories and Practices, 3*(1), 26–35.

Derrida, J. (1994/2006). *Specters of Marx: The state of the debt, the work of mourning, & the new international* (P. Kamuf, Trans.). New York, NY: Routledge.

Freire, P. (1992). *A pedagogy of hope: Reliving pedagogy of the oppressed.* Continuum.

Gilbert, J. (2016). Transforming science education for the Anthropocene—Is it possible? *Research in Science Education, 46*(2), 187–201.

Haraway, D. J. (2016). *Staying with the trouble: Making kin in the Chthulucene.* Duke University Press.

Higgins, M., Wallace, M. F. G., & Bazzul, J. (2019). Staying with the trouble in science education: Towards thinking with nature. In C. A. Taylor., & A. Bayley (Eds.), *Posthumanism and Higher Education: Reimagining Pedagogy, Practice and Research.* https://www.palgrave.com/gp/book/9783030146719.

Jamail, D. (2020). *The end of ice: Bearing witness and finding meaning in the path of climate disruption.* The New Press.

Liboiron, M., Tironi, M., & Calvillo, N. (2018). Toxic politics: Acting in a permanently polluted world. *Social Studies of Science, 48*(3), 331–349.

Mann, M. (2021). *The new climate war: The fight to take back our planet.* PublicAffairs.

Stengers, I. (2018). *Another science is possible: A manifesto for slow science.* Cambridge: Polity.

Tsing, A. L., Bubandt, N., Gan, E., & Swanson, H. A. (Eds.). (2017). *Arts of living on a damaged planet: Ghosts and monsters of the Anthropocene.* University of Minnesota Press.

Wohlleben, P. (2016). *The hidden life of trees: What they feel, how they communicate— Discoveries from a secret world.* Geystone Books.

Correction to: Redrawing Relationalities at the Anthropocene(s): Disrupting and Dismantling the Colonial Logics of Shared Identity Through Thinking with Kim Tallbear

Correction to:
Chapter 7 in: M. F.G. Wallace et al. (eds.), *Reimagining Science Education in the Anthropocene*, Palgrave Studies in Education and the Environment,
https://doi.org/10.1007/978-3-030-79622-8_7

The original version of this chapter was inadvertently published without an electronic supplementary material, which has been now updated. The correction to the chapter has been updated with the changes.

The updated version of this chapter can be found at
https://doi.org/10.1007/978-3-030-79622-8_7

Index

© The Editor(s) (if applicable) and The Author(s) 2022
M. F.G. Wallace et al. (eds.), *Reimagining Science Education in the Anthropocene*, Palgrave Studies in Education and the Environment,
https://doi.org/10.1007/978-3-030-79622-8

Printed in the United States
by Baker & Taylor Publisher Services